TP 1122 .C48 1990

Cheremisinoff, Nicholas P.

NEW ENGLAND INSTITUTE
OF TECHNOLOGY
LEARNING RESOURCES CENTER

Product Design and Testing of Polymeric Materials

NICHOLAS P. CHEREMISINOFF

MARCEL DEKKER, INC. New York and Basel

Library of Congress Cataloging-in-Publication Data

Cheremisinoff, Nicholas P.
  Product design and testing of polymeric materials / Nicholas P.
Cheremisinoff.
    p. cm.
  Includes bibliographical references.
  ISBN 0-8247-8261-5 (alk. paper)
  1. Plastics.   2. Engineering design.   I. Title.
TP1122.C48   1990
668.4--dc20   90-30844
                                                             CIP

This book is printed on acid-free paper.

Copyright © 1990 by MARCEL DEKKER, INC. All Rights Reserved

Neither this book nor any part may be reproduced or transmitted in any form or by any means, electronic or mechanical, including photocopying, microfilming, and recording, or by any information storage and retrieval system, without permission in writing from the publisher.

MARCEL DEKKER, INC.
270 Madison Avenue, New York, New York 10016

Current printing (last digit):
10 9 8 7 6 5 4 3 2 1

PRINTED IN THE UNITED STATES OF AMERICA

Preface

This volume is intended for the polymer-plastics product development engineer, who today is faced with the increasingly difficult task of designing polymeric materials for multifunctional end-use markets. The Edisonian approach to synthesizing a multitude of polymer candidates and testing each one in a variety of applications formulations is no longer an effective approach to product design and molecular optimization. Competition among worldwide polymer producers is too great and has reached levels of sophistication such that more detailed planning and execution of experimental programs as well as better understanding and application of polymer science principles are required to maintain competitive advantages in the technical performance of specialty rubbers and plastics. This book reviews many of the elements important to proper testing and evaluation of the physical and processing performance properties of polymeric materials, along with the designed experimental approach to planning, executing, and analyzing end-use applications responses.

The volume is organized into nine chapters. Chapter 1 covers the principles and methodology behind product testing. An overview of polymer science principles and general properties is given, along with the philosophy behind managing product development programs. Chapters 2 through 4 are devoted to important classes of test methods for base property and end-use performance characterizations. These chapters review the techniques and test procedures for molecular weight and molecular weight distribution measurement, thermal property analysis, stress-strain property characterization, and relaxation. Chapter 5 is devoted to laboratory scale processability testing. Considerable effort is spent in this chapter relating viscometric and small-scale testing to the optimization of molecular properties to achieve certain processing advantages, such as fast extrusion, swell and dimensional control of extruded articles, and surface quality. Chapter 6 makes an important departure from test methods and introduces the concepts of quality control from its importance in the laboratory to the actual marketplace where new products are introduced. This chapter sets the stage for the concepts of designed experimentation. Chapter 7 reviews the mathematics of statistics and probability needed for proper regression of experimental data. Chapter 8 deals with quality control, with emphasis given to production screening and sampling inspection techniques. Finally, Chapter 9 covers the mechanics of experimental design. Examples for application of these techniques in practical product optimization problems are given.

Nicholas P. Cheremisinoff

Contents

Preface   *iii*

CHAPTER 1   PRINCIPLES AND METHODOLOGY OF PRODUCT TESTING   1

    Introduction   1
    Principles of Polymer Science   2
    Characterization of Structural Properties   8
    Rheological Responses of Polymers   12
    General Properties and Uses of Elastomers   22
    Philosophy and General Concepts of
      Product Development   64
    Notation   78
    References   79

CHAPTER 2   CHARACTERIZATION OF MOLECULAR WEIGHT AND MOLECULAR WEIGHT DISTRIBUTION   80

    Introduction   80
    Determination of Number-Average Molecular
      Weight   82
    Determination of Weight-Average Molecular
      Weight   90

	Determination of Molecular Size	96
	Characterization of Molecular Weight Distribution	102
	Notation	113
	References and Suggested Readings	114
CHAPTER 3	THERMAL ANALYSIS OF POLYMERS	116
	Introduction	116
	Thermal Behavior Properties	116
	Measurement Techniques	122
	Differential Thermal Analysis	125
	Differential Scanning Calorimetry	129
	Measurement of Thermal Conductivity	141
	Thermogravimetric Analysis	143
	Notation	144
	References and Suggested Readings	146
CHAPTER 4	MEASUREMENT OF STRESS-STRAIN PROPERTIES	147
	Introduction	147
	Hardness	148
	Stress-Strain Properties and Testing	151
	Compression Stress-Strain, Shear Stress-Strain, and Flexural Stress-Strain Concepts	162
	Tear and Impact Testing	169
	General Comments on Short-Term Stress-Strain Properties	175
	Static and Dynamic Property Behavior	175
	Fatigue Testing	187
	Notation	194
	Suggested Readings	195
CHAPTER 5	PROCESSABILITY TESTING	197
	Introduction	197
	Viscoelastic Properties of Polymers	198
	Description of Conventional Viscometers	213
	Application of Capillary Rheometry to Processing Predictions	241
	Principles of Torque Rheometry	261
	The Oscillating Disk Rheometer	307

Contents   vii

	Notation	309
	References	311
CHAPTER 6	QUALITY IN PRODUCT DEVELOPMENT	313
	Introduction	313
	Philosophy of Quality and Its Relation to the Marketplace	314
	Market Research on Quality	317
	General Principles of Quality Control	332
	Closure	355
	References and Suggested Readings	360
CHAPTER 7	BASIC DEFINITIONS IN PROBABILITY AND STATISTICS	364
	The Importance of Frequency Distributions	364
	The Normal Frequency Distribution	368
	Basic Definitions and Theorems of Probability	368
	Principle of Duality, Random Experiments, and Probability	371
	Combinatorial Analysis, Counting, and Tree Diagrams	376
	Permutations, Binomial Coefficients, and Stirling's Approximation	378
	Definitions of Random Variables and Probability Distributions	381
	Important Theorems of Random Variables and Probability Distributions	390
	Principles of Geometric Probability	392
	Suggested Readings	393
CHAPTER 8	APPLICATION OF QUALITY CONTROL TECHNIQUES	394
	Frequency Distributions	394
	Sampling Inspection Philosophy	430
	Notation	436
	References	437
CHAPTER 9	DESIGN OF EXPERIMENTS AND APPLICATION TO PRODUCT DESIGN	438
	Basic Definitions and Terminology	438

	Statistics Basics for Experimentation Strategy	445
	Screening Designs	474
	Response Surface Concepts and Construction of Polynomial Models	481
	Application to Product Development	500
	Notation	511
	References and Suggested Readings	511
Appendix A	*Glossary of Plastics and Engineering Terms*	*513*
Appendix B	*General Properties and Data on Elastomers and Plastics*	*529*
Index		*561*

1
Principles and Methodology of Product Testing

INTRODUCTION

Ideas for new products generally develop in two ways. Researchers in product development can make technological forecasts of new developments in basic and applied research that could give birth to new products and processes. At the same time or separately, marketing and sales employ the tools of market research to identify unfulfilled needs in the marketplace or opportunities to compete with other products or applications. These two avenues often converge to embody the spirit of new product effort.

What ultimately ensures the financial success of any new product is the demand for it in the marketplace and its cost-effectiveness in manufacturing. Another way of saying this is that when a product is properly designed, it has several key features:

Has unique performance properties that demand a premium in the marketplace or at the very least provide competitive alternatives to existing materials or products

Can be manufactured to a specified consistency and in a cost-effective manner

Does not require alterations during or after production, which can increase the chances of inconsistency

Performs to the expectations and claims of the manufacturer under the environmental conditions for which it is intended to be used

It is a fair statement, then, that product design has a direct bearing on product quality. The poor design of any product, whether it is the raw plastic or elastomer material itself or the part into which it is ultimately fabricated, such as an insulating jacket for an electrical cable, is what limits 100% quality of conformance and is what will ultimately contribute to failure in the marketplace.

There are many factors that contribute to the success and failures of product development programs, which run the gambit from R&D philosophy, to management support, to the technical aspects of synthesis and product testing. It is beyond the scope of this volume to address the philosophy and management approach to proper product development, and indeed, beyond the author's level of expertise. Instead, this book is devoted to the technical aspects of plastics and elastomeric material testing, which plays a crucial role in polymer product development. The book addresses two primary areas of product testing: the principal test methods themselves, and the application of statistical techniques for implementing a proper laboratory program and analysis of test results. By understanding both the test methods and the regression and statistical approaches to planning experiments, the basis for establishing quality products and controlling their properties can be formulated. Along the way, important elements in establishing product development programs are pointed out. In this first chapter we provide terminology and fundamental concepts of polymer science and the philosophy of product development, to orient the reader for later discussions.

PRINCIPLES OF POLYMER SCIENCE

In this section we review the basic concepts and principles of polymer science. Its purpose is to orient the reader and to acquaint the less experienced user with concepts and terminology referred to in later discussions.

The word *polymer* comes from the Greek words *poly* (many) and *mer* (small units). In other words, many monomers are chained together to form a polymer. For example, polyethylene is formed by polymerization of many ethylene monomers. Polymers are synthesized by the process of polymerization, of which there are several.

Addition polymerization is also known as "radical-chain polymerization." The addition reaction starts with a free radical that is an initiator for the polymerization reaction. The free radical is usually formed by the decomposition of a relatively unstable component in polymer structures. In the reaction, repeating units add one at a time to the chain so that monomer concentration decreases steadily throughout the reaction.

$Mx^- + M \rightarrow Mx + 1$

High polymers can be formed at once by this type of polymerization. Long reaction times give high yields but do not affect molecular weight (Figure 1.1). Polymers made by chain reaction often contain only carbon atoms in the main backbone chain. These are called "homo chain polymers."

Condensation polymerization (also known as "step reaction polymerization") is analogous to the condensation of low-molecular-weight compounds. In polymer formation, condensation takes place between two polyfunctional molecules to produce one larger polyfunctional molecule with the possible elimination of a small molecule such as water. The reaction continues until almost all of one of the reagents is used up. The structural units of condensation polymers are usually joined by inter-unit functional groups. The types of products formed in a condensation reaction are determined by the functionality, that is, by the average number of reactive functional groups per monomer molecule.

$Mx + My \rightarrow Mx + y$

Generally, monofunctional monomers form low-molecular-weight products. Bifunctional monomers form linear polymers, and polyfunctional monomers give branched or cross-linked polymers. In this polymerization, molecular weight rises steadily throughout the reaction, as shown in Figure 1.2. Therefore, a long reaction time is essential for high-molecular-weight polymers. Polymers made by step reactions may have other atoms, originating in the monomer functional groups as part of the chain. These are called "hetero polymers."

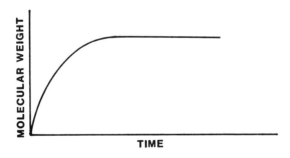

Figure 1.1 Molecular weight as a function of time in an addition reaction.

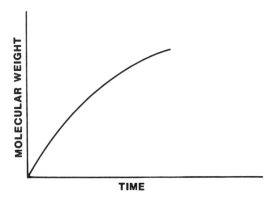

Figure 1.2 Molecular weight as a function of time in a condensation reaction.

The length of the polymer chain is specified by the number of repeat units in the chain. Polymers that have fully extended linear structure are unstable; however, polymers that have random structure (random coil) are stable. These are illustrated in Figure 1.3. Bifunctional monomers in condensation polymerization normally form linear polymers. Linear polymers can be dissolved, melted, molded, and have finite molecular weight. Polyfunctional monomers in condensation reaction normally form branch polymers or cross-linked polymers. Cross-linked polymers (network) are insoluble, infusible (i.e., have no melting point), nonmoldable, and have infinite molecular weight (Figure 1.4).

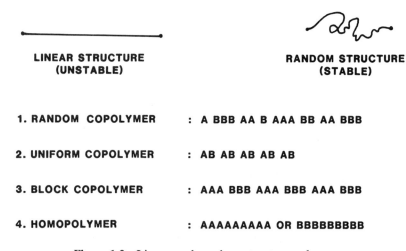

Figure 1.3 Linear and random structure polymers.

Principles and Methodology of Product Testing

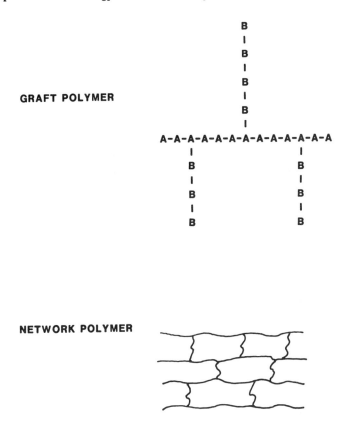

Figure 1.4 Graft and network polymers.

Most linear polymers can be made to soften and take on new shapes by the application of heat and pressure. These changes are physical rather than chemical. This type of material is referred to as a thermoplastic resin and is reprocessable.

Thermosets are materials that have undergone a chemical reaction (curing) by application of heat, catalyst, and so on. The cross-linked network extending throughout the final article is stable to heat and cannot be made to flow or melt. Curing of thermosets are described by A, B, and C stages upon degree of curing reaction. The A stage is the earlier stage, the B stage is an intermediate stage, and the C stage is the final stage of the curing reaction. This type of material is not reprocessable. Rubbers and elastomers are materials that deform upon application of stress and revert back to their original shape upon elimination of applied stress. They are capable of rapid elastic recovery. Materials are usually vulcanized by sulfur; however, peroxide cures are also common. Natural rubber is an elastic substance obtained by coagulating the milky juice of any of various tropical plants. In contrast, a synthetic rubber is one in which two or more monomers are artificially combined through chemical reaction. Therefore, many rubbers are copolymers.

A crystal is an orderly arrangement of atoms in space. The polymer crystalizes under a suitable condition and is called a crystalline polymer. Phase transition of the polymer occurs from solid to liquid. The transition temperature of this polymer is called the melting temperature and is often designated as T_m. All crystalline polymers contain noncrystalline portions as well as crystalline portions. There are many polymers that fall into the semicrystalline region.

An amorphous polymer does not crystallize under any condition. Phase transition of this polymer occurs from glassy state to rubbery state. The transition temperature of the polymer is called the glass transition temperature and is often designated as T_g. Most polymers have both T_g and T_m. The melting temperature and glass transition temperature can be found by measuring the specific volume or enthalpy of the polymer as a function of temperature. This subject is discussed in Chapter 3.

The glass transition temperature is the temperature of onset of extensive molecular motion. The slope of the temperature-specific volume curve above the glass transition temperature is characteristic of a rubber, and below the transition temperature is characteristic of a glass. T_g decreases with decreasing amorphous content. Therefore, T_g is sometimes difficult to detect in highly crystalline polymers.

Molecular weight ($\overline{M}_w$) is defined as the sum of the atomic masses of the elements forming the molecule. The structural formula of polyethylene is often expressed as shown in Figure 1.5. Therefore, the molecular weight of polyethylene can be calculated as n multiplied by the M_w of repeating units. As an example, suppose that we mix 1 g of polymer of 1,000,000 MW with 1 g of polymer of 1000 MW. What is the weight-average molecular weight ($\overline{M}_w$)? The molecular weight of 2 g of the polymer mix is 1,001,000. Therefore, $\overline{M}_w = 500{,}500$. The number-average molecular weight ($\overline{M}_n$) is calculated through Avogadro's number. If two molecules that have a ratio of 1:1000 molecules of polymer 1,000,000 to polymer 1000 mixed together, what is M_n? The solution is

1 molecule	1,000,000
1000 molecule	1,000,000
1001 molecule	2,000,000

Principles and Methodology of Product Testing

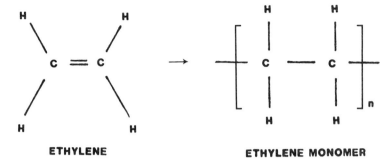

Figure 1.5 Structural formula of polyethylene n–Degree of polymerization.

Therefore, $\overline{M}_n = 1998$. Normally, $\overline{M}_w > \overline{M}_n$. The ratio of $\overline{M}_w$ and $\overline{M}_n$ is sometimes used as a measure of the breadth of the molecular weight distribution (MWD). The molecular weight distribution range of typical polymers is from 1.5–2.0 to 20–50. The MWD for the example above is

$$\text{MWD} = \frac{\overline{M}_w}{\overline{M}_n} = 25.6$$

The subject of molecular weight and molecular weight distribution is discussed in Chapter 2.

To obtain the best mechanical and physical properties from the fabricated products, various *additives* are often incorporated into polymer resins by formulation engineers. Color pigments are added to obtain suitable color, impact modifiers are added to improve brittleness, fillers are added to reach desired mechanical properties and reduce production cost, processing aids are added for ease of fabrication, and so on. The primary functions of *fillers* are cost-effectiveness and improved processing and end-product properties. They are also used to reduce the thermal expansion coefficient of the base polymer to improve its dielectric properties or to soften polymers. Some examples of fillers are $CaCo_3$, wood dust, talc, clays, silica mica, and carbon black. Lubricants are incompatible with the polymer and are used at all temperatures. They migrate to the melt-metal interface during processing, promoting effective slippage of the melt by reducing interfacial viscosity. This can prevent resin from sticking to the hot metal of the processing equipment by reducing friction between the two surfaces. Calcium stearate, zinc stearate, montainic acids, esters, and salts are examples. These types of lubricants are generally referred to as *external lubricants.* In contrast, *internal lubricants* are designed to make polymers compatible at processing temperatures. They reduce chain-to-chain intermolecular forces, thus decreasing melt viscosity. After reducing the cohesive forces between polymer chains, melt viscosity is lowered and flow properties (processing) are improved. Paraffin wax, stearate acid, esters, and polyethylene wax are examples.

Plasticizers are added to polymers to enhance flexibility, resistancy, and melt flow. Generally, high-boiling organic liquids are used as plasticizers. They function in reducing the glass transition temperature or brittleness of the plastics. Phthalates, epoxies, adipates, azelates, trimellitates, phosphates, polyesters, DOP, and DOA are examples.

Stabilizers are used to protect compounds from chemically breaking down under heat and to prevent discoloring or degrading. Tin and antimony compounds, phenols, epoxy compounds, organotin stabilizer, and barium/cadmium heat stabilizer are examples.

Reinforcements are used to improve the mechanical properties of the base polymers, primarily their strength and stiffness. Asbestos fiber, short and long glass fiber, graphite fiber, and carbon fiber are examples.

Cross-linking agents are used to convert a thermoplastic into a thermoset by increasing such properties as tensile strength and chemical strength. Organic peroxides, benzoyl peroxide, methyl ethyl ketone, and peroxides are examples.

A *blowing agent* improves stiffness, lowers labor cost, reduces energy usage, reduces resin consumption, and lightens the weight of the fabricated article.

Other additives of importance are *ultraviolet stabilizers* and *antioxidants*. The primary function of these additives is to prevent aging. Benzotnazole, hydroxybenzophenones, and benzoates are examples.

CHARACTERIZATION OF STRUCTURAL PROPERTIES

Physical and mechanical properties of polymers depend on their molecular weight and molecular structure. Properties such as tensile strength, elongation, modulus or stiffness, hardness, and processability have a direct relationship to molecular weight, with a certain minimum molecular weight required to achieve optimum performance characteristics. The effects of melting temperature (T_m) and glass transition temperature (T_g) and the increase of viscosity with molecular weight are illustrated in Figure 1.6. This plot helps to orient for us the regimes in which the properties of typical plastics, rubbers, viscous liquids, and other forms of polymeric materials may be found in terms of the variables of molecular weight and temperature. An increase in molecular weight generally results

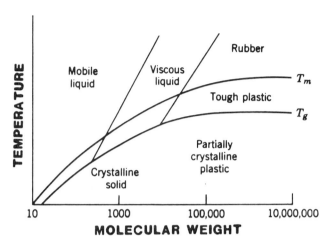

Figure 1.6 Relationships among molecular weight, T_m, T_g, and polymer properties.

in higher physical and mechanical properties, but this is achieved at the expense, for example, of ease of processability. This means that one must almost always compromise between optimum properties and optimum processing. This type of optimization problem requires that one synthesize polymers to within fairly specific molecular weight limits, as well as select appropriate molecular structures. Major variables determining the state of a polymer are the nature and magnitude of the restraints on the motion of its chain atoms (i.e., its molecular motion). The types of restraints and their effect on properties are illustrated in Figure 1.7, in which increasing restriction to molecular motion occurs as one moves on the plot either down or to the right, or both.

Properties often associated with long-chain polymers are those characteristic of the amorphous state of materials. These properties include low elastic modulus, high extensibility, and the polymer's viscoelastic behavior. The molecular basis for this behavior is related to the randomly coiled conformation of long chains in the amorphous state. When crystallinity is present, the material becomes stiffer, harder, tougher, exhibits higher melting, and is more resistant to solvents, has lower extensibility, and has improved mechanical strength. The ductility of crystalline polymers, for example, depends to a great extent on the size, perfection, and organization of the crystals. By synthesizing polymers whose architecture embodies the proper arrangement of crystalline and amorphous domains, a balance of engineering properties can be achieved.

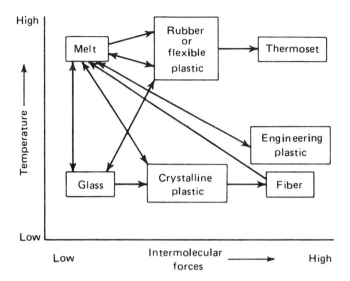

Figure 1.7 Interrelation of the states of bulk polymers.

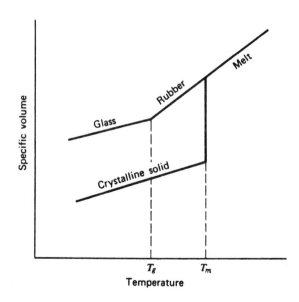

Figure 1.8 Regimes of bulk polymers in terms of volume and temperature.

Structure-property relationships can be understood better when we view polymers in terms of four characteristic regimes shown on the volume-temperature plot in Figure 1.8. All polymers can be described as exhibiting each of these states in varying amounts, but with a preponderance of one (with the exception of the liquid state). This implies that a homogeneous state or system is very rarely achieved. The characteristics of many polymers can therefore be described in terms of the amounts of these species that they contain.

In this chapter we provide a brief introduction to the behavior of polymers in their bulk states, functions of molecular structure, as determined by melt and concentrated-solution viscosity, dynamic electrical and mechanical behavior, stress-strain measurements, diffusion, and swell. Performance responses, with the exception of swelling and diffusion, can be characterized in terms of stimuli applied to the material and the subsequent response, as determined by its molecular and viscoelastic properties. Figure 1.9 illustrates conceptually the stimuli, responses, and

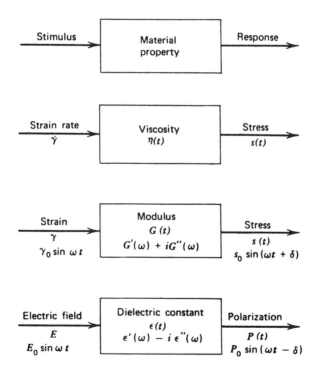

Figure 1.9 Stimuli, responses, and material properties for linear viscoelastic behavior.

material response for the linear viscoelastic behavior of polymers. As an example, a bulk-viscosity measurement is a shear strain applied at a constant rate. One can observe the response to this strain in the form of stress in the material, related to the strain through the property of bulk viscosity.

Any stimulus is essentially an interruption of the thermodynamic equilibrium of the system. The resulting response reflects the route by which equilibrium is again approached. The response always occurs later than the stimulus. This time delay is a phenomenon referred to as *relaxation,* in which the response continues after the stimulus is removed. It is a behavior that typifies polymers.

The nature of the stimulus can also establish the type of response. For example, if the stimulus is quite small, the response may exhibit a linear dependency. In this case, the total response to several stimuli is the sum of the responses to each one applied separately. This situation holds well for dynamic, electrical, and mechanical behavior, but not at all over practical ranges of measurements of stress-strain properties. In the latter case, the deformations are large and lead to such phenomena as yielding, permanent distortion, and failure.

In the balance of this chapter we review some of the important structure properties of all polymers in a very general way. The intent here is to provide a review of the basic property responses of polymers as well as to introduce terminology to the nonpolymer scientist. The importance of these properties to engineering applications of polymers will be discussed, making all too apparent the need for quality control testing in development activities and manufacturing operations.

RHEOLOGICAL RESPONSES OF POLYMERS

Polymers are composed of macromolecules that are similar in configuration but vary over a wide range of molecular weights. The varying types of macromolecules range from linear and flexible to branched and partially cross-linked to completely cross-linked systems, in which the molecular weight has increased without limit. Polymers that are strongly cross-linked can be mechanically dispersed in liquids. The flow properties of such systems are a function of the distribution of the sizes and shapes of the particles, and on particle-particle and particle-solvent interaction forces. The internal nature of the dispersed particles plays a minor role in establishing the flow properties of the dispersion. In contrast, upon dispersing linear and flexible, branched, or lightly cross-linked macromolecules in solution, macromolecular structure plays a dominant role in establishing flow properties. The polymer molecules can be conceptualized as bundles or coils through which solvent may drain, which gives rise to internal particle-solvent dynamic interactions. Since the segments of individual molecules are in constant agitation, the shape and conformation of each random coil undergoes constant change. For a constant temperature, a most probable extension of the individual molecule exists so that each coil may be depicted as a three-dimensional spring with a force of compression or extension arising when the equilibrium conformation is stimulated. Under a state of laminar shear, such a coil moves in the flow field with a rotational frequency equal to $\gamma/2$ (where γ = shear rate). In this rotation, segments are subjected to alternate extension and compression in which case the coil is constrained to vibrate with a frequency that is a function of shear rate.

When the polymer's molecular weight reaches or exceeds a critical value in a solution of fixed concentration, or when its concentration exceeds some critical range, flow characteristics can no longer be characterized by dilute solution properties. Most fabrication operations deal with concentrated solutions or forms of polymers as well as with polymers of high-molecular-weight ranges. In engineering applications such as those dealing in rubbers, the polymer can be highly loaded with a variety of fillers, reinforcing agents, and solid-liquid catalysts. The flow properties of reactive systems, let alone highly concentrated solutions, cannot be predicted from purely theoretical considerations. Therefore, for many practical processing operations, as well as for polymer process control and in the development of new polymers, the application of flow studies is essential.

At this point it is worthwhile to review some fundamental concepts regarding the flow behavior of polymers. Consider a thick layer of fluid sandwiched between two parallel plates, one of which is moved with respect to the other. This requires the application of a force, the shear stress s, which is proportional to the rate of change of flow rate with distance through the layer in the direction normal to the plates. The rate of change, denoted as $\dot{\gamma}$, is the time rate of change of shear strain (i.e., shear rate). The proportionality constant is the property known as viscosity, η:

$$s = \eta\dot{\gamma} \tag{1}$$

The material is Newtonian when $n \neq f(\dot{\gamma})$ and is non-Newtonian if $\eta = f(\dot{\gamma})$. In the latter case, the viscosity at specified values of s and $\dot{\gamma}$ is termed the apparent viscosity η_a.

Viscous behavior can be described by several types of flow curves. Figure 1.10 illustrates the classical responses. Curve N represents a Newtonian material, the slope of the straight line being the viscosity, η; curve P

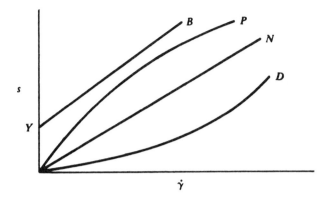

Figure 1.10 Types of viscosity behavior exhibited by polymer melts.

represents a material for which the viscosity decreases with increasing shear rate (i.e., a shear-rate thinning material); curve D represents a material for which the viscosity increases with increasing shear rate (i.e., a shear-rate thickening material); and curve B represents a material that behaves like a Newtonian material after a critical shear stress (called the yield stress), $s = Y$, is exceeded. For many materials the viscosity is also dependent on time of shear as well as the shear-rate history and the time of rest. Such materials are both shear and time dependent.

When a melt is subjected to shear, the flow curve is usually nonlinear. At low shear rates a constant viscosity, η_0, is exhibited. At some critical shear rate, the shear rate increases more rapidly than the shear stress. After this second flow regime, in which shear thinning takes place, the shear rate again becomes linear with shear stress, and an upper regime of flow ensues, with a viscosity designated η_∞. This behavior typifies the flow of concentrated systems and melts all kinds. It is characteristic of such diverse dispersions as clay suspensions, polyisobutylene in decalin, lubricating oils, aluminum soaps in toluene, and polyethylene. Figure 1.11 shows the responses. Fluid systems that exhibit the two viscosity regimes are also capable of storing energy when subjected to laminar shear. This phenomenon is evidenced by the display of normal stresses, that is, stresses in the direction perpendicular to the direction of applied shear. These are observed in several ways, including climbing phenomena on rotating elements. The stored energy may be in the form of energy of nonequilibrium orientation, of shear strain arising from particle deformation, or of dissociation of polymer molecules. All of these forms of stored energy vary in time of relaxation, according to the viscosity of the system and

Principles and Methodology of Product Testing

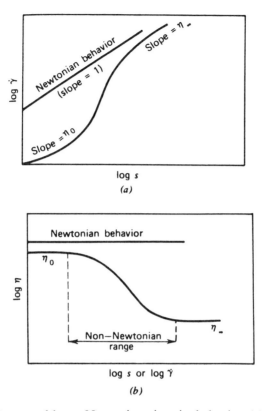

Figure 1.11 Upper and lower Newtonian viscosity behavior: (a) log shear rate versus log shear stress; (b) log apparent viscosity versus log shear rate.

the nature of the storing capacity. Stored energy of orientation may be dissipated rapidly in capillary viscometry by swelling and loss of velocity of the extrudate. The relaxation of shear stress is generally much more rapid than the relaxation of the stored shear-generated energy. Materials that are viscoelastic are capable of storing energy as recoverable shear strain, but the forms of energy storage in flowing systems are not limited to this. When the energy storage is reversible, normal forces may arise, but these will not be shown when the stored energy is dissipated as heat.

Flow behavior in the non-Newtonian region can be described by an empirical power law relationship:

$$s = K\dot{\gamma}^n \qquad (2)$$

where the exponent n is the non-Newtonian flow index. This type of expression is widely used, even though it may fit the data only over a narrow range of shear rate. Fortunately, it often describes flow data under polymer processing conditions reasonably well.

This approach can also be applied to the limiting zero-shear viscosity, η_0, to describe its dependence on concentration of polymer c and molecular weight M:

$$\eta_0 = K'c^n \tag{3}$$

The exponent n generally has a value near 5 for polymers with molecular weight above a critical value M_c. Fox et al. (1956) have shown that, below M_c, η_0 is proportional to M_w and above it, to M_w 3.4 (Figure 1.12). These two dependences can be combined in the equation, valid above $M = M_c$,

$$\eta_0 = K''c^5\overline{M}_w^{3.4} \tag{4}$$

The critical molecular weight M_c or the concentration at the break is attributed to the "entanglement transition," the point above which the vis-

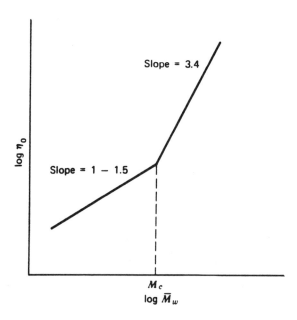

Figure 1.12 Flow behavior above and below critical molecular weight.

cosity is strongly influenced by polymer entanglement, polymer association, or polymer-polymer interaction. Some investigators (Chinai, 1965; Pezzin and Gligo, 1966) have not observed an abrupt change in the slope of the log zero-shear viscosity/log concentration curve, and slopes above the transition considerably greater than 5 (as high as 50) have been reported (Fujita and Kishimoto, 1961; Fujita and Mackawa, 1962; Hayahara and Takao, 1967). The observation of a power near 5 is quite general.

There are many different instruments for the measurement of melt viscosity [see Cheremisinoff (1987)]. The capillary viscometer is among the most common. In this instrument the polymer melt is forced from a reservoir through a capillary of known length and diameter, either at constant flow rate (shear rate $\dot{\gamma}$) or at constant load (stress s) by a plunger activated by either mechanical or pneumatic means. In the constant-load design, the extrusion plastometer measuring "melt index" (ASTM D1238) is an example. A fixed load is applied to the top of the plunger and the output rate is measured.

The constant-rate rheometer provides results more readily interpreted in terms of fundamental quantities. The Instron Capillary Rheometer, an accessory to the standard Instron tensile testing machine, and the Monsanto Processability Tester (MPT) are commonly used instruments. With the Instron, the crosshead drive of the testing machine advances the plunger at a constant rate and the load cell measures the force on the plunger. The precision-bore capillary is made from tungsten carbide or stainless steel. Temperatures from just above ambient to 340°C can be obtained, and steady-state temperature is reached within 2 to 3 min after loading the instrument.

The shear rate at the wall of the capillary, $\dot{\gamma}$, is proportional to the volume flow rate of polymer and inversely proportional to the cube of the capillary diameter, d_c. In terms of instrument variables, the flow rate is given as $\pi d_p^2 v$, where d_p is the diameter of the plunger and v is the velocity of the crosshead. If both diameters are measured in inches and v in in./min (any other unit of linear dimensions is correct), the flow rate in sec^{-1} is given by

$$\dot{\gamma}_w = \frac{2d_p^2}{15d_c^3} v \tag{5}$$

The fraction is a constant for a given instrument and capillary. The shear stress at the wall is proportional to the pressure drop across the capillary and its diameter, and inversely proportional to its length, l. In terms of instrument variables,

$$s_w = \frac{d_c}{\pi d_p^2 l} F \tag{6}$$

where F is the force on the plunger. Again, the fraction is an instrument constant.

The MPT operates on a similar basis with the added feature of having an interface with a computer so that force (or stress) can automatically be regressed to a power law relation. The coefficients of equation (2) are therefore automatically reported for a succession of preprogrammed shear rates to which the polymer sample is subjected. In addition, the instrument is equipped with an optical laser swell tester. The instrument is described in detail in Chapter 6.

A number of corrections can be made to the melt-viscosity data to obtain closer approximations to absolute values. For example, the Rabinowitsch correction compensates for non-Newtonian behavior in the calculation of the shear rate. The true shear rate is given in this correction by

$$\dot{\gamma}_{t,w} = \frac{\dot{\gamma}_w(3n + 1)}{4n} \tag{7}$$

where n is the non-Newtonian flow index defined in equation (2). This correction is on the order of 2 to 5%. Another correction, the entrance correction, is significant only if the ratio of length to diameter of the capillary is small, and is determined experimentally by taking measurements with capillaries of varying length. The procedure of Bagley (1957) can be applied. [See Cheremisinoff (1987) for illustrations pertaining to rubbers.] A correction due to a pressure drop across the plunger barrel can be significant for short, large-diameter capillaries (Metzger and Knox, 1965). These and several other corrections are described by Van Wazer et al. (1963).

Other useful observations during melt-flow determinations include measurement of the swelling of the polymer stream as it is extruded from the capillary, and qualitative observation of such surface characteristics as melt fracture and change in gloss. Die swelling, the ratio of the diameter of the extrudate to that of the capillary, is related to the elasticity of the melt. It generally increases with molecular weight, the presence of long-chain branching, and the breadth of the molecular weight distribution.

With concentrated solutions, viscosities and operating temperatures are lower and hence equipment for measuring viscosity is generally less complicated than melt rheometers. There are three rotational viscometers that are widely used: the cone-and-plate device, the coaxial cylinder, and the rotating cylinder.

The cone-and-plate viscometer has a significant advantage over a capillary viscometer in that in the former the shear rate is constant for small cone angles, whereas in the latter the shear rate varies from center to wall of the capillary. The Ferranti-Shirley cone-and-plate viscometer has a stationary flat lower plate and a rotating cone driven by variable-speed motor through a gear train. Shear rate is proportional to cone speed and is calculated simply as the angular velocity of the cone (rad/s) divided by the cone angle in radians. Shear stress is calculated from the torque of the spring, G, as

$$s = \frac{3G}{2\pi r^3} \tag{8}$$

where G is in dyn·cm and r is the radius of the cone in centimeters.

Typical data obtained with the Ferranti-Shirley viscometer for solutions of poly(vinyl chloride) in cyclohexanone are give in Figure 1.13. Extrapolated values of η_0 are plotted according to equation (3) in Figure 1.14. For the instrument used, $d = 2.0$ cm, the cone angle is 0.0055 rad, and scale divisions are converted to torque by means of a calibration constant of 1310 dyn·cm/div, determined by experiments with a standard oil of known viscosity. In Figure 1.13 the break in the visocisty-concentration curve at a critical concentration is clearly seen. For the sample used, the slopes below and above the break, respectively, are 1.4 and 4.2.

In a coaxial-cylinder (Couette) type of viscometer, the fluid is contained in a narrow gap between two coaxial cylinders, either of which can be rotated. The Haake Rotovisco is such an instrument, in which the inner cylinder is connected through a spring to a drive system. The operating principles of the Haake and Ferranti-Shirley instruments are similar. The shear rate in this instrument is simply calculated from the radii of the inner and outer cylinders:

$$\dot{\gamma} = \frac{2r_i r_o \Omega}{1/r_i^2 + 1/r_o^2} \tag{9}$$

where Ω is the angular velocity of the inner cylinder.

A modification of the couette-type viscometer leading to a very simple design is to eliminate the outer cylinder entirely. The inner (rotating) cylinder is simply immersed in a relatively large container of the fluid to be tested. The Brookfield viscometer and the similar Ferranti portable viscometer are of this type (referred to as rotating-cylinder viscometers).

The Brookfield instrument is one of the most widely used and popular industrial instruments for measuring viscosity because of its portability

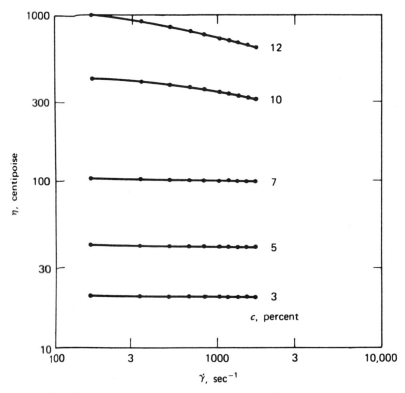

Figure 1.13 Concentrated solution viscosity as a function of shear rate and concentration.

and simplicity of operation. Data are obtained rapidly (less than 5 min) and a wide range of viscosities is covered (up to 80,000 poise) by the use of different-sized cylinders. The shear rate and shear stress are not readily calculated, but the simple approximation that the shear rate is approximately 0.2 times the revolutions per minute of the cylinder is useful. Conversion of dial readings to viscosities is normally made by the use of factors supplied by the manufacturer for specific combinations of speed and spindle.

Classically, the decrease in the bulk viscosity of polymers with increasing temperature is described by an equation of the Arrhenius type:

$$\eta = A \exp\left(\frac{-\Delta E}{RT}\right) \tag{10}$$

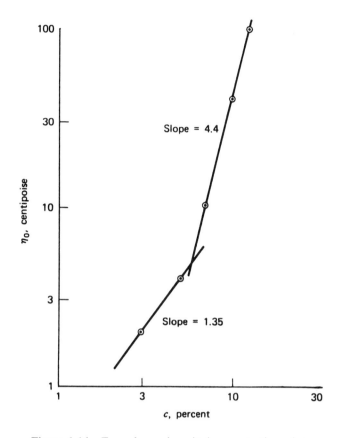

Figure 1.14 Zero-shear viscosity/concentration plot.

where A is a frequency factor depending on shear rate, shear stress, and molecular structure, and ΔE is the activation energy for viscous flow. The equation holds only over narrow temperature ranges, however, and in the non-Newtonian region the apparent activation energy depends on whether viscosity at constant shear rate or constant shear stress is considered. For many systems, ΔE decreases with increasing shear rate, and its value at constant shear rate is less than at constant shear stress. The two apparent activation energies are equal in the Newtonian region.

The temperature dependence of polymer melt viscosities can also be expressed in terms of the free volume using the Doolittle (1951, 1952) equation:

$$\eta = A \exp\left(\frac{BV_0}{V_f}\right) \qquad (11)$$

where the free volume V_f is the difference between the specific volume of the melt and its value V_0 extrapolated to the absolute zero of temperature without phase change. For polymer melts, the Doolittle equation often holds over a wider temperature range than the Arrhenius equation.

The change in viscosity of a material with temperature is associated with the concurrent change in its volume. This relationship is applied to polymers in the WLF equation,

$$\log \frac{\eta_T}{\eta_{T_g}} = \frac{17.44(T - T_g)}{51.6 + T - T_g} \qquad (12)$$

The WLF equation holds up to about $T_g + 100°$; above this, the Arrhenius equation usually represents the data better.

The ability to express the viscosity of a wide variety of polymer solutions and melts by a single universal equation permits the application of the time-temperature equivalence principle. This principle allows superposition of flow curves to produce a master curve from which flow behavior as a function of time can be predicted from data obtained in relatively short-time experiments at various temperatures (Leaderman, 1958; Tobolsky, 1958; Mendelson, 1968; Shenoy, 1988).

For low shear stress, equations (4) and (12) can be combined:

$$\log \eta = 3.4 \log \overline{M}_w + 5 \log C - \frac{17.44(T - T_g)}{51.6 + T - T_g} + k \qquad (13)$$

The subject of rheology and its importance to processability are treated in Chapter 6.

GENERAL PROPERTIES AND USES OF ELASTOMERS

Although it is the intent of this volume to provide general discussions and guidelines for characterizing and developing applications testing for all types of polymers, the best instruction can often be gained by specific examples. In this regard we will highlight some of the properties and characteristics of synthetic elastomers, such as ethylene-propylene polymers, to help illustrate some of the overall property advantages of elastomers.

Ethylene-propylene polymers are inherently stable rubbery materials, off-white to amber in color, and somewhat translucent. Although stable in

Principles and Methodology of Product Testing

normal storage, these materials, especially oil-extended polymers and very high diene polymers, must be protected from light-induced crosslinking before compounding and use.

Although compounding with carbon black and other ingredients plays a dominant role in processing and end-use requirements, the characteristics of the base polymer also establish final performances. The principal polymer characteristics of importance are:

Mooney viscosity (molecular weight)
Ethylene content (low-level crystallinity)
Diene content
Molecular weight distribution
Physical form

In terms of expected performance, compound properties best describe a rubber product for the applications engineer. Each compounding ingredient must be carefully selected to enhance the processing and performance properties of the base polymer in the compound.

Properly formulated, the ethylene-propylene polymer molecule imparts these significant features to products made from it:

Outstanding resistance to ozone attack
Excellent weathering ability
Excellent heat resistance
Wide range of tensile strength and hardness
Excellent electrical properties
Flexibility at low temperatures
Good chemical resistance, especially to polar media
Resistance to moisture and steam

Typical ethylene-propylene rubber compound properties are given in Table 1.1. Much of the environmental stability results from the fact that the side-chain double bonds, although reactive, do not act as sites for polymer chain breakdown. Table 1.2 compares EPDM properties to those of other commonly used rubbers containing main-chain double bonds.

E-P rubbers are practically impervious to ozone degradation. Vulcanizates are unaffected under the customary test conditions of 100 pphm ozone and 20% extension. Even after 15 months of exposure to this concentration and strained up to 60%, no cracking can be observed. When used in black compounds, this inherent resistance to ozone combined with superior resistance to degradation by ultraviolet radiation and extremes of atmospheric temperatures makes this rubber ideally suited for outdoor use. These properties are obtained without the use of an-

Table 1.1 Typical Properties of EPM/EPDM Compounds

Property	ASTM D2000 classification BA, CA, DA, BC, BE
Mechanical Properties (Reinforced)	
Hardness, Shore A	35 to 90
Tensile strength (MPa)	4 to 22
Elongation (%)	150 to 800
Tear strength (kN/m)	15 to 50
Compression set	15 to 35
Electrical Properties	
Dielectric constant	2.8
Power factor (%)	0.25
Dielectric strength (kV/mm)	26
Volume resistivity ($\Omega \cdot$ cm)	1×10^{16}
Thermal Properties	
Brittleness point (°C)	-55 to -65
Minimum for continuous service (°C)	-50
Maximum for continuous service (°C)	150
Maximum for intermittent service (°C)	175
Maximum theoretical temperature to break hydrocarbon bonds (°C)	204
Heat Capacity (J/kg · K)	
40 GPF/40 oil	1950
160 GPF/100 oil	1790
Dynamic Properties	
Resilience, Yerzley (%)	75
Elastic spring rate at 15 Hz (kN/m)	550
Loss tangent at 15 Hz (%)	0.14

Table 1.1 *(continued)*

Property	ASTM D2000 classification BA, CA, DA, BC, BE
Chemical Resistance	
Weather	Excellent
Ozone	Excellent
Radiation	Excellent
Water	Excellent
Acids and alkalis	Excellent to good
Aliphatic hydrocarbons	Fair to poor
Aromatic hydrocarbons	Fair to poor
ASTM oils	Good to fair
Oxygenated solvents	Good
Animal and vegetable oils	Fair
Brake fluid (nonpetroleum)	Excellent
Glycol-water	Excellent

tiozonants or waxes, which natural rubber, styrene-butadiene rubber, and chloroprene rubber compounds often require.

When properly compounded, EPM/EPDM vulcanizates are suitable for continuous service up to 150°C with extrusions up to 175°C. This performance is attributed to the completely saturated polymer backbone and isolation of the pendant olefinic sites from the main polymeric chain.

At high service temperatures, changes in the tensile strength of EPM/EPDM vulcanizates are minimal. The restriction on service life is usually caused by loss of elongation after aging at high temperatures. In EPDM, free-radical vulcanization with its carbon-carbon cross-links gives better heat aging than do the best monosulfidic sulfur cross-links. Further, as shown in Figure 1.15, the rate of elongation loss in EPM is better than in EPDM because there is no pendant unsaturation, which is considerably more reactive to free-radical oxidative attack than tertiary carbons on the main chain. In addition, the more rubber-rich compounds give the best results, since there is less localized strain amplification acting on the network.

Table 1.2 Comparison of Rubbers and Typical Compounds[a]

Property	Rubber type				
	EPDM	Natural	SBR	IIR	CR
Specific gravity (polymer)	0.86	0.92	0.94	0.92	1.23
Tensile strength (max.) (MPa)	22	28	24	21	28
Elongation (%)	500	700	500	700	500
Top operating temperature (°C)	150 to 175	75 to 120	75 to 120	120 to 180	90 to 150
Brittleness point (°C)	−55 to −65	−55	−60	−60	−45
Compression set (% in 22 hr at 100°C)	10 to 30	10 to 15	15 to 30	15 to 30	15 to 30
Resilience, Yerzley (%)	75	80	65	30	75
Tear strength (kN/m)	15 to 50	35 to 45	25 to 35	25 to 35	35 to 45
Dielectric constant	2.8	2.9	2.9	2.5	6.7
Volume resistivity ($\Omega \cdot$ cm)	10^{16}	10^{15}	10^{15}	10^{15}	10^{12}
Dielectric strength (kV/mm)	20 to 55	16 to 24	24 to 32	24 to 36	16 to 24

Resistance to:[b]	NR	SBR	IIR	CR
Weathering	E	F-G	F-G	G
Ozone	E	P	F	G
Acids and alkalis	G-E	G	G	E
Oils and solvents	P-F	P	P-F	G
Abrasion	G-E	G-E	G	G-E
Compression set	G-E	E	G	G
Tearing	G	E	F	G-E
Low temperature	G-E	G	F-G	F
Steam	G-E	P	P	G
Air permeation	F	F	F	F

[a] SBR, styrene-butadiene rubber; IIR, isobutylene-isoprene rubber (butyl rubber); CR, chloroprene rubber (neoprene).
[b] E, excellent; G, good; F, fair; P, poor.

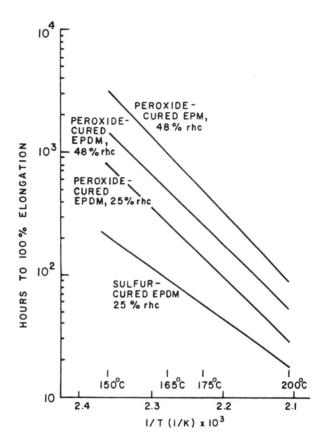

Figure 1.15 Typical data on heat resistance of EPM/EPDM.

Performance requirements such as extreme heat resistance or optimum cost in uncured applications may direct the choice toward copolymers. Especially with terpolymers, there are also situations where a blend of polymers will be best for a particular application.

The polymer or polymer blend finally selected will influence, to a great extent, the processing characteristics of the compound and also the level of physical properties achieved and, to a lesser extent, the compound cost. Property and processing benefits and limitations from increasing molecular weight, ethylene content, diene content, and molecular weight distribution and breadth are summarized in Table 1.3.

Table 1.3 Elastomer Selection: Benefit Summary

Property	As property increases: Benefits	As property increases: Limitations
Molecular weight (Mooney viscosity)	Tensile and tear strength increases; black/oil loading can increase for lower cost; hot green strength increases; collapse resistance improves; lower-pressure CV cures improve	More difficult to disperse; extrusion rate and ability to calender decrease
Ethylene content	Cold green strength increases greatly; flow at high temperature increases; tensile strength increases; higher filler/oil loadings possible; easier to pelletize raw polymer and compounds; peroxide cures improve	Tougher to mix; poorer low-temperature set and flex; higher tension set, less elastic recovery; higher shore hardness
Diene content	Faster cure rate; better low-pressure CV cures; more flexibility in selecting accelerators; compression set improves; modulus increases	Compound cost may increase; scorch safety decreases; shelf life decreases; elongation decreases
Molecular weight distribution		
Narrow MWD	Faster cure; faster extrusion speed; better extrusion smoothness; low die swell	Lower green strength; poorer extruder feed; poorer mill handling, adhesion to mill roll; chance of extrusion melt fracture; fracture calendering
Broad MWD	Improved calendering; improved mill handling; higher green strength especially when hot; improved extruder feeding at all temperatures; improved collapse resistance; less compound flow and tack	Higher die swell; slower cure rate and lower cure rate; chance of rougher extrusion surface

A brief description of the major neat polymer properties follows. To begin, neat polymer Mooney viscosity is a bulk viscosity measured in a shearing disk viscometer. There is no simple mathematical relationship between Mooney viscosity and any of the standard molecular weight averages for polymers. This is partly because of the deliberate changes in molecular weight distribution from polymer to polymer. However, higher Mooney viscosity is a general indication of higher molecular weight. Commercial elastomers fit into categories from low to very high Mooney. In general, the following compound performance observations in terms of processing and vulcanizate properties are made with increasing molecular weight.

Processing effects:

Increased hot green strength (improved collapse resistance; improved low pressure, continuous cures)
Potential for poor dispersion in mixing
Slower extrusion and poorer calenderability

Vulcanizate properties:

Increased tensile and tear
Lower compound cost as filler loading capability increases

Increasing ethylene content gradually introduces low-level crystallinity above 55 to 65% ethylene. This crystallinity usually melts in the range 30 to 90°C. Therefore, it has a considerable effect on the processability at temperatures below about 90°C; the upper temperature limit varies with molecular weight distribution also. General processing and vulcanizate effects with increasing ehtylene content are as follows:

Processing effects:

Increased cold and warm green strength
Easier pelletization of both polymer and compound (important to some feeding operations to end users)
Increased difficult in mixing and dispersion

Vulcanizate properties:

Increased tensile and crystallinity
Higher hardness
Lower compound costs through higher filler loading capability
Improved peroxide curves

Poorer low-temperature set and flex resistance
Higher tension set, less elastic recovery

Molecular weight distribution (MWD) primarily affects processing properties. Commercial polymers are offered in a wide range of MWD at different Mooney viscosity and ethylene levels. The general effects of narrowing MWD are as follows:

Processing effects:

Increased extrusion rate
Smooth extrusion surface
Lower die swell
Poorer extrusion feed
Greater chance of extrusion melt fracture and edge tearing
Poorer mill handling due to greater adhesion to the mill rolls
Poorer green strength
Poorer calendering
Greater compound flow and tack

Vulcanizate properties:

Increased cure rate
Higher cure state

The level of diene chiefly affects the rate and state of sulfur vulcanization. The effect of increasing diene level on processing and vulcanizate properties is as follows:

Processing effects:

Improved low pressure, continuous cures
Lower scorch safely
Increased flexibility in selecting accelerators
Shortened shelf life

Vulcanizate properties:

Faster cure rate
Higher modulus
Improved compression set
Decreased elongation
Decreased heat-aging resistance

The most widely used and best reinforcing fillers are the carbon blacks. The important properties are particle size and structures. The medium particle size and higher structure types, such as N550 FEF and N650 GPF, favor reinforcement and increase both hardness and stiffness. The fine-particle-size blacks (e.g., N330 HAF) are difficult to mix and disperse and at any given loading produce compounds with much higher viscosity. Coarser or low-structure blacks are suggested when high loadings are required for increased elongation and/or scorch time. The effect of carbon black structure and particle size is summarized in Table 1.4.

The prepration of carbon black/rubber "masterbatches" is common practice in industry. Masterbatching consists of adding the appropriate loading of carbon black in the form of a slurry to the rubber prior to coagulation. Table 1.5 provides a summary of the various common grades of blacks used in the synthetic rubber industry.

In plastic applications, medium- and high-color channel blacks are employed for tinting and to provide maximum jetness at low loadings. Carbon black is also used as antiphoto- and antithermal-oxidation agents in polyolefins. As an example, clear polyethylene cable coating is susceptible to crazing or cracking, accompanied by rapid loss in physical properties and dielectric strength when exposed to sunlight for extended periods

Table 1.4 Effect of Carbon Black Structure and Particle Size

		Particle size		Structure	
		Fine	Coarse	Low	High
Processing	Improves	→		→←	
Reinforcement	Increases	←		→	
Optimum loading (for tensile and tear) at increased oil loadings	Increases	→		→	
Mooney viscosity, modulus, and hardness	Increases	←		→	
Mooney scorch, elongation, and resilience	Increases	→		←	
Compound cost	Increases	←		←	

Best overall performance: N550 FEF, N683 APF, N650 GPF-HS
Best for cost compounding: N550 FEF, N650 GPF-HS/N762 SRF-LM blends

Principles and Methodology of Product Testing 33

of time. The carbon black essentially serves as a blackbody that absorbs ultraviolet and infrared radiations. It also serves to terminate free-radical chains and hence provides good protection against thermal degradation.

Carbon black is elemental carbon that is differentiated from commercial carbons such as coke and carcoals by the fact that it is in particulate form. Particles are spherical, quasi-graphic in structure, and of colloidal dimensions. Carbon blacks are manufactured either by a partial combustion process or by thermal decomposition of liquid or gaseous hydrocarbons. There are many commercial classifications of carbon black, but all blacks can generally be characterized by their method of production, as follows.

Lampblacks: produced via combustion of petroleum or coal tar residues in open shallow pans (typical sizes 600 to 4000 Å)
Channel blacks: produced by partial combustion of natural gas or liquid hydrocarbons in retorts or furnaces (400 to 800 Å), coarsest variety (400 to 4000 Å)
Thermal blacks: produced by the thermal decomposition of natural gas
Acetylene black: produced by the exothermic decomposition of acetylene

The various commercial grades and the specific properties imparted to end-use products via their incorporation depend on particle size (i.e., surface area), chemical composition and relative activity of the particle surface, and degree of particle-to-particle association.

Carbon black essentially remains as a discrete phase when dispersed in median such as rubber. The extent ot mixture reinforcement imparted to the rubber, and, for example, the intensity of color imparted to a coating formulation, depend primarily on the particle size or surface area.

The definitions used for characterizing average particle size vary. Arithmetic averages are defined as

$$D_n = \frac{\sum n_p d}{\sum n_p} \qquad (14)$$

where n is the number of particles of diameter d. A more meaningful definition is the surface-average diameter

$$D_a = \frac{\sum n_p d^3}{\sum n_p d^2} \qquad (15)$$

Table 1.5 Properties and Rubber Applications of Carbon Blacks

Type (designation)	Average particle size (Å)[a]	Typical applications
Channel blacks		
Easy processing (EPC)	290	Tire treads, heels, soles of shoes, mechanical goods products
Medium processing (MPC)	250	
Gas furnace blacks		
Semireinforcing (SRF)	800	Tire carcass/bead insulation, footwear and soiling, belts, hoses, packings, mechanical goods
High modulus (HMF)	600	Tire carcass/sidewalls, footwear, mechanical goods
Fine furnace (FF)	395	Truck tire carcass, breaker, and cushion
Thermal blacks		
Medium thermal (MT)	4700	Wire insulation/jackets, mechanical goods, blets, hose, packings, stripping mats
Fine thermal (FT)	1800	Natural rubber inner tubes, footwear uppers, mechanical goods, inflations

Oil furnace blacks		
General purpose (GPF)	510	Tire tread base, sidewalls, sealing rings, cable jackets, hose, extruded stripping products
Fast extruding (FEF)	400–450	Tire carcass, tread base, and sidewall, butyl inner tubes, hose, extruded stripping
High abrasion (HAF)	270	Camelback and tire treads, mechanical goods
High abrasion, high structure (HAF-HS)	250	Tire treads, products requiring high abrasion resistance
High abrasion, low structure (HAF-LS)	264	Channel black replacement in natural rubber applications
Intermediate superabrasion (ISAF)	230	Tire treads and camelback, mechanical goods
Intermediate superabrasion, high structure (ISAF-HS)	225	Tire treads, high abrasion service applications
Intermediate superabrasion, low structure (ISAF-LS)	227	Tire treads, high abrasion service applications
Superabrasion (SAF)	200	Tire treads, camelback, mechanical goods, heels/soles of shoes
Conductive (CF)	230	Antistatic and conductive rubber products, belts, hose, flooring material

[a] $D_n = \sum nd / \sum n$, where n = number of particles of diameter d.

Particle size distribution of commercial grades can vary from Gaussian to skewed.

Specific surface areas are evaluated by absorption techniques, and the arrangement of carbon atoms within a carbon black particle has been studied extensively by x-ray diffraction methods. Carbon black displays two-dimensional crystallinity, which is a structure defined as mesomorphic.

One of the most important properties of carbon black is its surface activity. Carbon itself has proved extremely useful as an adsorbent. All carbon blacks possess a distribution of energy sites, whereby those of highest energy are first occupied by adsorbate. As the adsorption process continues, fewer active sites become filled and the differential heat of adsorption decreases, eventually approaching the heat of liquefaction of the adsorbate at monolayer coverage. High-energy sites can be progressively reduced by heat treatment. At 3000°C all high-energy sites are reduced and the substance displays characteristics of a homogeneous surface of uniform activity.

Chain structure is another property of importance, particularly in terms of the properties imparted to rubber. In particular, extrudability, elastic modulus, and electrical conductivity of rubber vulcanizates are sensitive. As general comments, lampblack and acetylene black have a high degree of structure, whereas channel and low blacks are typically low in chain structure. Most commercial carbon black is available in bead or pellet form. The rubber industry typically employs carbon blacks having an average specific gravity of 1.8.

The relevant properties of carbon black to rubber compounding particle size/surface area, particle porosity, aggregate structure (i.e., bulkiness), the amount of carbon per aggregate, surface activity, and surface chemistry. Examples of widely used elastomers are SBR and BR.

Another important property is chemical resistance. These rubbers have excellent resistance to acids, alkalis, and hot detergent solutions. EPM/EPDM is also resistant to salt solutions, oxygenated solvents, synthetic hydraulic fluids, and animal fats. Chemical resistance was determined according to ASTM D471. The compound tested contained 25% Rockwell hardness C (RHC) and has optimized.

Although ethylene-propylene rubber compounds have only limited resistance when immersed in hydrocarbon solvents such as toluene and gasoline, they can be useful in a hydrocarbon atmosphere when exposure is intermittent or mild. This is evidenced by its successful use in automotive underhood applications.

Principles and Methodology of Product Testing

Other properties of importance include low-temperature performance and dynamic behavior. The resilience at room temperature is slightly less than that of natural rubber and generally equivalent to styrene-butadiene and polychloroprene elastomers. Although the low-temperature brittle point of ethylene-propylene (EP) rubber is about the same as that of styrene-butadiene rubber, it retains a greater percentage of its resilience at low temperatures. Figure 1.16 shows this in terms of stiffness. The dynamic response of EP rubber compounds is somewhat similar to that of natural rubber compounds, as indicated in the Yerzley test comparison illustrated in Figure 1.17. Ethylene-propylene rubber, however, is a more popular choice for dynamic parts because its age resistance better preserves initial design characteristics with time and environmental extremes. EPDM is a good first choice when high resiliency is desired. Where high damping is called for, butyl rubber is the proper choice, as shown in the Yerzley test.

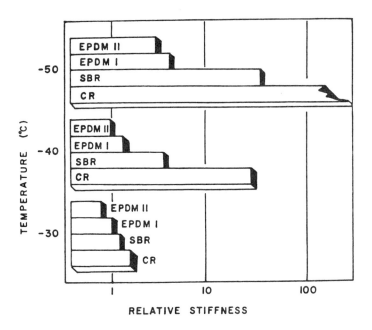

Figure 1.16 Relative stiffness in torsion of various rubbers.

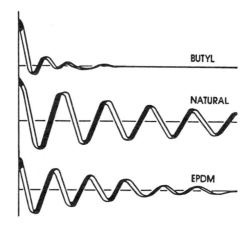

Figure 1.17 Yerzley test for various elastomers.

Rubber compounds contain a number of ingredients; generally, these fall into five categories:

1. Elastomers
2. Fillers
3. Plasticizers
4. Miscellaneous chemicals
5. Vulcanization system

Selection and application of these to the compound are determined by end-use requirements and ultimately represent a balance of the following four parameters.

1. Vulcanization (cure) rate
2. Vulcanizate physical properties
3. Cost
4. Processing behavior

As a general rule, ethylene-propylene elastomers are amorphous polymers. Unless the molecules are above approximately 60 wt% ethylene, they do not develop crystallization on stretching and therefore must be reinforced to achieve useful properties. Significantly large volumes of carbon black and other reinforcing fillers and oil can be incorporated—100 to 400 parts per hundred of rubber (phr) are common. This, in addition to the low specific gravity of ethylene-propylene rubber compared to other rubbers, allows the compound cost to be kept low.

Principles and Methodology of Product Testing

In many cases, proeprty requirements of the application and preference for sulfur cures will lead the polymer selection toward a terpolymer instead of a copolymer. However, certain acrylonitrile-butadiene copolymers and EPDMs are examples of polymers in which the molecular chains are not highly oriented when their gum vulcanizates are stretched. Figure 1.18 gives an example of an EPDM, showing the dramatic improvement in tensile and modulus properties with increasing carbon black loadings.

The preceding example helps to illustrate that the usefulness of elastomers depends greatly on the ability of fillers such as carbon black to impart reinforcement properties. The property of reinforcement is characterized in terms of the vulcanizate's stiffness, modulus, tear strength, rupture energy, cracking resistance, fatigue stress, and abrasion resistance.

It is important to note that some fillers are nonreinforcing, but are instead used as extenders in rubber formulations to reduce raw material

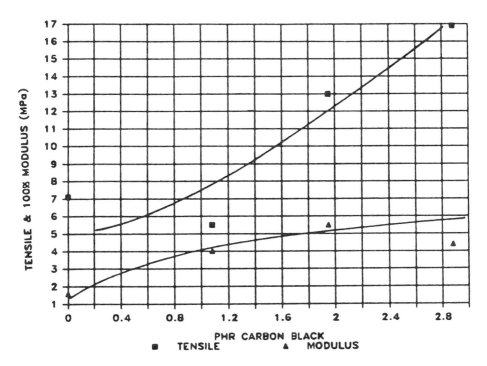

Figure 1.18 Tensile and modulus properties of sulfur-cured EPDM vulcanizate at different carbon black loadings.

costs. These fillers are usually chemically inert and have large particle sizes. Reinforcing fillers are the opposite; that is, typically, they are small in particle size and chemically active. Fully reinforcing fillers as those solid additives having mean particle sizes below 50 nm. The two principal additives are carbon blacks and silica. Solid fillers having sizes greater than 50 nm are clays, silicates, calcium carbonates, SFR carbon blacks, and a range of semireinforcing blacks that are usually derived by thermal oxidation.

It is important to make a distinction between the carbon black/rubber interactions of uncured and vulcanized materials. The effects of carbon black on such properties as resilience and stiffness will be different for these two states of the material. In the uncured state the incorporation of carbon black results in viscoelastic changes in the rubber system. In the vulcanized state, carbon black incorporation causes viscoelastic property changes, which in fact directly alter the rubber network. The specific properties affected in the vulcanizate are modulus, dynamic properties, and hysteresis.

Commercial carbon blacks used in rubber compounding applications are usually classified in terms of their morphology (i.e., particle size/surface area, vehicle absorptive capacity). In the ASTM nomenclature system for carbon black, which is strictly used in the United States, the first digit is based on the mean particle diameter as measured with an electron microscope. Other commonly used methods for classifying blacks by size are iodine number, nitrogen adsorption, and tinting strength. Important also are the shapes of individual particles and of aggregates. DBP absorption is the principal technique employed for measuring the irregularity of the primary aggregates.

The term *structure* characterizes the bulkiness of individual aggregates. Often when carbon black is mixed into rubber, some of the aggregates fracture. In addition to fracturing, the separation of physically attracted or compacted aggregates (microagglomerates) can occur. Aggregates can be held together fairly rigidly because of the irregular nature of their structures. Microagglomerates can also be formed during mixing. That is, high shear forces can cause microcompaction effects wherein several aggregates can be pressed together. Excessive microagglomeration can lead to persistent black network structures that can affect extrusion and ultimate vulcanizate properties.

The relationship of carbon black morphology to the failure properties of rubber vulcanizates has been studied extensively. Strength properties are usually enhanced with increasing black surface area and loading. (References at the end of the chapter provide detailed discussions.) The upper limit of black loading for maximum tensile strength and tear resistance depends on carbon black fineness and structure, with the former usually having the greater effect. Coarser, lower-structure blacks generally show peak strength properties at higher loadings. In terms of the ultimate level of strength reinforcement for different blacks, structure is significant as a dispersing aid, which may be attributable to better bonding between black and polymer. High structure is also important at lower black loadings, particularly for tear resistance.

There is some evidence that the strength reinforcing properties of fillers are directly related to modulus development. These properties are derived from the large stresses held by the highly extended polymer chains attached to the immobile particles. Boonstra (1973) has related high tensile strength to energy dissipation by a slippage mechanism at the filler surface. This is supported by the fact that high-surface-area, inactive (partially graphitized) carbon blacks give very high tensile strength under standard testing conditions. Under more severe conditions, however, adhesion between black and polymer becomes important.

Andrews and Walsh (1958) have shown that the path of rupture through a filled vulcanizate passes from one filler aggregate to another, which are sites of high stress concentration. One approach to increasing strength is to lengthen the overall rupture path. The finer the filler and the higher the loading, the greater the effective increase in total cross section. There is a limiting point, however, when the packing of the filler aggregates becomes critical and they are no longer completely separated by the polymer. Preferential failure paths between aggregates may then be formed. These, combined with the fact that a smaller amount of polymer is being strained, can result in failure. The level of filler-polymer adhesion is important, especially as the upper limits of loading (aggregate packing) are reached.

Filler-polymer adhesion is important to the amount of elastic energy released by internal failure. Strongly adhering fillers enlarge the volume of rubber that must be highly strained during the process of rupture. It is worth noting that the relative importance of different carbon black parameters that influence vulcanizate properties greatly depends on the conditions of testing.

Carbon black fineness is the dominant factor governing vulcanizate strength properties. Tensile strength, tear resistance, and abrasion resistance all increase with decreasing black particle size. Cyclic processes such as cracking and fatigue are more complex because they may involve internal polymer degradation caused by heat buildup. Beatty (1964) points out that it is important to note whether or not the end-use application represents constant energy input or constant amplitude vibrations. The latter mode of service greatly reduces the fatigue life of rubber compounds containing fine or high-structure blacks. Beatty has reported no significant particle size effect on fatigue on the basis of equal energy input.

The relationship of carbon black fineness to failure properties is not straightforward since it is difficult to separate the effects of aggregate size and surface area. It is likely that both are important, since aggregate size relates to the manner in which the surface area is distributed. In general, decreasing aggregate size and increasing black loading reduce the average interaggregate spacing, thereby lowering the mean free rupture path. Increasing black fineness and loading can be viewed in terms of either increased interface between black and polymer, or increased total black cross section (lower aggregate spacing). Both show a strong effect on tensile strength up to the maximum tensile value that can be achieved.

The ultimate value for tensile strength across all blacks appears to be determined by either aggregate size or specific surface area. Tensile strength can be improved to a limiting value by increasing the loading. It is not possible, however, to match the ultimate tensile of a fine black by increasing the loading of a coarse one, at least not without other modifications (e.g., improving the ultimate dispersion or the bonding between black and polymer).

Blends of two or more elastomers are used in a variety of rubber products. The compatibility of elastomers in terms of their relative miscibility and response to different fillers and curing systems is of great importance to the rubber compounder. From the standpoint of carbon black reinforcement, certain combinations of polymers can give less than optimum performance if the black is not proportioned properly between them. Note that equivalent volume proportionality of the black is not always desirable. This depends on the nature of the polymers and their relative filler requirements in terms of strength reinforcement.

The compatibility of elastomers blends has been studied by a variety of different techniques, including optical and electron microscopy, differential thermal analysis (DTA), gel permeation chromatography (GPC), solubility, thermal and thermochemical analysis, and x-ray analysis. These and other methods have been discussed by Corish and Powell (1974).

Microscopie methods are especially useful in studying the phase separation (zone size) of different polymer combinations and in determining the relative amounts of filler in each blend component. Walters and Keyte (1962) studied elastomer blends by means of phase-contrast optical microscopy, where the contrast mechanism is based on differences in the refractive indexes of the polymers. Their work covered blends that included NR/SBR, NR/BR, and SBR/BR. Their results indicated that few, if any, elastomers can be blended on a molecular scale. They also provided direct evidence that fillers and curing agents do not necessarily distribute proportionately.

The term *compatibility* can be defined in terms of the glass transition temperature (T_g). Blends that have T_g values of the individual polymer components are considered incompatible, while those that result in a single, intermediate T_g can be viewed as compatible. Also, solubility parameters are considered a prerequisite to compatibility. In the category of compatible blend combinations Carish (1967) lists NR/SBR, NR/BR, and SBR/BR. Incompatible blends are NR/NBR, NBR/BR, and NR/CR.

The compatibility of elastomer blends can also be defined from the standpoint of interfacial bonding between the different polymer phases. The techniques of differential swelling, using solvent systems and temperatures such that one elastomer is highly swollen and the other is below its temperature with the polymer chains remaining tightly coiled, can be used. Thus the polymer below its temperature could be treated in the same manner as a filler with respect to the way it restricts the swell of the other polymer. A high degree of swelling restriction is indicative of interfacial bonding between phases, while the reverse is analogous to the dewetting that occurs with nonreinforcing mineral fillers in elastomer networks. Also, the type of curing system is important in achieving interfacial bonding.

The importance of the proper curing system in blends of low-functionality rubber with more unsaturated or polar elastomers must also be emphasized. Large amounts of polymer can be extracted from blends when unsatisfactory curing systems (e.g., sulfur-ZnO alone) are employed. This can be attributed to curative diffusion because it occurs whether or not the curatives are added to the blend or to the low-functionality rubber alone.

Good blending is favored by a similarity of solubility parameters and viscosity. For polymers varying in unsaturation or polarity, cure compatibility must be considered in terms of both properly cross-linking each polymer component and achieving satisfactory interfacial bonding. In contrast, the addition of fillers to elastomer blends can significantly alter the state of the polymer phases. Where large differences in unsaturation exists between the polymers in a blend, staining methods may be used to render the high-unsaturation polymer more opaque to provide contrast for either light- or electron-microscopic analysis.

Factors that contribute to reinforcement include dispersion, aggregate breakdown and interactions, and distribution between the separate phases of a polymer blend. For blends of dissimilar elastomers (e.g., in unsaturation, polarity, crystallization, or viscosity), problems can arise in achieving optimum black distribution in the end product. When fillers are added to elastomer blends, one very obvious change is a reduction in the size of the separate polymer zones. In other words, there is a preferential location of the black in one of the blend components. This phenomenon may be related to both molecular structure and relative viscosity. The great mobility and possibly more linear structure of one of the blend component molecules apparently enables better wetting of the black surface. This characteristic is often observed in SBR/BR blends. The lower initial viscosity of the BR would also tend to favor the acceptance of the black. It is, however, possible to alter the normal pattern of distribution in terms of how the filler is added, along with mixing conditions. Other polymer factors that favor preferential filler adsorption in elastomer preblends are higher unsaturation and polarity.

Another phenomenon contributing to reinforcement is carbon black transfer. This is the ability of a carbon black to migrate from one polymer to another during mixing. Transfer is favored by a low heat history and high extender-oil content. Both these parameters minimize interaction between black and polymer. Hence solution or latex masterbatches tend to favor transfer, while hot mixing in a Banbury does not. Black transfer also occurs when a masterbatch of a low-unsaturation rubber is cut back with a high-unsaturation polymer. Polarity is also a factor.

As noted above, carbon black is widely used as a reinforcing filler; however, the unique property of improving tear and abrasion resistance in rubber vulcanizates, is not fully understood. It is likely that carbon black surface and rubber interact both chemically and by physical adsorption. The parameters important to this interaction are capacity, intensity, and geometry. The total interface between polymer and filler is expressed in units of square meter per cubic centimeter of vulcanizate or compound. The intensity of interaction is determined by the specific surface activity per unit area. It should be noted that adsorptive energies vary greatly in different locations on the black surface, and it is likely that this distribution is responsible for the variability in properties among different types of carbon blacks. Geometrical properties are characterized by the structure, which is basically anisometric, and particle shape and porosity. Greater anisometry results in looser particle packing. The void volume is used as a measure of the packing density.

Carbon black particles form irregular structures that tend to break down during intensive mixing. It is therefore a combination of carbon black reactivity, rubber and black chemical, physical, and rheological properties, as well as the conditions of mixing that establish the final strength properties of the vulcanizate. As a rule of thumb, high-structure blacks (e.g., ISAF) impart a higher modulus (at 300% elongation) than does a corresponding normal black. The high modulus is determined by both the anisometry and surface activity. The separate influence of these properties can be observed by heat treating these blacks as treatment and properties approach these of graphite. Through recrystallization, highly active sites on the carbon black surface lose their high activity. In this situation the entire surface area becomes homogeneous and adsorption energies approach their lowest state. Figure 1.19 shows the effect of surface activity on vulcanizate performance. Heat treatment typically results in a minor decrease in surface area and further reductions in tear resistance. Modulus and elongation can be decreased by a factor of 3 or 4. The decrease in properties can be attributed to the removal of highly active sites from the surface. Boonstra (1967) shows that this is accompanied by a decrease in water adsorption and propane adsorption.

Microscopic examinations of thin sections of vulcanizate in which carbon black or other fillers and polymer were mixed for a short time reveal only that this additive exists as coarse agglomerates in an almost pure rubber matrix, without almost any colloidal dispersion of the additive. Only after mixing has been continued at greater intensity and/or for longer periods can one observe how these agglomerates gradually disappear, making room for an increasing amount of additive. In all cases the initial product of mixing is comprised of an agglomerate that is formed by the penetration of rubber in the voids between carbon black particles or other additives, such as mineral fillers (e.g., clays) under the pressure and shear that builds up during Banbury mixing and on a roll mill. Once these voids are filled, the additive is incorporated but not yet dispersed. As soon as the agglomerates are formed, continuous mixing exposes them to shearing forces that break them up and eventually disperse them. In other words, agglomerate formation and breakdown take place almost simultaneously.

Inorganic fillers are used in white or colored compounds or as extenders in carbon black stocks to reduce cost. Silicas, clays, talcs, and whitings are examples of the classes of mineral fillers most commonly used. Frequently, a combination of fillers is used to obtain the required balance of reinforcing properties, processability, and economics for the particular application.

Fine-particle-size silicas are the most reinforcing of the mineral fillers. However, they are difficult to disperse, retard cure, and produce boardy, hard stocks. Hard clays and talcs are easier to mix, provide some reinforcement, and retard cure to a lesser extent than do silicas. To overcome the retarding effect, small amounts of diethylene glycol or polyethylene glycols are commonly added.

Talcs, whitings, and soft clays impart little reinforcement but aid pro-

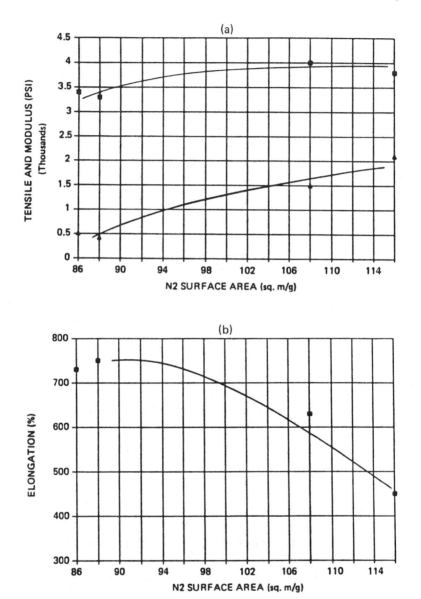

Figure 1.19 Effect of graphitization of black on vulcanizate properties: (a) physical properties; (b) vulcanizate elongation; (c) scorch.

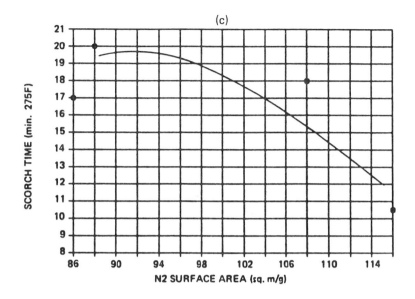

cessing by reducing compound viscosity. Table 1.6 provides guidelines on the effects of different mineral fillers. Ethylene-propylene rubber as well as some other synthetic elastomers are compatible with paraffinic or naphthenic process oils. Aromatic oils should be avoided because they have a detrimental effect on cure. Viscosity is an important consideration

Table 1.6 Effect of Mineral Fillers on Compound Properties

	Compound
Good reinforcement	Silica, hard clay, platy talc
Good mill mixing	Silicate, calcined clay, platy talc
Good extrusion	Silica, platy talc, calcined clay
Good internal mixing	All except ground whiting
Lowest viscosity	Platy talc, whiting
	Vulcanizate
Highest hardness	Silica, silicate
Highest modulus	Silica, platy talc
Highest tear	Silica, silica
Good compression set	Whiting, soft clay
Good water resistance	Calcined lay, platy talc, soft clay

when selecting an oil plasticizer. In general, the higher viscosity oils enhance physical property development, improve heat resistance, and minimize shrinkage. Lower-viscosity oils improve both resilience and low-temperature flexibility and tend to reduce compound viscosity more but are more fugitive.

Because of the very high filler loadings often employed in EPDM compounds, the oil performs several important functions:

Improves wetting and incorporation of fillers
Reduces power consumption of the mix equipment
Lowers batch temperature
Reduces the risk of compound scorch
Improves extrusion and all other shaping operations

As an extender, oil almost always lowers compound cost. Table 1.7 provides a guide to the selection of process oil type.

A contour map is a useful method of presenting rubber compounding data. It is similar to a line of constant elevation on a topographical map, representing a single value over the range of the grid coordinates. In rubber technology the coordinates are usually expressed in loading levels of fillers and plasticizers in the compound. The family of curves so generated is called a contour plot and can be used in two different ways.

1. To determine the general effect of the compounding variables on a property of interest
2. To select a combination of filler and plasticizer loading to produce a desired property value

A typical contour map for a polymer is shown in Figure 1.20. Rubber manufacturers will typically supply these processing maps for different grades of their polymers.

Table 1.7 Process Oil Types Typically Used

For:	Use:
Best physical properties	High-viscosity oils
Maximum heat resistance	High-viscosity paraffinic oils
Highest resilience and best low-temperature flex	Low-viscosity oils
Lowest compound viscosity	Low-viscosity oils
Peroxide cures	Only paraffinic oils

Principles and Methodology of Product Testing

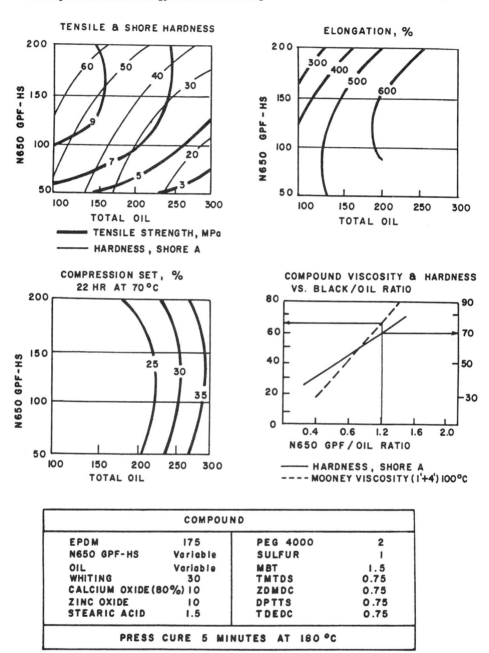

Figure 1.20 Construction and use of typical contour map for a particular polymer prepared in a sulfur cure compound.

Vulcanization is primarily a cross-linking process in which chemical bridges, characteristic of the cure system employed, form a three-dimensional network with the polymer chains. The chemical method of vulcanizing EPDM depends on the chemical structure of the polymer. Ethylene and propylene are the basic monomeric units of EPM/EPDM. The structures are more complex than described earlier in that we have not attempted to represent the probable distributions of ethylene and propylene units determined by the ratio of the monomers employed, the catalyst system, and the polymerization conditions. Among the commercial EPM/EPDM producers, many products are significantly different, despite commonality of Mooney viscosity or monomer ratio.

EPM copolymer is cross-linked exclusively via a free-radical mechanism, usually accomplished by the decomposition of organic peroxides; initaition with electron-beam radiation is also possible. The tertiary hydrogen on the EPM main chain is abstracted by a peroxy radical to form a polymer radical. This subsequently combines with adjacent radicals to complete the cross-link. In general, free-radical vulcanization is enhanced by polyfunctional coagents such as ethylene dimethacrylate, triallyl cyanurate, and the like. These reagents increase the cross-linking efficiency of the peroxide and minimize undesirable side reactions. However, the decomposition rate of the peroxide, which is a function of temperature is not altered. Acidic compounding ingredients such as stearic acid promote ionic breakdown of peroxides and should be avoided. Calcium stearate can be used in place of stearic acid with peroxide cures.

One undesirable side reaction with free-radical vulcanization is the competing chain breakdown at the tertiary hydrogen site. For this reason, higher ethylene grades of EPM, which have fewer such sites, are preferred for vulcanization whenever rheological or end-use considerations permit.

The peroxide cure system can also be used with EPDM; the principal vulcanization site is at or near the side-chain double bond on the diene. Reactivity rates are significantly greater than for EPM copolymers due to the presence of allylic hydrogen on the diene. Peroxides are used with EPDM when the ultimate in heat and compression set resistance is required. Generally, these are high-quality compounds in which the plasticizer level is minimized to avoid interference with the cure. If a higher plasticizer level is necessary, additional peroxide can be added to compensate. The peroxide decomposition rate, which is a function of temperature, is a key factor in cure development. The onset and development of cure are dependent on the rate of free-radical generation as well as the structure of a given peroxide.

Convenient sulfur vulcanization through the diene is used in the majority of EPDM applications. The type of cross-link is related to the specific cure system used. High-sulfur systems favor the formation of a polysulfidic bond. These vulcanizates offer high stress-strain properties and good flexibility, but thermal stability is marginal compared to systems generating monosulfide bonds.

Vulcanization rate is related to the amount and type of diene present in the polymer, as well as to the cure system. When properly boosted with ultra accelerators, a nominal 5-wt% ENB grade can be vulcanized in fast cure cycles. However, the amount of diene is not the only controlling factor: Narrow MWD polymers attain a higher cure rate and state than do broad MWD polymers. This is illustrated in Figure 1.21. The basic sulfur cure systems for EPDM consist of:

Cross-link agent: sulfur, sulfur donor
Primary accelerator: thiazole
Ultra accelerator: thiurum, dithiocarbomate

Compared to more highly unsaturated rubbers, very large amounts of thiuram and dithiocarbamates are used in EPDM; up to 2 to 3 phr of each can be used in some applications.

Bloom resistance is also a factor in the choice of a sulfur system. Bloom is a gradual migration to the surface of a partly insoluble byproduct of sulfur vulcanization, usually a zinc dialkyl dithiocarbamate. Bloom is especially sensitive in press cures; limited amounts of mixed accelerators give less bloom than equivalent amounts of a single accelerator. Polyetehylene glycol (e.g., PEG 3350) inhibits the formation of bloom by

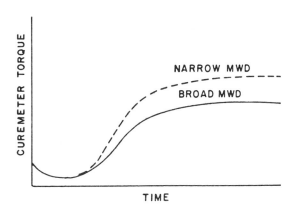

Figure 1.21 Effect of polymer MWD on cure state.

increasing the solubility of the accelerators. Some sulfur cure systems that bloom in press cures do not give bloom in steam cures.

There are a variety of methods used to achieve vulcanization. In *compression molding* the uncured stock is placed directly in the mold cavity. The mold is then assembled and placed in the hydraulic press. The press closes the mold and causes the stock to fill the cavity. When cure is effected, the mold is removed from the press, disassembled, and the rubber parts removed.

Transfer molding is a refinement of the compression molding process in which the mold is tightly closed before the cavity is filled. Stock is "transferred" from a holding pot to the cavity under pressure.

Injection molding is similar to transfer molding except that the stock is metered into the mold cavities through runners fed by an injection head. On some machines one injection head may serve several presses and molds. This higher degree of automation permits short, high-temperature cures with short change times.

Open steam in an autoclave has long been used to cure extruded and calendered products. With this method, the article may be in direct contact with the steam or wrapped with fabric tape.

Continuous vulcanization combines processing and curing steps into one continous operation on many extruded profiles. Traditionally, extruded profiles and hoses have been separately processed and vulcanized. During the processing step, the profile is extruded in a continuous operation, but the extrudate is accumulated for later batch vulcanization, in a steam autoclave, for example.

Basically, there are three stages to the continuous vulcanization process:

1. Formation of the profile
2. Heating to curing temperature
3. Curing

Heat may generally be inside the extrudate by dielectric heating (ultrahigh frequency, UHF), by friction (shear head), or by heat transfer from the outside [hot air, fluid bed, liquid curing media (LCM)]. Curing is achieved by maintaining the temperature for the time needed to cross-link the rubber fully.

Typically, some of these operations may be combined as shown in Figure 1.22. For example, shear head or UHF heating may be followed by one or more hot-air curing stages. Shear head extrusion followed by one-stage hot-air curing is an efficient method of continuous vulcanization.

Principles and Methodology of Product Testing 53

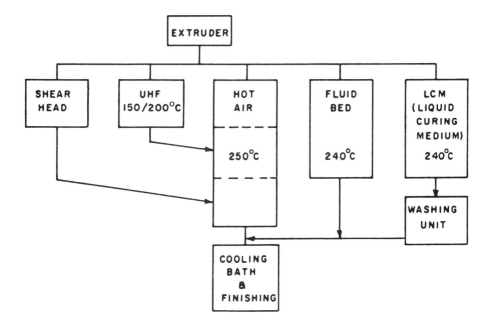

Figure 1.22 Continuous vulcanization process.

In a shear head, the compound can be heated from 150 to 220°C. Extrudability can be good, even for somewhat complex profiles, since the compound viscosity decreases with shear rate. Narrow-MWD Vistalon grades give higher heat buildup because of their lower dependence of viscosity on shear rate.

The general effects of major variations in carbon black and oil content are illustrated by the processing map in Figure 1.23. Intrinsic properties of the polymer determine shape and location of the "optimum processing zone" for any particular polymer. A typical processing map for a polymer compounded EPDM is shown in Figure 1.24. The boundaries are not strict limits but only indications of usual performance. Within the optimum processing envelope, simple compounds usually process well. Outside the envelope more compound development is frequently required for acceptable processability.

In the following paragraphs, the various processing operations used for compounding and forming rubber articles are described qualitatively.

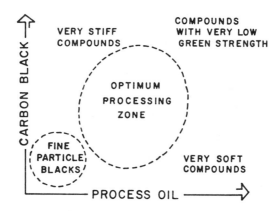

Figure 1.23 Features of a processing map.

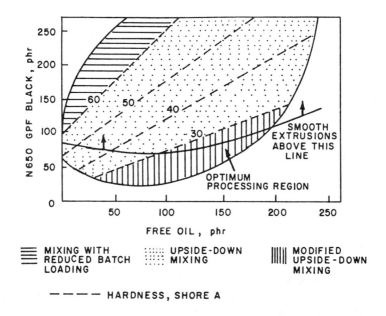

Figure 1.24 Typical processing map for a compounded polymer.

Detailed treatments of each of these process operations are given in the references cited at the end of this chapter.

The first operation of importance is mixing. The prime objective in mixing is to obtain a homogeneous dispersion of carbon black, oil, and polymer (i.e., a compound). There are various types of mixer configurations employed throughout the rubber industry. Cheremisinoff (1987) describes several of these. Figure 1.25 illustrates the most common internal mixer, called a Banbury.

With elastomers, the mixer should be charged with a 10 to 15% overload for the most rapid and efficient mixing. Most compounds having a cured hardness of 40 to 80 Shore A are thoroughly mixed after 3 or 4 minutes in a large internal mixer at low speed (20 rpm). For high loadings, the "upside-down" addition of ingredients to the mixer is generally most effective. This involves the addition of all the fillers and oil prior to the addition of polymer. However, soft sponge stocks, compounds with very high oil levels, and low-viscosity mineral-filled stocks should be mixed conventionally or "right side up."

Dump temperatures in the range 125 to 150°C are suggested for the N762 SAF and N550 FEF loaded stocks. Accelerated compounds or those containing blowing agents should be dumped at temperatures below 110°C. Highly loaded, high hardness, high Mooney compounds should be mixed with no overload and include 10-phr microcrystalline wax for the best dispersion. Often, these tough compounds require two-pass internal mixing to obtain optimum vulcanizate properties.

The term "incorporation" refers to the wetting of carbon black with rubber. During this operation, entrapped air is squeezed out between the voids of the rubber and carbon black particles. Early stages of mixing reveal that as the carbon black becomes incorporated, relatively large agglomerates (on the order of 10 to 100 µm) form. Cotten (1984, 1985) suggests that the time required for full carbon black incorporation can be determined by measuring the time required to reach the second power peak during a mixing cycle. Cotten (1985) performed mixing studies in a Banbury mixer, observing from microscopic examination of rubber/carbon black compounds the progression of black dispersion at different times. The rates of carbon black dispersion in this study were computed from the maximum torque data. A typical power curve generated by the author in reproducing Cotten's work is shown in Figure 1.26 using an oil-extended EPDM. Following Cotten, the rubber was first masticated for about 2 minutes and then the rotors were stopped. Carbon black was charged in the chute, the mixer ram inserted, and the rotors started again. Mixing times were measured from the instant when the rotors were restar-

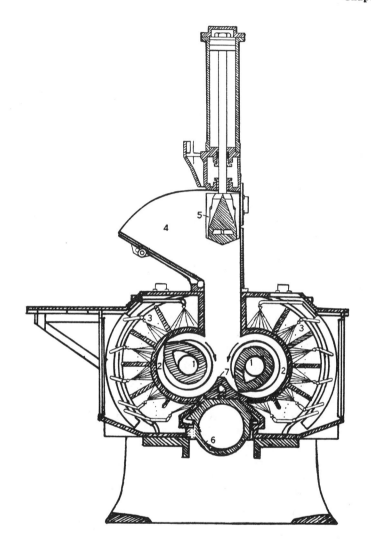

1. ROTORS OR AGITATORS
2. MIXING CHAMBER
3. COOLING SPRAYS
4. FEED HOPPER
5. FLOATING WEIGHT
6. SLIDING DISCHARGE
7. SADDLE DISCHARGE OPENING

Figure 1.25 Cutaway view of a Banbury mixer.

Principles and Methodology of Product Testing

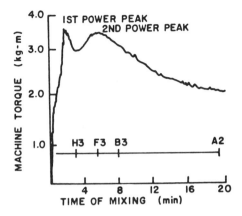

Figure 1.26 Typical torque response curve for rubber mixing.

ted. The carbon black incorporation time was taken to be the time required to attain the second power peak shown in Figure 1.26.

Bound rubber in studies such as this can be measured by standard solvent extraction techniques. For example, toluene at room temperature or boiling heptane or hexane extractions can be used with a known aliquot of rubber sample. From a measurement of the residue weight, one can calculate the percentage of insoluble polymer.

The percentage (by volume) of unincorporated (nonwetted) carbon black can be estimated from density measurements. The volume of air in the batch (V) at time t can be computed from the difference in densities (ρ' and ρ) at time t and the time at the termination of mixing when carbon black dispersion is better. The formula used for this calculation is

$$V = \frac{B}{\rho'} - \frac{B}{\rho}$$

where B is the batch weight (phr).

The first power peak corresponds to the ingestion of the batch into the mixing chamber. This coincides with the instant that the ram in the loading chute reaches the bottom of its flight and thus removes any additional hydrostatic pressure from the mixing chamber. Cotten has found that the fraction of undispersed carbon black decreases linearly with the time of mixing at the point where the second power peak occurs. At shorter mixing times the compound can appear inhomogeneous and crumbly. The mixing times can be observed to be inversely proportional to the rotor speed. When normalized, data tend to collapse to a single linear correlation. Cotten's work shows that there is a strong correlation between incorporation time and the rate of incorporation as computed from the slope of regression lines for change of densities with time for carbon black compounds in a single polymer.

The existence of a double power peak depends in part on the properties of the carbon black itself. Using pelletized carbon black will result in an absence of the first power peak because no additional hydrostatic pressure is applied when the mixer chute is lowered (hence the measured torque is no longer affected). The time to reach the second power peak, however, remains the same. Cotten notes that with fluffy carbon black, the time to reach the second power peak increases if no pressure is applied. Large agglomerates of carbon black form during the initial mixing stages regardless of the type of carbon black used.

Other studies by Turetsky et al. (1976) and Smith (1964) have shown good correlation between the measured black incorporation time (BIT) and weight-average molecular weight, M_w, of polymers. Incorporation times will increase sharply with increasing polymer molecular weight. This effect tends to mask expected decreases in incorporation times when the Mooney viscosity is decreased through the addition of oil. Decreasing concentrations of carbon black and increasing oil loading tend to reduce incorporation time. This is generally thought to be related to a lowering of the polymer viscosity.

Typical filler-reinforced elastomers contain curatives, plasticizers, and stabilizers in the rubber filler. After mixing and curing a masterbatch, both physical and chemical cross-linking processes transform the system into a network structure. The structure is composed essentially of two networks: a stationary and a transient network. The stationary network is comprised of chemical cross-links that are formed by both curatives and fillers. This network also includes permanent crystalline structures from crystallizable polymers. In contrast, the transient network is comprised of trapped entanglements and transient ordered structures. Payne and Watson (1965) and Lee (1985) have proposed two mechanisms that explain the reinforcing effects of filler in elastomers. The first of these mechanisms can be described as a hydrodynamic interaction in which the filler particles are responsible for the reinforcement. The second mechanism is that of strong chemical interactions between the filler and the matrix. Both of these reinforcing interactions contribute to the mechanical properties of filler-reinforced polymers. As noted earlier, the mechanical properties of engineering importance are the tensile modulus, the tensile strength, and the ultimate stretch ratio. Tensile modulus is mathematically defined as the slope of the stress-strain curve at zero strain for a given masterbatch. Tensile strength is defined as the force at break per

Principles and Methodology of Product Testing

unit area of the original sample. The ultimate stretch ratio is the uniaxially fractured gauge length divided by the original gauge length of the filler-reinforced polymer specimen.

The milling of properly compounded rubber stocks can easily be accomplished on cool or warm roll mills. Hot stock freshly dumped from the internal mixer or cool slabs will band immediately and stay on the cooler roll. Sheet-off operations are fast and trouble-free. Mill roll sticking of soft stocks can be avoided by warming the mill rolls slightly. Figure 1.27 illustrates the operation of milling.

Low-to-intermediate Mooney and oil-extended polymers can be mixed on an open mill. The polymer has an initial tendency to resist banding, which is overcome after a few passes through a tight nip. Once a band is formed, the stearic acid, zinc oxide, and filler should be added as rapidly as the polymer will incorporate these ingredients, without concern about large amounts of unmixed filler on top of the mill. Process oil should be added with the carbon black to avoid mill surface lubrication and loss of bandage.

Often, to save time, a manually prepared premix of black and oil may be made. Such a premix consists of all the oil plus sufficient black to make a sort of dry paste—usually about equal weights of black and oil. This premix may not be necessary if a mill apron is used. Leaving a small amount of stock on the mill, or "seeding" the next batch to be mixed, can also save

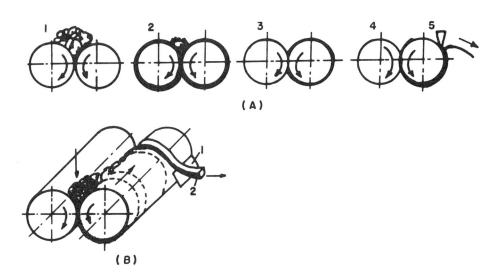

Figure 1.27 Milling and mill mixing.

mixing time. During mixing the batch may flip to the faster roll. Since a temperature differential is maintained between the rolls, the batch should return to the cooler, slower roll when the addition of compounding ingredients is complete.

The operation of extrusion is discussed in several chapters in this volume. So only some general comments are made here. An extruder is a standard piece of equipment used in rubber and plastic processing for such operation as injection and blow molding, for processing thermosetting resins, and for the hybrid process of injection blow molding. The apparatus is fed room-temperature resin in the form of beads, pellets, or powders, or if rubbers are being processed, the feed material may be in the form of partuculates or strips. The unit converts the feedstock into a molten polymer at sufficiently high pressure to enable the highly viscous melt to be forced through a nozzle into the mold cavity (injection-type molding) or through a die (e.g., blow molding or continuous extrusion of articles). In the initial portion of the extruder the polymer is conveyed along the extruder barrel and is compressed. The material is then heated until soft, eventually reaching a molten state. As fresh feed material enters, heat transfer takes place between the molten fluid and solid polymer. Once in the molten state, the extruder acts like a pump, transferring the molten polymer through the extruder channel, building up pressure prior to flow through the discharge nozzle or die. The principal components of a single-screw extruder are illustrated in Figure 1.28. The machine has a motor drive, a gear train, and a screw that is keyed into the gear-reducing train. The fluid layers between the screw flights and barrel wall maintain the screw balanced and centered.

Units are equipped with continuous variable speeds and barrels are usually electrically heated, usually by band heaters with either "on-off" or proportional control. The barrel can be "zoned according to the number of controllers on the heater bands. Depending on the application and type of service, the screw may be cored to allow for heating or cooling.

A die can be attached at the end of the extruder. Die designs range from very simple geometries such as an annulus for pipe and tubing profiles, to very complex faces such as the rubber seals used as glass run channels around the windows of automobiles.

Older extruders used in manufacturing various consumer articles are equipped with minimal instrumentation. Standard instrumentation usually consists of a pressure gauge at one point along the barrel (usually at the head) and a thermocouple in the hot-melt region. In noncritical operations, the operator will monitor pressure, temperature, and screw speed. Mass flow rates are typically monitored by the sample weight-time method.

Principles and Methodology of Product Testing

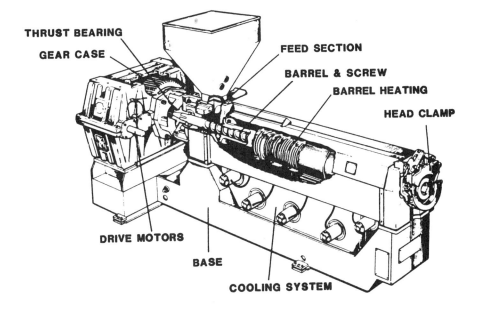

Figure 1.28 Features of a single-screw extruder.

In operations requiring close tolerance on extruded articles, a greater degree of instrumentation is employed. Usually, several pressure transducers and thermocouples along the barrel are used to ensure uniform extrusion and to control barrel and stock temperatures. Some designs may include thermocouples on the screw to monitor and control conveying flights.

There are numerous types of extruders, the most common of which is screw extrusion, which can be of the single or twin type. Another common type is the plunger or ram-type extruder. Materials such as TFE Teflon and ultrahigh-molecular-weight polyethylene are normally handled by ram extrusion. The melting temperatures of these polymers are very high; hence these materials cannot be pumped readily as in screw extrusion.

Many commercial extruders plasticate and pump materials in the range 10 to 15 lb/hr-hp. However, pumping capacity is a relative quantity that depends on the material. In adiabatic mixing, machine capacities can be as low as 3 to 5 lb/hr-hp. Also, a machine that handles a thermoplastic elastomer could show as much as three- to four-fold increase in mass throughput when switched to a low-melt-viscosity material such as nylon.

Screw configuration depends to a large extent on the properties of the material being processed. Figure 1.29 illustrates several common arrangements. A constant-pitch metering screw is usually employed in applications not requiring intensive mixing. Where mixing is important, say for color dispersion, a two-stage screw equipped with a let-down zone in

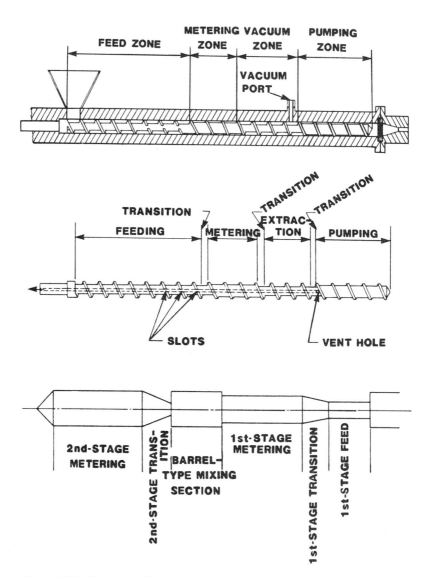

Figure 1.29 Screw profiles and cross section of vented barrel extruder.

the center of the screw is appropriate. Turbulence promoters can also be included at or near the tip of the screw. In some applications with two-stage screws, venting at the let-down section may be needed.

The screw section immediately ahead of the gear train acts as a solids metering or feed zone. It is characterized by a deep channel between the root of the screw and the barrel wall. The plasticating zone is ahead of the solids metering zone. This is a transition region where the channel narrows. The purpose of this zone is to provide intense friction between solids and a region for melting of the polymer to take place. Near the tip of the screw is the melt metering zone, where pressure builds up. In this region the polymer melt is essentially homogenized and raised to the proper temperature for extrusion of the article.

The action of an extruder is analogous to that of a positive-displacement pump. The flight depth along the screw (i.e., the ratio of the solids metering channel depth to the melt pump channel depth) is known as the compression ratio. The purpose of screw flights is to enable the screw to transport polymer down the barrel. The pitch angle of the flights again depends on the type of material handled. Many elastomer applications employ a general-purpose screw that is of a constant pitch (i.e., flight equals the diameter). The pitch angle of this single-flighted screw is usually 17.61°. Typical extruder specifications are as follows:

Compression ratio: 2:1 to 6:1 (for materials ranging from LDPE to some nylons)
Pitch angle: 12 to 20°
Length to diameter (L/D): 16:1 to 36:1 (low 40's typically for easily melting/ flowing polymers requiring high mixing and venting)
Extrusion pressures: 10,000 to 300,000 psi

For extrusion pressures, low ratings ($\leqslant 100,000$ psi) are usually sufficient for many thermoplastic materials. The upper limit typifies FEP Teflon.

The lead on a screw is defined as the distance between the flights. As an approximation, it is equal to the ID of the barrel for a single-flighted screw. The radial clearance between the flight tip and the barrel is tight (usually 0.001 in./in. of barrel ID). The reason for such a tight clearance is that if the gap is too great, the material may flow back along the barrel, resulting in a loss of melt pumping zone capacity.

The initial region of the extruder plays an important role in the machine's overall operation. It consists of a hopper or feed arrangement and a solid feedstock conveying region. Its purpose is to transfer the cold polymer feed from the feed hopper into the barrel, where it is initially compressed. This compression forces air out between the interstices of resin pellets or rubber chunks (air being expelled back through the hopper) and breaks up lumps and polymer agglomerates. This action creates a more homogeneous feedstock that can readily be melted.

There are a variety of twin-screw extruder designs employed throughout the polymer industry, with each type having distinct operating principles and applications in processing. Designs can be generally categorized as corotating and counterortating twin-screw extruders. Eise et al. (1981) have classified twin-screw extruders in terms of the mechanisms of operation. Molding operations are the next group of processing methods, generally divided into three techniques: compression, transfer, and injection. Each case requires a different compounding approach to achieve efficient mold fill. Although all EPDMs may be used in these molding operations, the low Mooney fast-curing (high and very high diene) EPs are preferred in the more sophisticated transfer and injection processes. References at the end of the chapter provide detailed discussions on the various topics overviewed.

PHILOSOPHY AND GENERAL CONCEPTS OF PRODUCT DEVELOPMENT

The current environment of intense international competition among polymer manufacturers requires that an organization be capable of developing quality products quickly and efficiently. For this reason it is worthwhile for companies to make a critical review of the product development process to assess the areas of needed improvement. In this section, the author's views on the general philosophy and key elements of an efficient new products development process are highlighted.

In many organizations, the development process is performed by a series of specialists who contribute to the product development effort in sequence, passing it from one function to the next. Each functional group makes its major contributions to the product only once, and there is little dialogue between them.

In contrast, the parallel approach brings together everyone responsible for any aspect of the project at the beginning, in the same room, to cooperate on the development strategy. Furthermore, it stresses their continued interaction throughout the process. Having technology (product/process/finishing group members), marketing, manufacturing, and quality assurance personnel work together accomplishes several things. First, it helps departments with contradictory goals understand each other's concerns and align with each other's goals, and it brings the people responsible for the project closer, ultimately inspiring cooperation. Second, it generates a larger number and wider diversity of ideas. Through this process, the product development engineer can start to understand the constraints of the manufacturing engineer, who must control costs, and that of the marketer, who must introduce a new product in an expedient manner. In the parallel system, everyone's opinion is heard on all aspects of the project because everyone stays with the project from beginning to end. Examples of programs containing elements of the series and parallel approaches are shown in Figures 1.30 and 1.31. The parallel approach essentially brings together a broad range of talent, trying things out, learning from each attempt, and then trying again. Through a recycling and interactive approach, it enables more reliable products to be developed in a shorter time.

Management's strategic role in new product development is to provide the initial kickoff to the development process by signaling a broad strategic direction or goal for the company. Ideally, it should not restrict creativity by defining a specific work plan, but rather allow room for discretion and local autonomy to those in charge of the development project. Management should decide on broad strategic directions or goals by constantly monitoring the external environment (i.e., competitive threats and market opportunities and in evaluating company strengths and weaknesses).

Management can implant a certain degree of tension within the project team by giving it a wide degree of freedom in carrying out a project of strategic importance and by setting challenging parameters. The creation of controlled tension helps to cultivate a "must-do" attitude and a sense of cohesion among members of the project team.

One appraoch to new product development is to establish program task forces consisting of members with diverse backgrounds and temperaments. In theory, each team should be given unconditional backing from top management so that it may begin to operate like a corporate entrepreneur and engage in strategic initiatives. This type of team approach is well suited to serve as a motor because of its visibility, its authority over the direction of development, and its sense of mission.

Three elements or qualifications emerge from task forces. First, the group is autonomous. This autonomy enables the second desirable element to emerge—the group comes up with extremely challenging goals. It does not seem to be content with incremental improvements alone and is in constant pursuit of a quantum leap. Third, the group is composed of members of diverse functional specializations, thought processes, and behavioral patterns. The total becomes more than the sum of its parts when members assemble and interact with each other. Variety is amplified and new ideas are generated as a result. In essence, cross-fertilization takes place.

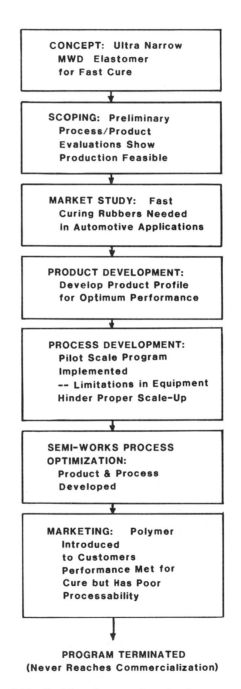

Figure 1.30 Serial path to product development.

Principles and Methodology of Product Testing

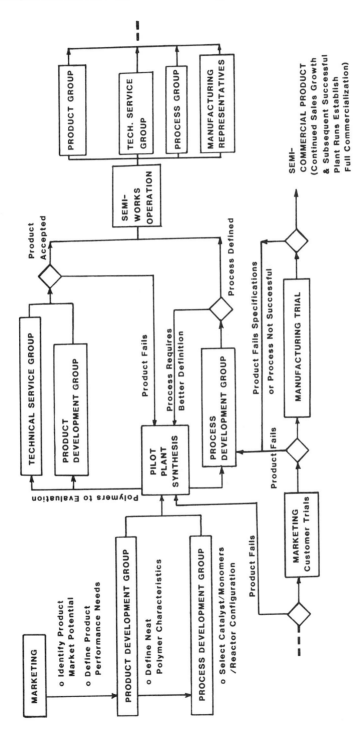

Figure 1.31 Example of a product development program having elements applied.

The group dynamics of the task force team approach influence the manner in which the development project proceeds. The autonomous, cross-fertilization nature of the group produces a unique set of dynamics; for example:

Cohesion is promoted as team members face challenging objectives and goals.
Some ambiguity is tolerated since there are diverse backgrounds in the group.
Overspecification is avoided since it may impede creativity.
Sharing of information is encouraged.
Decision making is intentionally delayed to extract and assemble as much up-to-date information as possible from the marketplace and various technical departments.
Sharing of responsibility is accepted as the group embarks on a risk-taking mission.

The end result is an iterative process of experimenting even at a very late phase of the development process. This leads to some redundancy or repetition of effort, which in fact is good from the overall quality and assurance of final recommendations of the product. This style is in fact closer to the Japanese system of product development. In contrast, a NASA-type program is more sequential in nature, designed to have distinct and separate program phases. A comparison of the approaches is illustrated in Figure 1.32.

Management must take the responsibility of implanting the seeds of control in the team approach by selecting the right personnel for the project team. Further, it should not walk away and expect the team to return with a new product after a certain period. Instead, it should monitor the

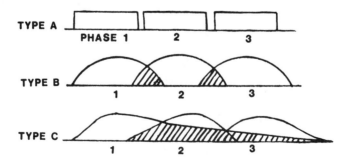

Figure 1.32 Sequential (A) versus overlapping (B and C) phases of development.

Principles and Methodology of Product Testing

balance in team membership, and add or delete specific members if deemed necessary. Also, team members should be encouraged to extract as much information from the field—from customers and competitors—and to bounce them off other members. This sharing of information helps to keep everyone up to date and to build cohesion within the group.

Many R&D organizations in fact have elements of the general methodology outlined above already in their product development programs. Often, however, the approach and methodology are not applied to every project, resulting in inefficiencies in terms of the overall product development effort. Figure 1.33 illustrates by way of a fishbone chart the various aspects of product development along with the factors that can affect the overall success. This logic diagram outlines a sequential approach; however, as illustrated earlier, a parallel process can in fact be adopted along each phase of the development effort. Key elements that emerge in the process are:

1. *Problem types.* All problems and programs fall into three generic categories:
   a. *Present market pull:* short term need for a new product or grade either to match competition and/or capture a market opportunity.
   b. *Crisis problems:* analytical technical servicing and quality control problems requiring technology support, most often customer specific and market pull in nature. In this category the plant can become a customer; longer-term development programs can be based on "market pull" or "product push" criteria.
2. *Focal point leadership.* Essential to any program is a central figure of authority or organization such as a product planner(s) who serves as the focal point for all aspects of the program execution. Key elements that must be addressed and coordinated by a market planner are:
   a. Establish the basis for a product's market potential.
   b. Provide ownership of a program, once it has been initiated. By this it is meant that the market sectors must be able to demonstrate the need for a new product development program in terms of adequate venture analysis. Upon sponsorship of a program through management, the particular market sector becomes the customer and technology provides the service of development.
   c. Organize task force membership for product development (i.e., the present mechanism of small teams of persons selected for problem solving is adequate). Elements from manufacturing, technology, and marketing should be included.
   d. Manufacturing must be given sufficient lead time to participate, react, scope-out, and implement necessary process modifications. Input early in a program could prevent effort being expended on an impractical and/or cost-ineffective approach.
   e. Clearly defined time schedules and sequence of development

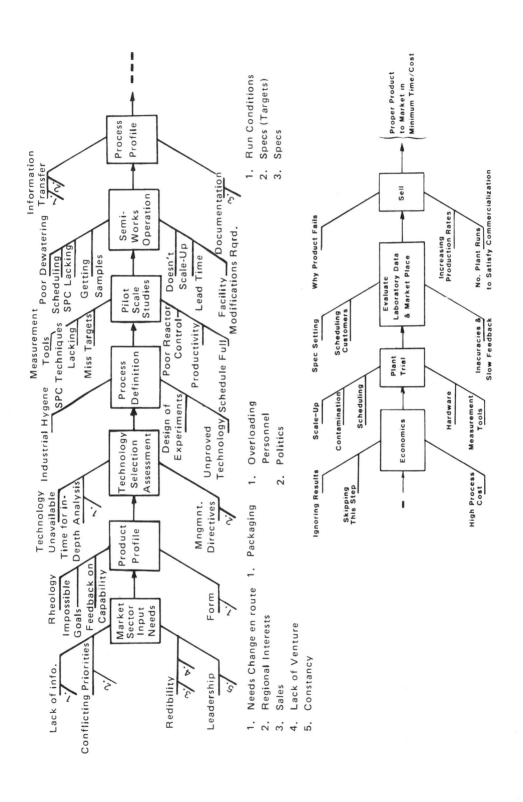

Principles and Methodology of Product Testing

steps should be outlined at inception and updated routinely. PERT diagrams are a useful tool.

The project planner can be fulfilled by several persons in any one program. Logically, the responsibility should lie with an appointed member of the market sector who is given the assignment of task force leader.

Criteria for Success

Establishing a yardstick for what is and is not a successful program is essential not only from the standpoint of individual product development activities, but in the evolution of R&D philosophy as well. The elements contributing to success are sales volume, plant productivity and process-product reproducibility, margin and profitability, and the degree of process robustness. The criteria for success in a general sense can be defined as the ability to *provide the proper product to the marketplace within a minimum time frame*. The following are specific yardsticks or criteria that should be tested for each product development program.

1. Manufacturing achieves success during its first pass.
2. A robust process and product is provided.
3. The product meets sales projections and satisfies market needs.
4. Product and process are optimized at the time of sales (i.e., the product and process must be capable of achieving production rates and economics).
5. Both the polymerizing and finishing departments are capable and comfortable in producing the product.
6. Time schedules are met.
7. Results are communicated in a timely fashion between regions to enable grade interchangeability. Documentation should be aimed to both manufacturing sites from a process and product quality standpoint and to all marketing sectors, so that other potential marketing opportunities can be identified from a particular product development.

Effective Prioritization

For product development to be efficiently implemented within the guidelines of success, the organization must be capable of making decisions concerning prioritization of programs. To make such decisions, manage-

---

Figure 1.33 Fishbone chart showing phases of product development from inception to commercialization, together with factors that could affect success.

ment must be able to quantify or rate the potential value of each product development proposal. This should be derived from a proper material balance based on existing resources; that is:

[product value (present and future)] = [market value] − [investment]
          (I)   (II)   (III)

[actual effort expended] = [total resources available] − [effort required
       (i)   (ii)   for success]
                                                                   (iii)

The actual data required to provide material balances that enable prioritization to meet market strategy are as follows:

Marketing venture analysis:

Volume/price/profitability
Timing to capture market potential
Competitive status (both in terms of competing manufacturers and potential for material (other products, not just the products made in the company); displacement for market application
Customer commitment

Technology input:

Probability of success
Resources required
Availability of resources
Practicality of timing in relation to existing work load

Manufacturing input:

Potential capital investment
Resources required
Lead time required
Resources (manpower) available

   To assist in assessing the potential value of a new product development effort, early interaction between marketing, technology, and in some cases manufacturing is encouraged. The use of venture analyses, particularly at the initiation of a program is important. Table 1.8 provides an illustrative form that could be used as a basis for assessing whether R&D efforts should be expanded for a particular program. This form can be prepared by various marketing sectors with assistance from technology and reviewed by management, as to the potential value of a product development program.

Principles and Methodology of Product Testing

Table 1.8 New Grade Economic Potential Assessment

I. *Market assessment*

    Industry potential _____ Mlb in 5 years _____ Mlb now

    This grade (Vistalon)

        Potential _____ Mlb in 5 years

    Other Vistalon and competitive grades used in the specific application now

Grade	Approximate volume	Approximate price

    Basis for expected penetration of this grade to the volume level shown above

    _____

    _____

    Projected price for this grade _____ ¢/lb (1987¢)

    Sources of value over competitive materials _____

II. *Cost assessment*   Current prod. rate   Current total cost*

    Most similar Vistalon existing grade ____   ____ klb/D   ____ ¢/lb

    New grade

        Approx. expected diene % _____ oil phr _____

        Friable bale _____ Dense bale _____ New facilities needed _____ k$

        Expected production rate after 4 runs _____ klb/D

        Reasons for unusual operations/costs _____

    Expected Δ total cost from most similar Vistalon grade   ____ ¢/lb

        Variable expected lined-out cost   ____ ¢/lb

        Total expected lined-out costs[a]   ____ ¢/lb

III. *Margin (lined out)*   Total cost   Variable cost

    Projected price   ____ ¢/lb   ____ ¢/lb

    Projected cost   ____ ¢/lb   ____ ¢/lb

    Margin   ____ ¢/lb   ____ ¢/lb

IV. *Cost to develop the new grade*

    Research/CPU   ____ k$

    VSU run   _50_ k$

    Plant run costs   ____ k$ 15% off grade at 20¢/lb realization + 10% scrap first run

Plant investment _____ k$
ATS/introductory costs _____ k$
Total development costs _____ k$
Biggest risks/uncertainties in development _____
_____
_____

V. *Economics*

Sales volume in 5 years _____ M/lb (from I above)
× Total margin _____ ¢/lb (from III above)
Projected IBIT _____ M$

$$\frac{\text{Projected IBIT}}{\text{Total development costs}} = \frac{M\$}{M\$} = \begin{matrix} 1.0 = \\ 0.6 = \\ 0.3 = -15\% \text{ DCF} \\ 0.2 = \\ 0.1 = \end{matrix}$$

Any key choices with major economic impact?
Properties versus costs? _____
Properties versus quality? _____
Targets versus time to develop? _____

[a]Should include nonmanufacturing costs.

The development of new products can be organized into five states of flow: an exploratory research phase, feasibility research, development, market test phase, and a customer introduction phase. Each of these phases is recommended to contain parallel elements of phase development. The key subelements in each phase are outlined in Table 1.9.

Table 1.9 States of Flow in New Product Development

Activity	Suggested Responsibility
Exploratory Research Phase	
1. Establishment of corporate and business group objectives	Sector managers and product VPs
2. Continuous surveillance of marketing situation	Sector managers
3. Analysis of corporate and product divisions' strengths and resources	Product VPs
4. Characterization of the business of the industry sector and product divisions	Sector managers, marketers, product VPs
5. Specifications of criteria for new product fields	Regional sectors and product groups
6. Generation of product and market ideas	Everyone
7. Screening, selection, and preliminary validation of new product and market ideas	Regional product management
8. Feasibility authorization; decision point	Regional product management
Feasibility Research Phase	

1. Market research
  a. Market feasibility of the concept
  b. Characteristics of market, its size and trends
  c. Nature of competition
  d. Strategy of product placement on form, price, quality
	Industry sectors' marketing specialists or consultants or product network

2. Experimental technical research
  a. Establishment of performance specifications
  b. Design studies of basis technical alternatives
  c. Feasibility of manufacturing
  d. Estimate of development time and costs
	Producer and process technology groups

(continued)

Table 1.9 *(continued)*

	Suggested Responsibility

3. Analysis and integration of findings
   a. Time
   b. Costs
   c. Personnel required
   d. Commercial potential
	Product management committee

4. Development project authorization	Industry sector and product group management

Development Phase

1. Technical development
   a. Leading to product prototype
   b. Laboratory and use testing
   c. Preliminary design finalization
	Product technology group

2. Production costing and planning
   a. Materials
   b. Labor
   c. Equipment and space
	Product management and product manufacturing

3. Market evaluation
   a. Product prototype
   b. Marketing mix plans
	Management and industry sectors

4. Market forecasting
   a. Demand analysis
   b. Cost analysis
   c. Price analysis
   d. Break-even analysis
	Management and industry sectors

5. Management review; decision point	Polymers group product management

Market Test Phase

1. Planning
   a. Selection of geographic area and accounts
   b. Complete schedules and budgets
   c. Establishment of standards to test performance
	Product management and industry sectors

2. Experimental production for market test	Product management and manufacturing

3. Final production planning	Product management, product manufacturing
4. Execution of market test	Industry sectors

5. Analysis and review
  a. Product modifications/ marketing modifications
  b. Package modifications
  c. Price modifications
  d. General performance of test against forecast and standards
	Industry sectors

6. Final plans for commercialization
  a. Budget
  b. Critical actions, responsibilities, checkpoints
	Venture assessment, product management

7. Final management review; decision to launch	Polymer group management

Customer Introduction Phase

1. Prepare inventory
  a. Production run
  b. Statistical analysis of production for suitability
  c. New product program introduced to all sales personnel
  d. Training of salespeople and order personnel
  e. Distribution of product line to outlets
  f. Public showing and exhibits and announcement to trade
  g. Distribution of promotional materials
	Manufacturing and regional product network and industry sectors

2. Launch
  a. First calls, customer programs launched
  b. Filling of initial orders, technical service
  c. Field reports and information feedback
	Industry sectors and product groups

3. Commercialization
  a. Absorption by established functions
  b. Product 3-year plan
	Existing organizations

NOTATION

A	frequency factor in Arrhenius equation
B	batch weight
c	concentration
D_a	surface-average particle diameter
D_n	arithmetic-average particle size
d	diameter
d_c	capillary diameter
d_f	plunger diameter
E	activation energy
F	force
G	torque
K	power law consistency index
$K'\ K''$	zero-shear viscosity coefficients
l	capillary length
M	molecular weight
M_c	critical concentration molecular weight
n	power law exponent
n_p	number of particles
R	universal gas law constant
r	radius
s	shear stress
T	absolute temperature
T_g	glass transition temperature
T_m	melting temperature
t	time
V_f	free volume
V_0	volume at absolute zero of temperature
v	velocity
Y	yield stress

Greek Symbols

$\dot{\gamma}$	shear rate
η	visocisty
η_a	apparent viscosity
η_0	Newtonian plateau viscosity
η_∞	constant viscosity at high shear
ρ	density
Ω	angular velocity

REFERENCES

Andrews, E. H., and A. J. Walsh, *J. Polym. Sci., 33,* 39 (1958).
Bagley, E. B., *J. Appl. Phys., 28,* 624 (1954).
Beatty, J. R., *Rubber Chem. Technol., 37,* .1341 (1964).
Boonstra, B. B., in *Rubber Technology,* M. Morton, ed., Litton, New York, 1973.
Cheremisinoff, N. P., *Polymer Mixing and Extrusion Technology,* Marcel Dekker, New York, 1987.
Chinai, S. N., and W. C. Schneider, *J. Polym. Sci., A3,* 1359-1371 (1965).
Corish, P. J., *Rubber Chem. Technol., 44*(3), 814 (1971).
Corish, P. J., and B. D. W. Powell, *Rubber Chem. Technol., 47*(3), 481 (1974).
Cotten, G. R., *Rubber Chem. Technol., 57,* 118 (1984).
Cotten, G. R., paper presented at the meeting of the Rubber Division, American Chemical Society, Los Angeles, Apr. 23-26, 1985.
Doolittle, A. K., *J. Appl. Phys., 22,* 1471-1475 (1951).
Doolittle, A. K., *J. Appl. Phys., 23,* 236-239 (1952).
Eise, K., M. Werner, H. Herrmann, and U. Burkhardt, in *Advances in Plastics Technology,* Vol. 1, No. 2, Van Nostrand Reinhold, New York, 1981.
Fox, T. G., S. Gratch, and S. Loshaek, Chapter 12 in *Rheology: Theory and Applications,* Vol. 1, F. R. Eirich, ed., Academic Press, New York; 1956.
Fujita, H., and A. Kishimoto, *J. Chem. Phys., 34,* 393-398 (1961).
Fujita, H., and E. Maekawa, *J. Phys. Chem., 66,* 1053-1058 (1962).
Hayahara, T., and S. Takao, *Kolloid Z. Z. Polym., 225,* 106-111 (1967).
Leaderman, H., Chapter 1 in *Rheology: Theory and Applications,* Vol. 2, F. R. Eirich, ed., Academic Press, New York, 1958.
Lee, M. C. H., *J. Appl. Polym. Sci., 29,* 499 (1984).
Mendelson, R. A., pp. 587-620 in *Encyclopedia of Polymer Science and Technology,* Vol. 8, H. F. Mark, N. G. Gaylord, and N. M. Bikales, eds., Wiley, New York, 1968.
Metzger, A. P., and J. R. Knox, *Trans. Soc. Rheol., 9,* 13-25 (1965).
Payne, A. R., and W. F. Watson, *Rubber Chem. Technol., 36,* 147 (1963).
Pezzin, G., and N. Gligo, *J. Appl. Polym. Sci., 10,* 1-19 (1966).
Shenoy, A. V., Chapter 10 in *Encyclopedia of Fluid Mechanics,* Vol. 7, N. P. Cheremisinoff, ed., Gulf Publishing Co., Houston, Tex., 1988.
Smith, B. R., *Rubber Chem. Technol., 49,* 278 (1964).
Tobolsky, A. V., Chapter 2 in *Rheology: Theory and Applications,* Vol. 2, F. R. Eirich, ed., Academic Press, New York, 1958.
Tobolsky, A. V., A. Mercurio, and K. Murakami, *J. Colloid Sci., 13,* 196-197 (1958).
Turetsky, S. B., P. R. Van Bushirk, and P. F. Gurberg, *Rubber Chem. Technol., 49,* 1 (1976).
Van Wazer, J. R., J. W. Lyons, K. Y. Kim, and R. E. Colwell, *Viscosity and Flow Measurement: A Laboratory Handbook of Rheology,* Wiley, New York, 1963.
Walters, M. H., and D. N. Keyte, *Trans. IRI, 38,* 40 (1962).

2
Characterization of Molecular Weight and Molecular Weight Distribution

INTRODUCTION

Methods of measuring the molecular weight and molecular weight distribution of polymers are reviewed. As the result of random processes at some stage in polymerization, all synthetic polymers, and all naturally occurring ones except perhaps a few of biological interest, have a distribution of molecular weights. Figure 2.1 illustrates the typical features of a molecular weight distribution (MWD) curve. Shown are the approximate positions of several important average molecular weights. These averages are discussed throughout the chapter. It should be noted that the z-average molecular weight is directly accessible only from ultracentrifuge measurements, which are not discussed here. Like all the other averages, however, it can be calculated from fractionation data. Also, the value of the viscosity-average molecular weight and its relation to the other modes vary with the polymer-solvent interaction forces in the particular measurement in which it is determined.

The significance of the absolute and relative values of the various average molecular weights must be viewed in relation to polymer properties. The properties of strength, toughness, cure characteristics, and low sensitivity to chemical attack characteristic of polymers as a class of materials are typically not well pronounced until a molecular weight of around 10,000 is reached. Second, it is possible to obtain information on the

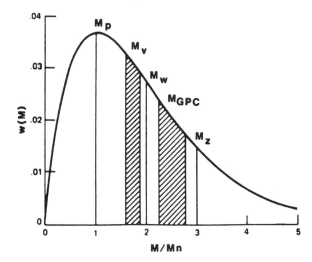

Figure 2.1 Distribution of molecular weights in a typical polymer showing the positions of important averages.

breadth of the molecular weight distribution from the relative values of two different averages.

Because of their accessibility, the ratio of the weight average, $\overline{M}_w$, to the number average, $\overline{M}_n$, is often used. The various averages may be highly sensitive to presence of a small fraction of material in an extreme range of molecular weight. Specifically, $\overline{M}_n$ is highly sensitive to the presence of a small number fraction of low-molecular-weight material such as oil or low-molecular-weight species referred to as oligomers; and $\overline{M}_w$ is similarly sensitive to small amounts by weight of high-molecular-weight polymer. These sensitivities can become important limitations in some instances.

With random-coil polymers, it is necessary to distinguish between their size, or dimensions, and their mass. The two are not necessarily simply related, and their relative magnitudes can change with the structure of the molecule as well as with experimental variables such as polymer-solvent interaction forces. It is expedient to distinguish between experiments measuring size, which may be correlated empirically with molecular weight in restricted circumstances, and those yielding directly one of the average molecular weights.

To obtain information about the molecular weight distribution, it is necessary to separate the molecular species in a sample by a fractionation process, and to determine the amounts and molecular weights of all the fractions. From these data the distribution of molecular weights can be computed. The separation and determination steps may be carried out independently, or be combined, as in a chromatographic process with suitable calibration such as gel permeation chromatography (GPC) or in an ultracentrifuge. The separation may result from the molecular weight dependence of solubility, as in classical fractionation; of molecular size, as in GPC; or in rare cases, of some other property. Interpreting the results of fractionation experiments directly in terms of molecular weight can be difficult unless it can be demonstrated that the separation step is independent of other variables that might alter the relationship between the driving force for separation and molecular weight.

DETERMINATION OF NUMBER-AVERAGE MOLECULAR WEIGHT

The number-average molecular weight can be determined from one of two types of measurements: chemical or physical methods based on endgroup analysis, and those based on measurement of one of the colligative properties: vapor-pressure lowering, freezing-point depression (cryoscopy), boiling-point elevation (ebulliometry), and osmotic pressure.

In end-group analysis, the chemical reactions used to count chain ends are either identical with or closely related to those utilized to follow the course of polymerization. These and some of the physical methods that can be applied are discussed in standard polymer chemistry texts. The major requirements for the application of the method are that it be known that the polymer is linear, that is, each molecule has only two ends; whether the groups to be determined occupy both ends or only one end of the chain; and that all of the end groups are accounted for. The calculations relating the count of end groups to the molecular weight are straightforward.

Among the colligative properties, the osmotic pressure provides the largest and most precisely measured effect. Limitations of available osmotic members are such, however, that this technique (referred to as membrane osmometry) is accurate only for polymers with rather high molecular weight. To cover the range of interest, another method must be used for samples with low molecular weights. The traditional technique used is vapor-phase osmometry, in which vapor-pressure lowering is measured indirectly. Low-angle laser light scanning (LALLS) is discussed later.

It may also be recalled that the number-average molecular weight $\overline{M}_n$ is the simple counting average in which the mass of the sample, expressed in atomic mass units or dalts, is divided by the number of molecules it contains: $\overline{M}_n = w/N$. Expressing N as the sum over all species of the number N_i of molecules of the ith kind, and w similarly as w_i, where $w_i = N_iM_i$, the defining equation for the number-average molecular weight is

$$\overline{M}_n = \frac{\sum N_i M_i}{\sum N_i} \qquad (1)$$

Membrane osmometry is one of the few techniques with which values of $\overline{M}_n$ higher than a few tens of thousands can reliably be measured. The upper limit of the method, corresponding to the smallest osmotic height that can be measured, is in the range 500,000 to 1,000,000. The lower limit is set by the permeability of the available membranes, as described below, at a rather indefinite 10,000 to 50,000, depending on the sample and membrane.

There are a variety of materials that can be used as the semipermeable barrier in membrane osmometry, the most popular being modern cellulosic membranes. One of the most widely used of these materials is gel cellophane, the modifier indicating material taken from the production line in a water-soaked condition, free of plasticizer and other additives. Depending on the speed of equilibration and retentiveness to low-molecular-weight species desired, it may never be allowed to dry, or may be dried and resoaked. Other commercially available membranes include cellulose regenerated by the deacetylation of acetylcellulose, and bacterial celluloses.

The membrane is usually obtained water-wet, and must be conditioned if it is to be used in an organic solvent. In this procedure, the water is replaced by the solvent to be used; if the latter is not miscible with water, the conditioning is carried out in two steps using an intermediate solvent miscible with both water and the final solvent. A common choice for the intermediate solvent is a lower alochol such as ethanol or isopropanol, or it can be acetone.

Standard conditioning procedures involve soaking the membrane in a series of mixtures of water and intermediate solvent, the latter in increasing amounts until usually by the fourth mixture it is present alone. A similar series of soakings in mixtures of the intermediate and final solvents follows. Suggested times for each stage vary widely, from a few minutes up to 24 hr. The preparation procedures are empirical, and few systematic studies of the processes have been made.

Automatic osmometers have the major advantages of achieving equilibration within 5 to 10 min, compared to several hours for simpler osmometers, and automatic indication and recording of the osmotic pressure. There are several designs that have evolved over the last two decades, many of which are still in use in industrial operations. One technique is illustrated in Figure 2.2. This design uses a horizontal membrane that separates the solvent (below) and solution compartments. Both are open to the atmosphere, eliminating all valves except that draining the solution compartment. Below the solvent compartment, a section of glass capillary connects, through a flexible tube, to a solvent reservoir in a tower at the left. An air bubble is introduced into the capillary before the unit is assembled, and placed so that its meniscus is between a light source and a detector. During operation, movement of the meniscus in response to solvent flow through the membrane alters the amount of light reaching the photodetector. This produces a signal which operates a servo system moving the solvent reservoir to the point which restores the meniscus to its original position. The difference in reservoir levels with solution and pure solvent in the upper (solution) compartment is the osmotic pressure of the solution. Figure 2.3 illustrates another design first introduced by the Shell Development Laboratories. The system employs a horizontal membrane to separate the solvent and solution compartments, but here the solution compartment (below the membrane) is completely closed by valves. It has a flexible metal diaphragm as one wall. Transfer of solvent through the membrane deflects this diaphragm. A deflection of approximately 25 nm

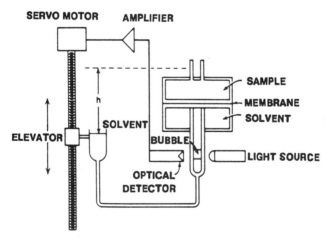

Figure 2.2 Major components of one type of automatic membrane osmometer.

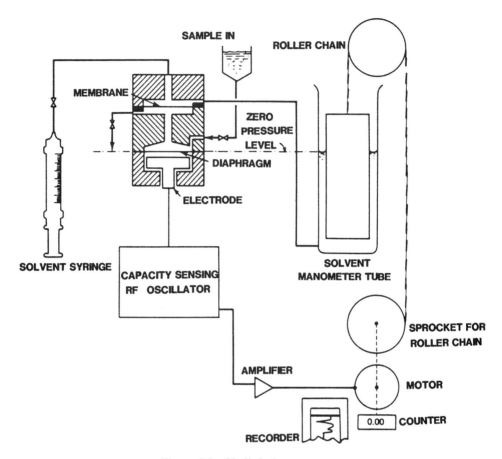

Figure 2.3 Shell design osmometer.

(1 μ in.) is sufficient to cause a change in frequency of the circuit, generating a signal that operates a servomotor to apply an appropriate hydrostatic pressure (negative in this case) to the solvent side of the membrane.

In osmotic measurements, with solvent and solution separated by a membrane permeable to the solute only, the chemical potential μ_1 of the solvent on the solution side is lower than that of the pure solvent because of the presence of the solute molecules. This chemical potential can be increased by the application of pressure to the solution, and thermodynamics identifies the rate of change of μ_1 with pressure as the partial molar volume of the solvent. For dilute solutions, this is nearly equal to the actual molar volume of the solvent V_1 and is independent of pressure in the range of osmotic pressures. One can then write the difference in solvent chemical potential across the membrane as

$$-\Delta\mu_1 = \pi V_1 = -RT \ln a_1 \tag{2}$$

where π is the osmotic pressure. This is the pressure difference between solution and solvent. For dilute solutions, application of Raoult's law allows the activity to be replaced with the mole fraction of solvent, n. Expansion of the logarithm gives

$$-\ln n_1 \simeq 1 - n_1 = n_2 = c\frac{V_1}{M} \tag{3}$$

At infinite dilution all the approximations become exact, and van't Hoff's limiting law results:

$$\left(\frac{\pi}{c}\right)_0 = \frac{RT}{M} \tag{4}$$

The concentration dependence of the osmotic pressure can be written by expanding equation (4) in a power series in c,

$$\pi = RT(A_1 c + A_2 c^2 + A_3 c^3 + \cdots) \tag{5}$$

where the A's are known as virial coefficients. Comparing equations (4) and (5) it can be seen that $A_1 = 1/M$. M is the number-average molecular weight $\overline{M}_n$. If, as is often the case, A_3 and all higher virial coefficients are negligibly small, equation (5) can be rearranged to

$$\frac{\pi}{RTc} = \frac{1}{\overline{M}_n} + A_2 c \tag{6}$$

The second virial coefficient A_2 is a measure of the polymer-solvent interactions.

$$A_2 = \frac{\rho_1}{M_1 \rho_2^2}\left(\frac{1}{2} - \chi_1\right) \tag{7}$$

Parameter A_2 is a predictor of the thermodynamic "goodness" of a solvent for a given polymer. When A_2 is high, the thermodynamic drive for solution to take place is also high. As A_2 decreases, the solvent becomes "poorer" until, at $A_2 = 0$, polymer of infinitely high molecular weight just precipitates from the solution. Molecules of lower molecular weight are more soluble, however, so that with real samples one can experience negative values of A_2 when precipitation is imminent. In the usual case, A_2 decreases with decreasing temperature, but the opposite may be true. The second virial coefficient also decreases slowly with increasing polymer molecular weight.

Data obtained from osmotic experiments are the osmotic heights, measured in centimeters of solvent, at a series of polymer concentrations. To utilize these data in equation (6), it is easiest to plot π/c versus c and obtain the intercept $(\pi/c)_0$, using π and c in the units given. Then $\overline{M}_n$ can be found from the following relationship:

$$\overline{M}_n = \frac{RT}{(\pi/c)_0} \tag{8}$$

The second virial coefficient can be obtained from values of π/c at any two concentrations:

$$A_2 = \frac{(\pi/c)_{c_2} - (\pi/c)_{c_1}}{RT(c_2 - c_1)} \tag{9}$$

A serious limitation of membrane osmometry is the diffusion of low-molecular-weight species (which may or may not be polymer) through the membrane. If any species is present which diffuses through the membrane slowly enough that its concentration on the two sides of the membrane has not reached the same value during the time of the experiment, the apparent osmotic pressure observed will not be the true osmotic pressure of the sample. Even though the osmotic pressure changes only a few percent per hour, serious errors can result. Also, it is not accurate to extrapolate a changing osmotic pressure back to teh time of the start of the experiment in an attempt to correct for diffusion. One must abandon the experiment and try again with a more retentive membrane if this occurs or to start with a sample from which the diffusible species have been removed by fractionation or extraction.

We now briefly examine vapor-phase osmometry. These instruments are referred to as "vapor-pressure osmometers." Figure 2.4 illustrates one design that consists of two parts: a thermally insulated measurement chamber and an electronics unit. Within the measurement chamber, two matched thermistors are suspended in a temperature-controlled cell that is saturated with solvent vapor. A drop of polymer solution is placed in

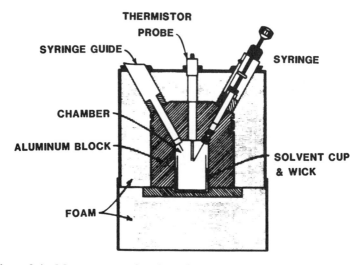

Figure 2.4 Measurement chamber of a typical vapor-phase osmometer.

one thermistor, and a drop of pure solvent on the other, by menas of hypodermic synringes. The vapor pressure of the solvent in the solution drop is lowered by the presence of the solute. This results in an excess condensation of solvent from the vapor in the chamber onto this drop, causing the release of heat of vaporization and raising the temperature of this drop slightly. Since the thermistors are electrical elements having a high-temperature coefficient of resistance, the temperature difference between the solution and solvent drops can be detected as an imbalance of a Wheatstone bridge circuit in the electronics portion of the system. The theory behind the operation is as follows.

From Raoult's law, which states that in an ideal solution the partial vapor pressure of each component is proportional to its mole fraction, it can be shown that the vapor-pressure lowering is

$$p_1^0 - p_1 = p_1^0 n_2 = p_1^0 c \frac{V_1}{M} \tag{10}$$

Superscript zero denotes pure solvent. Thermodynamics relates the vapor-pressure lowering to the temperature difference observed by means of the Clapeyron equation. The temperature difference is proportional to the difference in electrical resistance, Δr, between the two thermistors. If all the excess heat of vaporization were utilized to produce the observed

Characterization of Molecular Weight

temperature difference, these relations would allow a limiting law, analogous to van't Hoff's law in membrane osmometry, to be stated explicitly. This is not the case, however, since some heat is lost, for example by conduction along the thermistor leads. This is a constant fraction of the total, and typically, 70 to 80% of the theoretical temperature difference is achieved. Furthermore, the temperature coefficient of resistance of the thermistor, at the temperature of the measurement, is usually not precisely known. Usually, one combines all the necessary constants and proportionality factors into a single calibration constant k, such that

$$\left(\frac{\Delta r}{c}\right)_0 = \frac{k}{\overline{M}_n} \tag{11}$$

Virial expansion of equation (11) gives

$$\frac{\Delta r}{kc} = \frac{1}{\overline{M}_n} + A_2 c \tag{12}$$

To determine k, one measures $(\Delta r/c)_0$ for a low-molecular-weight substance of known molecular weight and applies equation (11).

Values of $\Delta r/c$ may scatter or exhibit curvature as a function of c instead of falling on a straight line. This is usually attributed to solvent effects and suggests that a plot of Δr versus c may yield a straight line that does not pass through the origin. If so, the intercept of this line at $c = 0$ may be used as a small correction to values of Δr. It is often found that, with this correction, values of $\Delta r/c$ follow equation (12) more satisfactorily, and values $\overline{M}_n$ so obtained are correct for known samples.

In this technique, all solute molecules that are nonvolatile under the conditions of the measurement contribute to $\overline{M}_n$. This can be viewed as an advantage, as the method is useful for low-molecular-weight solutes in providing a rapid method of measuring the molecular weight of rather small quantities of sample. On the other hand, it represents a limitation, since a very small weight percent of low-molecular-weight impurity, which would rapidly equilibrate in membrane osmometry and have an effect, can lead to serious errors in $\overline{M}_n$ determined by this technique. In fact, very few polymers with $\overline{M}_n > 50{,}000$ by membrane osmometry can be obtained pure enough to give their true molecular weights in vapor-phase osmometry, even though the sensitivity of research instrumentation is adequate to measure them. Other limitations, resulting from the mode of operation of the instruments, require careful control of such variables as drop size and measurement time.

DETERMINATION OF WEIGHT-AVERAGE MOLECULAR WEIGHT

The weight-average molecular weight is defined as follows:

$$\overline{M}_w = \sum w_i M_i = \frac{\sum c_i M_i}{c} = \frac{\sum N_i M_i^2}{\sum N_i M_i} \qquad (13)$$

This mode of the MWD can be measured by either light scattering or equilibrium ultracentrifugation experiments. Ultracentrifugation is the method of choice for many biological materials; however, it is seldom used for random-coil synthetic polymers and therefore is not discussed here.

In light-scattering systems, measurement is made of the difference in scattered-light intensity between a polymer solution and its solvent. This scattered intensity depends on both concentration and the angle between the incident and scattered light beams. The second requirement sets the major design features of light-scattering photometers.

The importance of proper preparation of both the solvent and polymer solutions cannot be overstated. Any foreign matter, particularly dust and dirt, in a liquid can be expected to scatter light. The scattered intensity is proportional to the square of the mass of the scattering particle and to the square of its difference in refractive index from the liquid. Large dust particles can be expected to scatter far more light than can polymer molecules. It is essential that all foreign material be removed completely from the solutions and solvent to be measured.

There are several techniques available to clarify solutions and solvent, usually followed by more than ordinary care in subsequent handling. The filtration of solutions through ultrafine filters, with pore sizes of a few tenths of a micrometer, usually suffices to remove superficial dust and foreign matter. The filters are conveniently supported in holders and the filtered liquid is led directly into the scattering cell. The solutions should never be forced through filters by pressure, but allowed to flow by gravity or with very slight vacuum on the exit side. In some cases, the solution may be centrifuged in a laboratory centrifuge or preparative ultracentrifuge.

The clarified solutions, in the scattering cell, may be observed with a strong light beam. Dust particles can best be seen at small angles of observation. When properly prepared, the solutions will not exhibit any bright particles.

The essential components of a light-scattering photometer are a light source, cell and holder, and detector mounted so as to view the cell over a range of angles. Figure 2.5 illustrates the basic arrangement. The light

Characterization of Molecular Weight

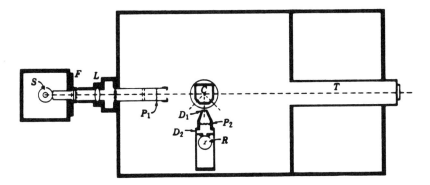

Figure 2.5 Major components of a light-scattering photometer.

source S can be a laser. A lens L provides an approximately parallel beam of light passing into the large sample compartment, A polarizer P_1 may be used. It can be assumed that the light is vertically polarized (with respect to the plane of the cell and detector, which is conventionally horizontal).

The scattering cell at C is usually cylindrical except for plane entrance and exit windows for the primary light beam. (Figure 2.5 illustrates observations at 45°, 90°, and 135° only.) The cell is centered on the axis of rotation of the receiver or detector R and a photomultiplier tube. Diaphragms D_1 and D_2 define the field of view seen by the detector, typically encompasing 2 to 5° of scattering angle. A second polarizer P_2 is provided and is also normally set to pass vertically polarized light. The receiver arm can be rotated manually from outside the cell compartment. After passing through the cell, the primary light beam is caught in a trap T.

In other commercial photometers, a small scattering cell is immersed in a larger bath filled with a liquid having the same refractive index as the contents of the scattering cell. A prism attached to and rotating with the phototube assembly views the scattering cell from inside this bath. This arrangement effectively eliminates reflections from the surfaces of the cell.

Since the ratio of intensities of the incident and scattered beams is typically of the order of 10^6, it is inconvenient to measure both beams with the full sensitivity of the detector. At some stage a calibration of the instrument with some material of known scattering power is required. This is usually carried out in two stages:

1. A primary standard is used to calibrate a permanent working standard. This can be an opal glass or plastic, perhaps combined with a

neutral filter to reduce the light intensity, which is permanently mounted in the instrument or replaces the scattering cell.
2. The working standard is measured from time to time during the measurement of the polymer solutions to provide the required calibration.

A second parameter in the light-scattering determination is the specific refractive increment, dn/dc, which is the concentration dependene of the refractive index of the polymer solution. This is a constant, so that a single determination of the difference in refractive index between a solution of known concentration and the solvent suffices for its determination. Since this difference is in the fifth or sixth decimal place of refractive index in the usual case, conventional refractomers are not sensitive enough. The use of a differential refractometer is required. This instrument uses a simple double-prism cell in which the deviation in direction of a light beam is directly proportional to the desired refractive-index difference. Figure 2.6 illustrates the operation.

The determination of the weight-average molecular weight by light scattering is based on the Debye (1947) theory:

$$\frac{Kc}{\Delta R_\theta} = \frac{1}{\overline{M}_w P(\theta)} + 2A_2 c \qquad (14)$$

The scattered intensity is expressed as the Rayleigh ratio, R_θ, which is the ratio of scattered-light intensity, per unit volume of scattering solution and unit solid angle at the detector, to the incident light flux per square centimeter. The Rayleight ratio, which is a function of the scattering angle θ, has the dimensions of cm^{-1}. ΔR_θ denotes the Rayleigh ratio of the solution less that of the solvent. The constant K is defined by

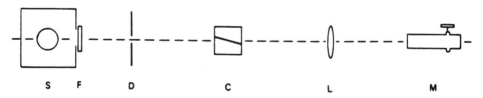

S   F   D   C   L   M

Figure 2.6 Optics of a differential refractometer: S, source (mercury arc lamp); F, monochromatizing filter; D, slit; c, cell; L, lens focusing image onto M, microscope with traveling cross hair and scale.

Characterization of Molecular Weight

$$K = \frac{2\pi^2 n^2 (dn/dc)^2}{N_0 \lambda^4} \tag{15}$$

where

n = refractive index
dn/dc = specific refractive increment
N_0 = Avogadro's number
λ = wavelength of the incident light as measured in air

$P(\theta)$ is a particle scattering factor that defines the angular dependence of the scattered light and relates it to particle size for a given model such as a sphere, random coil, or other type. Limiting cases are $P(\theta) = 1$ at $\theta = 0$, and for small θ,

$$\frac{1}{P(\theta)} = 1 + \frac{16\pi^2}{3\lambda_s^2} s_z^2 \sin^2 \frac{\theta}{2} \tag{16}$$

where s_z^2 is the z-average mean-square radius of gyration for random-coil polymers and λ_s is the wavelength of light in the solution: $\lambda_s = \lambda/n$. $P(\theta)$ can therefore be considered as a correction factor to the scattered intensity to reduce it to its value at $\theta = 0$.

The analysis of light-scattering data to yield the weight-average molecular weight, second virial coefficient, and radius of gyration is sufficiently complex that a computer program is often developed to handle computations in routine work.

Data regression can be organized into three parts. In the first, the instrument readings are converted into values of ΔR_θ, the Rayleigh ratio for (solution less solvent). Next, the data are prepared for the extrapolations to $c = 0$ and $\theta = 0$. These extrapolations are made graphically. Finally, $\overline{M}_w$, A_2, and $(s_z^2)^{1/2}$ are calculated.

An initial correction to the data is required by the fact that the Rayleigh ratio is defined per unit volume of scattering liquid, whereas with the usual cell geometries the scattering volume observed is not constant, but varies inversely as $\sin \theta$. This dependence is illustrated in Figure 2.7. Next, the data are corrected for slight variations in the reference intensity, as determined by reading the secondary standard one or more times during the determination with each solution. At this point a refraction correction might be applied to the data.

A refraction correction is required if the primary calibration was made using a solvent with a different refractive index from that of the solvent used in the determination. The relation used is

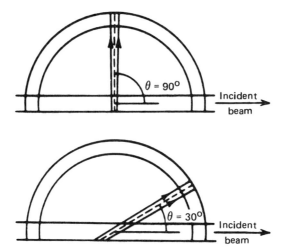

Figure 2.7 Geometry of a light-scattering cell, showing the difference in observed volume at θ = 30° and θ = 90°. [Adadpted from Billmeyer et al. (1971).]

$$R_\theta = k_c \left(\frac{n_s}{n_c}\right)^2 \times \text{(intensity ratio)} \qquad (17)$$

where the subscripts s and c stand for the solvent and control, respectively.

The extrapolations to $c = 0$ and $\theta = 0$ required by equation (14) are usually carried out by the graphical method developed by Zimm (1948). Use is made of the expansion of $1/P_2(\theta)$ for random-coil polymers [equation (17)]. The parameter $\sin^2(\theta/2)$ is the appropriate variable for the extrapolation as a function of angle. Zimm's analysis involves a plot of the left-hand side of equation (17) against $\sin^2(\theta/2) + Kc$, where K is an arbitrary constant selected to provide a conveniently spaced plot. The rectilinear grid characteristic of the Zimm plot is illustrated in Figure 2.8.

$\overline{M}_w$ is computed from

$$\overline{M}_w = \frac{1}{K(c/\Delta R_\theta)_{c=0,\theta=0}} \qquad (18)$$

A_2 is the slope of the $\theta = 0$ line and the radius of gryation $(s_z^2)^{1/2}$ from the slope of the $c = 0$ line using equation (16). The second virial coefficient is obtained from

Characterization of Molecular Weight

$$A_2 = \frac{(1/2)K[(c/\Delta R_\theta)_{c_2} - (c/\Delta R_\theta)_{c_1}]}{c_2 - c_1} \tag{19}$$

The mean-square radius of gyration, on combining equations (14) and (16) and noting that $\lambda s = \lambda/n$, is given by

$$s_z^2 = \frac{3\lambda^2}{16\pi^2 n^2} \times \frac{\text{slope}}{\text{intercept}} \tag{20}$$

where slope and intercept refer to the Zimm plot.

The light-scattering method is applicable over a wide range of molecular weights, as long as there is sufficient difference in refractive index between the polymer and the solvent. Its major limitation is the requirement that solutions be free from all extraneous scattering material. Special treatment is required for some systems, such as copolymers and very large molecules.

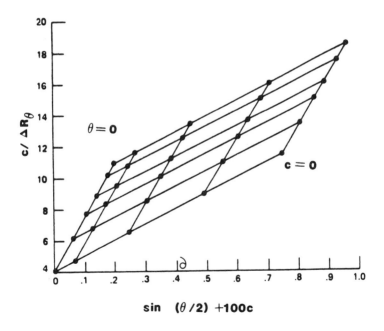

Figure 2.8 Zimm plot.

DETERMINATION OF MOLECULAR SIZE

The term *size* generally refers to the description of the amount of space taken up by the molecule. It can be expressed either as a linear dimension or as a volume. The size of a single molecular coil varies with time as a result of conformational changes due to Brownian motion. Similarly, the molecular size varies from molecule to molecule of identical mass and structure. Size can therefore be described only in terms of average properties. Two such average-sized parameters describing linear dimensions of polymers are the root-mean-square end-to-end distance $(r^2)^{1/2}$ and the radius of gyration. The latter is defined as the root-mean-square distance of the segments of the chain from its center of gravity. When polymer solutions are referred to, the common volume parameter is the hydrodynamic volume, which refers to the volume that the chain appears to occupy based on a specific property (e.g., viscosity increase it imparts to the solution). For linear polymers, all these quantities are uniquely related; for example, $r^2 = 6\,s^2$.

Light scattering yields the size parameter $(s_z^2)^{1/2}$; unfortunately, other methods do not yield the same z-average radius of gyration. As such, that it is difficult to compare light-scattering results directly to others. Gel permeation chromatography separates molecules according to their size, not mass, but the technique is used primarily as an approach to measuring the molecular weight distribution.

The method most often used to measure molecular size is dilute-solution viscosity measurements. The terminology within this subject category often leads to confusion even among the most experienced. Before describing the measurement technique, the following nomenclature of solution viscosity terms is summarized:

Relative viscosity (viscosity ratio):

$$\eta_r = \frac{\eta}{\eta_0} \simeq \frac{t}{t_0} \tag{21}$$

Specific viscosity:

$$\eta_{sp} = \eta_r - 1 = \frac{\eta - \eta_0}{\eta_0} \simeq \frac{t - t_0}{t_0} \tag{22}$$

Reduced viscosity (viscosity number):

$$\eta_{red} = \frac{\eta_{sp}}{c} \tag{23}$$

Inherent viscosity (logarithmic viscosity number):

$$\eta_{inh} = \frac{\ln \eta_r}{c} \qquad (24)$$

Intrinsic viscosity (limiting viscosity number):

$$[\eta] = \left(\frac{\eta_{sp}}{c}\right)_{c=0} = \left(\frac{\ln \eta_r}{c}\right)_{c=0} \qquad (25)$$

The relation for relative viscosity is defined as the ratio of the efflux time for the solution, t, to that of the solvent, t_0. The viscosities of the solution (n) and solvent (n_0) are related to the corresponding efflux times by the following relations:

$$\eta = Ct\rho - \frac{Ed}{t^2} \qquad (26a)$$

$$\eta_0 = Ct_0\rho_0 - \frac{E\rho_0}{t_0^2} \qquad (26b)$$

where ρ is the density and C and E are constants for viscometer. For dilute solutions, $\rho \simeq \rho_0$. Also, viscometers are designed such that for efflux times > 100 sec, the second terms can be neglected.

The specific viscosity is the relative increment in viscosity of the solution over that of the solvent, and the reduced viscosity is this quantity taken per unit concentration. This is still dependent on c, however, so extrapolation to $c = 0$ is required. It is convenient to extrapolate not only the reduced viscosity but also the inherent viscosity to $c = 0$. By expanding the logarithm in the defining equation for η_{inh} and observing the behavior as $c \to 0$, it is easy to show that both quantities extrapolate to the same intercept, denoted the intrinsic viscosity (η). The units of η_{red}, η_{inh}, and $[\eta]$ can be seen to be reciprocal to those of concentration and are dl/g since η_r and η_{sp} are dimensionless ratios.

A family of straight lines is obtained in these extrapolations for a series of polymer samples differing only in molecular size. These lines have slopes that are not constant but vary in a regular way, increasing as $[\eta]$ increases. They are described by

$$\frac{\eta_{sp}}{c} = [\eta] + k'[\eta]^2 c \qquad (27)$$

$$\frac{\ln \eta_r}{c} = [\eta] + k''[\eta]^2 c \qquad (28)$$

where k' and k'' are constants. For all but unusual cases, k'' is negative, and often $k' - k'' = 0.5$. Since k'' is usually smaller in absolute value than k', reflecting a smaller change with concentration in the inherent viscosity compared to the reduced viscosity, the value of the inherent viscosity at a fixed concentration, 0.2 or 0.5 g/dl, is sometimes taken as an easily obtained approximation to the intrinsic viscosity.

Viscometry generally requires relatively simple apparatus; for example, one or more glass capillary viscometers, a constant-temperature bath, and a timer. Accurate temperature control is critical and variations of solution temperature during the measurements must be kept below 0.01 to 0.02°C. The measurement of efflux times can normally be carried out by visual observation of the passage of the liquid meniscus past two lines marked on the viscometer, at which times automatic timers can be used. In these instruments, photocells mounted on the viscometer actuate an electric or electronic timer.

Figure 2.9 illustrates two common types of viscometers for solution viscosity measurements. The Ostwald viscometer is a constant-volume device, whereas in the Ubbelohde instrument, the effluent from the capillary flows into a bulb separate from the main liquid reservoir, so that the viscometer operates independent of the total volume of liquid present. This allows the convenience of preparing and measuring solutions that have a range of concentrations without the need of transferring them. Both types of viscometer are available with a variety of capillary diameters giving a selection of efflux times.

There are several theories relating the intrinsic viscosity to molecular size. Staudinger (Staudinger and Heuer, 1930) was the first to note the viscosity of polymer solutions as evidence for their long-chain, high-molecular-weight character, stating that $\eta_{sp} = KcM$, where K is a constant that can be evaluated from measurements on oligomers of known molecular weight. This expression implies that the reduced viscosity is independent of concentration, which is simply not the case. In addition, later work has demonstrated that Staudinger's equation must be modified by raising M to a power a, usually between 0.5 and 0.8, yielding the expression

$$[\eta] = k\overline{M}_v^a \tag{29}$$

This relation is attributed to Mark, Houwink, and Sakurada. Here the viscosity-average molecular weight is

$$\overline{M}_v = \left(\frac{\sum N_i M_i^{1+a}}{\sum N_i M_i}\right)^{1/a} \tag{30}$$

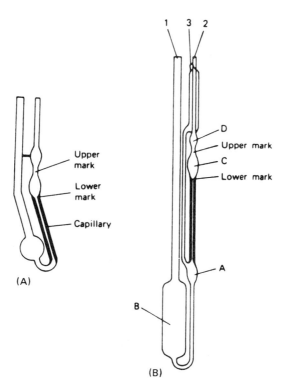

Figure 2.9 Capillary viscometers used for measurement of dilute-solution viscosity. (A) Ostwald or Cannon-Fenske type. (B) Ubbelohde viscometer.

The value of this average and its relation to others depends on a, a measure of polymer-solvent interactions. Since $\overline{M}_v$ is not available from other experiments, one often substitutes $\overline{M}_w$ for it, as the nearest directly measurable average. Of course, for a monodisperse polymer all averages would be identical, allowing K and a to be determined with ease, but such samples are not usually available. Extensive tables of K and a are available [e.g., Kurate et al. (1966)], but caution is required in using these sources since they are not selective, although reasonable comprehensive.

The empirical correlation between [η] and molecular weight is subject to several restrictions. This is understandable in light of the fundamental dependence of the intrinsic viscosity on molecular size rather than mass. Only if there is a unique relationship between size and mass can equation (29) be expected to apply. This requires that the polymer be linear, not branched; that the polymer solutions be measured in the same solvent and at the same temperature for which K and a were determined; and that the polymer be of the same chemical type as that used in determining K and a. It should be noted that since $[\eta]$ depends on the solvent and temperature, its specification is incomplete unless these quantities are given as well as the units.

The most useful theory relating the intrinsic viscosity to molecular size is that of Flory (1949, 1950, 1951):

$$[\eta]M = \Phi_0 V_h = \Omega_0 \xi^3 (s^2)^{3/2} \tag{31}$$

where V_h is the hydrodynamic volume, and Φ_0 is a universal constant with the value 2.8×10^{21} if V_h is expressed in milliliters and $[\eta]$ in dl/g. To relate the product $[\eta]M$ to a more familiar size parameter such as the radius of gyration $(s^2)^{1/2}$, we must introduce another parameter, ξ, which depends on solvent power. Newman et al. (1954) showed that for a polydisperse system, equation (31) is correct with $M = M_n$, and a number-average size parameter, such as $(s_n^2)^{1/2}$, results.

Perhaps the greatest value of Flory's viscosity theory lies in the interrelations with his theories of the configuration of real polymer chains and the thermodynamics of polymer solutions. The dimensions of polymer chains are determined both by short-range interactions, which can be predicted by calculation (Flory, 1969), and by long-range interactions. The latter contributions depend on the polymer-solvent interactions and are described by an expansion factor a', the ratio of the chain dimension in a given solvent and at a specified temperature to its value in the absence of such interactions (unperturbed dimensions): $a' = (s^2/s_0^2)^{1/2}$, for example. The chain assumes its unperturbed dimensions at a special temperature (for a given solvent) called the Flory temperature θ'. Thermodynamic considerations relate θ' to the temperature of phase separation at infinitely high molecular weight and to the temperature at which $A_2 = 0$. The unperturbed dimensions of polymers vary in a known way with molecular weight, such that quantities of type s_0^2/M depend only on chain structure (chemical type), not on solvent or temperature.

Use can be made of these facts by replacing s^2 by $a'^2 s_0^2$ in equation (31) and isolating the terms in s_0^2/M:

$$[\eta] = \Phi_0 \left(\frac{s_0^2}{M}\right)^{3/2} M^{1/2} \xi^3 a^3 = K M^{1/2} \xi^3 a^3 \tag{32}$$

where $K = \Phi_0 (s_0^2/M)^{3/2}$ is a constant for a given polymer, independent of solvent, temperature, and molecular weight.

The dependence of ξ on solvent power is given by Ptitsyn (1959) as

$$\xi^3 = \xi_0^3(1 - 2.63e + 2.86e^2) \tag{33}$$

where e is a parameter that can be related to other measures of solvent power: for example, the exponent a in equation (29): $e = (2a - 1)/3$. The value of e_0, referring to the unperturbed state of the chain at $T = \theta'$, is independent of solvent power and can be incorporated into K. The result is that at the θ' temperature, where $a = 1$, a is predicted to be exactly ½.

The solution properties are important, particularly for rubbers manufactured in solution. Polymer characterization relies heavily on solution properties, and certain applications involve solutions. For characterization of molecular weight, molecular weight distribution, long-chain branching, and compositional distribution, solution measurements and fractionation procedures are the methods of choice. Intrinsic-viscosity measurements offer the simplest method to estimate molecular weight; even without calibration, reasonable estimates can be made. In good solvents using structure correlations, a Mark-Houwink (M-H) relationship can be estimated. The M-H relationship is assumed to be a function of composition, which perhaps has contributed to the vast number of relationships that have been published. Ver Straate (1986) reports a correlation for EPM based on a series of polymers of about 50 wt % ethylene:

$$[\eta] = 2.47 \times 10^{-4} \overline{M}_w^{0.759} \quad \text{dl/g} \tag{34}$$

based on the solvent decalin at 135°C and

$$[\eta] = 2.92 \times 10^{-4} \overline{M}_w^{0.726} \quad \text{dl/g} \tag{35}$$

in 1,2,4-trichlorobenzene at the same temperature. The polymers were prepared with $VOCl_3$ and ethylaluminum sesquichloride in a continuous backmixed reactor. The polymer samples were believed to be essentially linear and of most probable molecular weight distribution (i.e., weight to number polydispersity of 2.0).

Data for the compositional dependence of the M-H exponents are not available over the entire range of molecular weights or ethylene and diene content. An approximate relationship for K variation with ethylene content is given in Figure 2.10, where the exponent has been fixed. Other data support a nearly linear interpolation between polyethylene and polypropylene.

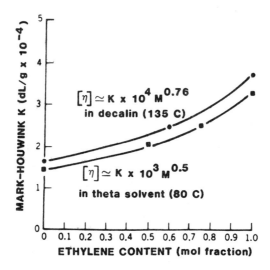

Figure 2.10 Dependence of intrinsic viscosity on composition.

CHARACTERIZATION OF MOLECULAR WEIGHT DISTRIBUTION

The principal methods for measuring molecular weight distribution are gel permeation chromatography (GPC) and turbidimetric titration. As noted earlier, the separation that is an essential part of any study of molecular weight distribution is often based on change in another property of the sample with molecular weight, rather than directly on molecular weight itself. Only in ultracentrifugation is the driving force for separation directly the molecular weight. Most common fractionation methods involve separation due to differences in solubility. But solubility is affected primarily by the chemical nature of the polymer and its relation to that of the solvent, and reflects differences in molecular weight to a lesser extent. Similarly, the separation in GPC is based on differences in molecular size.

The term *gel permeation chromatography* has its origins in the biological field as gel filtration. There has been muchd ebate on whether either term is appropriate for describing the process, in which molecules are separated according to their hydrodynamic volume, as this property influences their ability to permeate the internal pore structure of microporous "gel" particles. Both names are well established in the literature.

Chromatographic processes refer to those operations in which the solute is transferred between two phases, one of which is stationary and the other moving, often traversing a long tube called a column. In GPC, both phases are liquid, but in contrast to most liquid-liquid chromatography, where the two liquid phases are immiscible, the two phases in GPC are the same liquid (solvent) and are differentiated only in that the stationary phase is that part of the solvent which is inside the porous gel particles, while the mobile phase is outside.

The transfer of solute between the two phases takes place with the driving force of diffusion, resulting from a difference in concentration of solute between the two liquid phases, and restricted by the molecule's ability (depending on its size) to penetrate the pore structure of the gel. This process is that of permeation. Other separation processes applicable to chromatographic systems are those based on solubility differences (partition, as in gas-liquid chromatography), adsorption of a layer of molecules onto a surface (as in gas-solid chromatography), and ion exchange.

The term *gel* is misleading in application to GPC because of its common connotation of a soft, compressible material. The cross-linked dextrans used in gel filtration do have these characteristics, but GPC gels are typically hard and incompressible. This permits the use of high pressures in the liquid transport system, up to 1000 psi or more with correspondingly high flow rates. The original gels used in this technique consisted of polystyrene cross-linked with divinyl benzene. They are polymerized in a suspension system, with large amounts of an inert solvent-nonsolvent mixture present in the monomer, selected so that the polymer when formed is on the verge of precipitation from the liquid phase. The internal pore structure results. Porous glass can also be used in place of cross-linked polystyrenes.

The gel particles are about 100 µm in diameter. Examination of thin sections with transmission electron microscopy shows that the sizes of the internal pores are satisfyingly compatible with those calculated for polymer molecules, which they are known to separate.

The basic flow system is illustrated in Figure 2.11, consisting of a solvent reservoir, pump, and associated devices. The stream is split, one half going through a dummy column to provide pressure drop and into the reference side of the detector, the other through the column or columns for the separation and to the sample side of the detector. The sample is injected into the sample stream, either through a septum or with an injection loop and valve arrangement.

A differential refractometer is customarily used as the detector, but ultraviolet, infrared, or even flame-ionization detectors can also be used. A

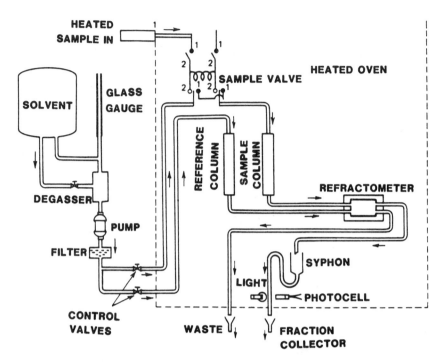

Figure 2.11 Flow arrangement of a gel permeation chromatograph. [See Maley (1965), for principles of operation.]

number of options are available, such as automatic injection of a series of samples for unattended operation, sample collection with a fraction collector, and recycle capability to improve resolution by cycling that portion of the mobile phase containing the sample through the columns more than once. High-temperature operation can be carried out for polymers soluble only at elevated temperatures.

GPC columns typically have a diameter of 3/8 in. and are 3 to 4 ft long. Three or four are commonly used in series, packed with gels having different pore sizes to ensure good separation capability over a wide range of molecular weights. A flow rate of 1 to 2 ml/min is common, but the volume of the columns is such that the time for elution of all components is 2 hr or more.

Peak broadening is more serious in liquid than in gas chromatography. It originates in part from the flow properties of the viscous mobile phase through the packed column, and partly from nonuniformities in the internal pore structure of the gel. It is customary to test GPLC

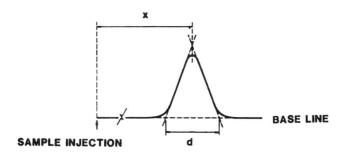

Figure 2.12 Calculation of plate count in GPC.

columns for satisfactory resolution and a low-level peak broadening by the plate-count method. A monodisperse sample is injected, usually a low-molecular-weight liquid, and the plate count is calculated. This is illustrated in Figure 2.12. Several hundred plates per foot is required to give adequate resolution for polymer systems.

Polymer molecules are separated by size in the GPC experiment because of their ability to penetrate part of the internal volume of the gel particles, taht is, the stationary phase. As the sample moves along the column with the mobile phase, the largest molecules are almost entirely excluded from the stationary phase, while the smallest find almost all the stationary phase accessible. The smaller the molecule, the more of the stationary-phase volume is accessible to it and the longer it stays in that phase. Small molecules thus fall behind larger ones and are eluted from the column later. For large molecules completely excluded from the gel, the retention volume V_r is equal to the interstitial or mobile-phase volume V_0, whereas for very small molecules, for which the entire stationary phase is accessible, $V_r = V_0 + V_i$, where V_i is the internal pore volume of the gel, that is, the stationary-phase volume. Intermediate species have retention times $V_r = V_0 + K_d V_i$, where K_d is the ratio of pore volume accessible to that species, to the total pore volume. K_d is a separation constant varying from 0 to 1, and in GPC all species are eluted with retention volumes between V_0 and $V_0 + V_i$. In partition chromatography the corresponding partition coefficient can have values much greater than unity.

V_r cannot yet be predicted from molecular parameters. Calibration curves are prepared by running standards of known molecular weight and narrow molecular weight distribution, and preparing a plot of log molecular weight versus retention volume as shown in Figure 2.13. This plot is often nearly linear over a range of molecular weights.

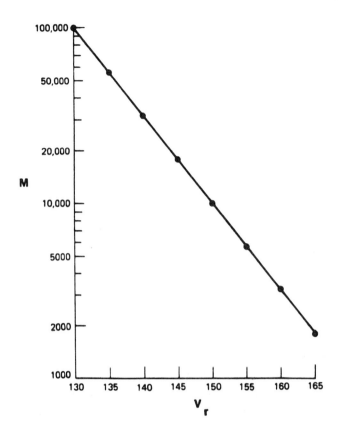

Figure 2.13 Typical GPC calibration curve.

Note that the true nature of the separation is based on hydrodynamic volume, not molecular weight. The restrictions given earlier apply when the attempt is made to substitute the latter variable for the former. So if a calibration is made with samples of narrow-distribution linear polystyrene in tetrahydrofuran at 25°C, it will not apply to a polymer of another composition, or a branched polystyrene, or to measurements in another solvent or at another temperature. Since narrow-distribution polymers other than polystyrene are not generally available, this becomes a serious problem. An approximate correction (Q factor), which takes account of differences in molecular weight per unit chain length but ignores the effects of solvent power, is often applied. Exact treatment requires that the calibration be made directly in terms of hydrodynamic volume.

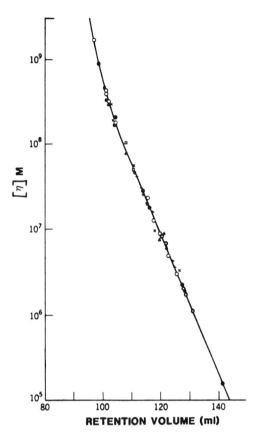

Figure 2.14 Universal GPC calibration curve based on hydrodynamic volume, developed by Grubisic (1967).

To do this, the calibration curve is plotted as log $[\eta]M$ versus V_r, as in Figure 2.14. If the solvent used is one in which polystyrene is soluble, the narrow-distribution samples serve very well to provide this curve. A variety of linear and branched polymers will fall on the curve. For linear polymers, interpretation in terms of molecular weight is straightforward: If the Mark-Houwink-Sakurada constants K and a are known, for example, log $[\eta]M$ can be written $\log M^{1+a} + \log K$, and V_r can be directly related to M. It should be noted that a new and different average molecular weight is defined by this process:

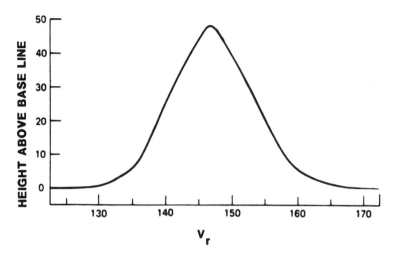

Figure 2.15 GPC chromatogram for poly(methyl methacrylate).

$$\overline{M}_{\text{GPC}} = \frac{\sum_{i=1}^{\infty} w_i M_i^{1+a}}{\sum_{i=1}^{\infty} w_i M_i} \tag{36}$$

The GPC average molecular weight is equal to $\overline{M}_z$ for $a = 1$ and is somewhat greater than $\overline{M}_w$ if $a = 0.5$. With branched polymers the material eluting at any value of V_r consists of a mixture of species having different molecular weights and degrees of branching but constant hydrodynamic volume. When narrow-distribution samples of known molecular weight and intrinsic viscosity are not available to provide a universal calibration curve for the solvent used, more complex analyses are possible.

The basic data from GPC are values proportional to the amount of material, and molecular weights. These are related through corresponding values from sample chromatogram and calibration curve, respectively; V_r serves no other purpose. Recorded measurements from a differential refractometer are generally given in chart divisions, corrected for a baseline. A typical chromatogram is illustrated in Figure 2.15. These readings represent total concentration of polymer at that retention volume and are proportional to the quantities $N_i M_i$ appearing in equations (1) for $\overline{M}_n$, (13) for M_w, and elsewhere. By the use of corresponding values of M_i from the

Characterization of Molecular Weight

calibration curve, it is easy to calculate N_i and $N_iM_i^2$ and to obtain the sums necessary to calculate $\overline{M}_w$ and $\overline{M}_n$ from the equations indicated. One can also find the necessary data to calculate $\overline{M}_v$ and $\overline{M}_{GPC}$ from equations (30) and (36) if a is known, and the z-average molecular weight

$$\overline{M}_z = \frac{\sum_{i=1}^{\infty} N_i M_i^3}{\sum_{i=1}^{\infty} N_i M_i^2} = \frac{\sum_{i=1}^{\infty} w_i M_i^2}{\sum_{i=1}^{\infty} w_i M_i} \qquad (37)$$

It must be remembered that the quoted molecular weights are not correct, and even the ratio $\overline{M}_w/\overline{M}_n$ is in error. The first approximation to correcting the data is to consider the fact that, neglecting differences in polymer-solvent interactions, equal numbers of repeat units in the chain should make the same contribution to its hydrodynamic volume. This leads to the concept of the "Q factor," the weight in daltons per angstrom of chain length.

On a relative basis, GPC data can be used to provide graphs of the molecular weight distribution. Two types are useful: The first is the cumulative distribution curve, obtained by accumulating and then normalizing the GPC readings. The resulting data are plotted against M (Figure 2.16A), or as is useful for broader distributions, against $\log M$ (Figure 2.16B). In considering the logarithmic plot, it is useful to note that

$$w(\log M) d(\log M) = \frac{2.3 M_w(M)\, dM}{2.3 M} = w(M)\, dM \qquad (38)$$

The second useful curve is the differential distribution, obtained by differentiating the smooth curve drawn through the points of Figure 2.16A. This is shown in Figure 2.17. An approximate relative curve can be obtained by omitting the smoothing and plotting the GPC readings directly against M. Note that the chromatogram itself is not unlike the differential distribution curve in general features, differing primarily in that the scale of V_r is not linear but approximately logarithmic in M. This is shown in Figure 2.18.

As emphasized earlier, many of the limitations of the GPC method stem from the fact that the separation is based on molecular size, although it is common practice to interpret the results in terms of molecular weight. For linear homopolymers, this leads to difficulties in absolute calibration techniques, which are partially overcome by making the calibration in

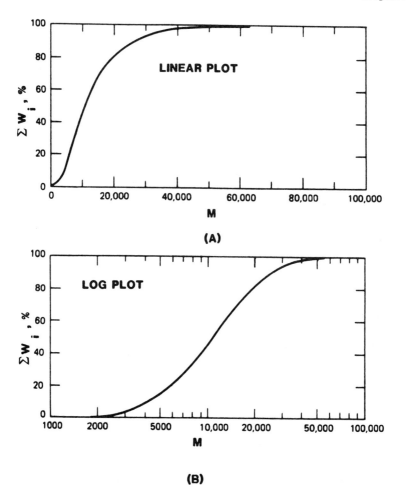

Figure 2.16 Cumulative molecular weight distribution curve.

terms of hydrodynamic volume. The problem of branched polymers can be simplified by use of the universal calibration techniques, but no complete interpretation in terms of molecular weight distribution is possible. Copolymers present a different problem in that (1) V_r is determined at least in part by chemical composition, which influences molecular size because of polymer-solvent interactions, and (2) refractive index, and thus detector response per unit concentration, also depend on chemical com-

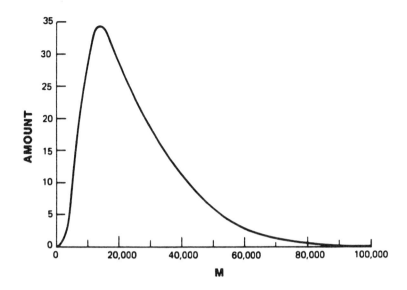

Figure 2.17 Differential molecular weight distribution.

position. If the refractive index of a homopolymer depends on molecular weight (which it does at sufficiently low molecular weight), the detector response is no longer constant over the entire chromatogram and a separate calibration must be constructed. Other detectors can be used, such as infrared or ultraviolet absorption or flame ionization. Useful information about the chemical composition of the eluate can be derived by combining results obtained with two different detectors, especially if one is sensitive to an absorption band specific to one of the species in a copolymer. It is important to note that with existing columns, peak broadening can seriously affect the distribution curve for samples with M_w/M_n less than about 2.

An alternative method to GPC is turbidimetric titration, which is an analytical fractionation technique, in which polymer is precipitated continuously by decreasing its solubility in the solvent. Early work was performed by titrating a nonsolvent into the polymer solution, but the technique of cooling a solution to cause precipitation is also used. The amount of polymer precipitated is measured by the turbidity of the solution. Critical to this analysis is the establishment of a suitable solvent-precipitant system in which the precipitated phase remains stably suspended rather than settling rapidly as is usually desirable in other fractiona-

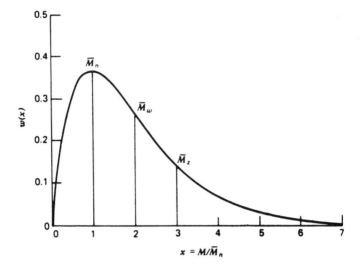
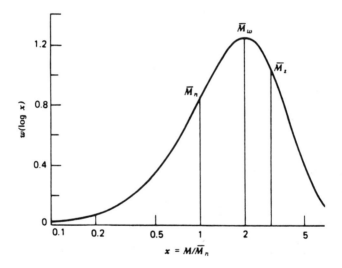

Figure 2.18 Most probable molecular weight distribution curve and corresponding GPC chromatogram.

Characterization of Molecular Weight

tion methods. Since turbidity is uniquely related to concentration only at constant size and mass of the scattering particles, these parameters should be maintained constant throughout the titration, new particles forming as more polymer is precipitated, rather than existing particles growing or agglomerating. Also, the particle size should be relatively independent of the molecular weight and the solvent composition. Since the turbidity depends on the difference in refractive index between the two phases, the solvent and nonsolvent must have similar refractive indices to minimize the change in the refractive index of their mixture as the titration proceeds.

Usually, turbidity itself is not measured, but rather, scattering at a convenient angle. Equipment ranges from conventional or modified light-scattering photometers to analytical spectrophotometers. Most often scattered-light intensity is measured at 90° angle. It is also convenient to measure the transmitted light intensity in this case. Because of difficulties in readily establishing a suitable solvent-precipitant system for new polymers, this method is not widely used.

NOTATION

$A_{1,2,3}$	virial coefficients
a	Mark-Houwink exponent
a_1	solvent concentration
C	viscometer constant
c	concentration
E	viscometer constant
e	solvent parameter
K	constant
k	calibration constant
k', k''	constants
$\overline{M}_n$	number-average molecular weight
$\overline{M}_w$	weight-average molecular weight
N	number of molecules
n	mole fraction of solvent
p	vapor pressure
R	universal gas law constant
R_θ	Rayleigh ratio

r	electrical resistance; molecular distance
s_z^2	z-average mean-square radius of gyration for random-coil polymers
T	absolute temperature
V_1	molar volume
V_h	hydrodynamic volume
w	mass or weight

Greek Symbols

η	viscosity
θ'	Flory temperature
θ	scattering angle
λ	wavelength of incident light
μ_1	chemical potential
ξ	coefficient of intrinsic viscosity
π	osmotic pressure
ρ	density
Φ_0	universal constant for intrinsic viscosity

REFERENCES AND SUGGESTED READINGS

Billmeyer, F. W., *Textbook of Polymer Science*, 3rd ed., Wiley-Interscience, New York, 1984.
Billmeyer, F. W., H. I. Levine, and P. J. Livesey, *J. Colloid Interface Sci.*, 35, 204–214 (1971).
Debye, P., *J. Phys. Colloid Chem.*, 51, 18–32 (1947).
Ferry, J. D., *Viscoelastic Properties of Polymers*, 3rd ed., Wiley, New York, 1980.
Flory, P. J., *J. Chem. Phys.*, 17, 303–310 (1949).
Flory, P. J., *J. Polym. Sci.*, 5, 745–747 (1950).
Flory, P. J., *J. Am. Chem. Soc.*, 73, 1904–1908 (1951).
Flory, P. J., *Statistical Mechanics of Chain Molecules*, Wiley, New York, 1969.
Graessley, W. W., and M. J. Struglinski, *Macromolecules*, 19, 1754 (1986).
Grossiord, J.-L., and G. Couarraze, *C.R. Acad. Sci., Ser. 2*, 302, 615 (1986).
Grubisic, Z., P. Rempp, and H. Benoit, *J. Polym. Sci.*, B5, 753–759 (1967).
Kurata, M., I. Masamichi, and K. Kamada, pp. IV-1 to IV-72 in *Polymer Handbook*, J. Brandrup and E. H. Immergut, eds., Wiley, New York, 1966.
Maley, L. E., J. Polymer Sci. C8, 253–268 (1965).
Malkin, A. Ya, and A. E. Teishev, *Vysokomol. Soedin., Ser. A*, 30, 175 (1988).
Newman, S., W. R. Krigbaum, C. Laugier, and P. J. Flory, *J. Polym. Sci.*, 14, 451–462 (1954).

Staudinger, H., and W. Heuer, *Berichte, 63B,* 222–234 (1930).
Van Straate, G., in *Encyclopedia of Polymer Science and Engineering,* Vol. 6, 6th ed., Wiley, New York, 1986.
Zimm, B. H., *J. Chem. Phys., 16,* 1099–1116 (1948).

3
Thermal Analysis of Polymers

INTRODUCTION

A major property of any polymer is its thermal behavior. Knowledge of this behavior is essential not only for the selection of proper processing and fabrication conditions, but also for the full characterization of the material's physical and mechanical properties, as well as for the selection of appropriate end uses. The temperature-dependent properties of polymers undergo major changes at one of two transition points. For crystalline polymers, the point of greatest importance is the crystalline melting point, while for amorphous polymers, it is the glass transition temperature.

In this chapter we consider methods for measuring thermal properties, and their dependence on molecular-structure parameters. The discussion is limited to the properties of glass transition temperature, the crystalline melting point, specific heat and enthalpy, thermal conductivity, and thermal degradation.

THERMAL BEHAVIOR PROPERTIES

Among the main properties characterizing the thermal behavior of polymeric materials is the galss transition temperature. The glass transition temperature, T_g, is defined as that temperature below which the polymer is glassy and above which it is rubbery. These descriptions apply strictly only to amorphous polymers, and in the presence of substantial crystallinity, the change of properties at the glass transition may be obscured. To simplify the discussion, we refer only to amorphous polymers for now.

We are concerned with the molecular interpretation of T_g as the temperature of the onset of large-scale motion molecular chain segments. At very low temperatures (near absolute zero) chain atoms undergo only low-amplitude vibratory motion around fixed positions. As temperature increases, both the amplitude and the cooperative nature of these vibrations among neighboring atoms increases until at a well-defined transition, at temperature T_g, segmental motion becomes possible and the material becomes leathery or rubbery. Above T_g, the chain segments can undergo cooperative rotational, translational, and diffusional motions, and as the temperature is raised sufficiently (e.g., $T_g + 100°C$) the material behaves like a high-viscosity liquid.

Let's consider two well-known polymers, polystyrene and natural rubber (cis-polyisoprene). At room temperature, polystyrene is brittle (in the glassy state), while the rubber exhibits typical elastomeric properties. But when rubber is cooled below its T_g ($-72°$) (e.g., by immersion in liquid nitrogen), it becomes both rigid and brittle. Tapping it on a hard surface shatters the piece by brittle or glassy fraction. In contrast, when polystyrene is heated above $T_g = 100°C$, it is no longer brittle but has typical, if poor elastomer properties. This example illustrates both the nature of the glass transition and its value as a reference temperature for comparing the physical and mechanical behavior of materials.

An important aspect to consider is the effect of molecular structure on T_g. The basis for considering the effects is the onset, as T_g is approached, of cooperative motion of groups of atoms larger than, say, the monomer. As the temperature increases, these groups become larger until above T_g the entire polymer coil is the elastic unit (as stipulated by the kinetic theory of rubber elasticity). This means that any molecular parameter affecting chain mobility can be expected to influence T_g. Among the variables that are important are chain microstructure, including chemical type of monomer units, copolymerization effects, and tacticity; molecular structural parameters such as molecular weight and MWD, branching, and cross-linking; and the presence of low-molecular-weight constituents such as plasticizers, oligomers, and diluents. Also, among the major parameters influencing T_g are chain stiffness, bulkiness and flexibility of side chains, and the polarity of the chain. The latter parameter is expressed in terms of the cohesive-energy density or solubility parameter.

The influences on the glass transition of molecular weight, branching, cross-linking, copolymerization, and diluents or plasticizers can be accounted for by the free-volume theory. Quantitative relations can be derived with the assumption that chain ends contribute more free volume to the system than do similar segments along the chain, that the volumes of the components are additive, and that both components contribute to the thermal expansion of the polymer by their glassy-state expansion coefficients below T_g and their rubbery-state coefficients above T_g. For example, the influence of molecular weight can be expressed as

$$T_g = T_{g\infty} - \frac{K}{\overline{M}_n} \tag{1}$$

where $T_{g\infty}$ is the glass transition temperature for polymer of infinite molecular weight and K is a constant, both being characteristic of a given polymer type.

Amorphous random copolymers exhibit a single glass transition temperature which lies between the values of T_g for the two homopolymers. For predicting this glass temperature,

$$T_g = \frac{w_1 T_{g1} + K w_2 T_{g2}}{w_1 + K w_2} \tag{2}$$

where T_{g1} and T_{g2} refer to the two homopolymers present in weight fractions w_1 and w_2 (where $w_1 + w_2 = 1$) and K is given by

$$K = \frac{(\alpha_r - \alpha_g)_2}{(\alpha_r - \alpha_g)_1} \tag{3}$$

where the α's are the thermal-expansion coefficients of the homopolymers, the subscript r referring to the rubbery state and g to the glassy state. It is often better to determine K experimentally; however, the equation fits experimental data well. In contrast, block and graft copolymers exhibit two glass transitions if the blocks are long enough to cause phase separation. Equation (2) can also be applied to polymer-diluent systems; since the value of T_g for a low-molecular-weight diluent or plasticizer is very low, T_g is always decreased.

Branching tends to increase free volume and decreases T_g, whereas cross-linking decreases free volume and increases T_g. The extent to which tacticity influences T_g varies with the type of polymer. The lower value of T_g for isotactic poly(methyl methacrylate) compared to the syndiotactic polymer is explained in terms of easier rotation about chain bonds in the isotactic structure. Similarly, increasing the syndiotacticity of poly(vinyl chloride) raises T_g substantially.

Thermal Analysis of Polymers

In crystalline polymers, the crystallites tend to reinforce or stiffen the structure, increasing T_g proportional to the amount of crystallinity. The effects of structural variables on T_g are summarized in Figure 3.1.

We now direct attention to the crystalline melting point. For crystalline polymers, this constitutes the most important thermal transition. The property changes at T_m are far more drastic than those at T_g, particularly if the polymer is highly crystalline. These changes are characteristic of a thermodynamic first-order transition and include a heat of fusion and discontinuous changes in heat capacity, volume or density, refractive index, birefringence, and transparency. Any of these parameters can be used to detect T_m.

Crystalline melting usually occurs over a broad temperature range, since the exact melting point of each crystalline region depends on both

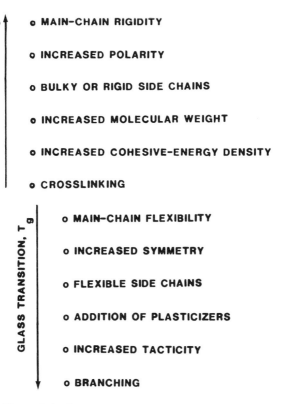

Figure 3.1 Structural factors affecting T_g.

its size and its perfection, larger and more perfect crystallites having higher melting points. The value of T_m is usually taken to the melting point of the highest-melting crystallites, that is, the point of disappearance of the last traces of crystallinity. This interpretation is adhered to, for example, in the determination of T_m by optical microscopy described later.

Since crystal perfection and crystallite size are influenced by the rate of crystallization, T_m depends to some extent on the thermal history of the specimen. The rate of polymer crystallization increases as the temperature is dropped below T_m [more properly, the melting point of the largest and most perfect crystals that can be formed by slow crystallization or annealing (i.e., the equilibrium melting point)]. At some temperature a maximum rate will occur, which may or may not be realizable experimentally;' at lower temperatures, the rate again decreases and reaches zero at T_g. This behavior is illustrated in Figure 3.2 for natural rubber. It is possible to quench the polymer in the amorphous state to below T_g; crystallization does not take place until the temperature is increased continuously, crystals will first form and then melt. For many other polymers, notably polyethylene and polypropylene, the entire crystallization curve cannot be observed, the rate of crystallization being so great that the material cannot be quenched to the amorphous state.

Values of T_m and T_g for some common polymers are listed in Table 3.1. The next property of importance is the material's thermal conductivity. Knowledge of the thermal conductivity is essential for any application involving the conversion, exchange, or transport of thermal energy.

Thermal conductivity K is defined as the ratio of the heat flow across

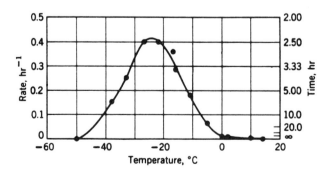

Figure 3.2 Rate of crystallization of natural rubber as a function of temperature.

Table 3.1 Typical Glass-Transition and Melt Temperatures for Common Polymers

	Glass transition		Melt temperature	
	°C	°F	°C	°F
Poly(α-methylstyrene)	175	347	—	—
Polycarbonate	150	302	220–267	428–513
Polymethacrylontrile	120	248	130–150	266–302
Poly(acrylic acid)	106	222.8	—	—
Poly(methyl methacrylate)	105	221	>200	392
Polyacrylonitrile	104	219.2	317	602.6
Polystyrene	100	212	—	—
Poly(vinyl chloride)	83	181.4	220–240	428–464
Poly(ethylene terephthalate)	69	156.2	267	512.6
Poly(ethyl methacrylate)	65	149	—	—
Polycaproamide (nylon 6)	50	122	215–225	419–437
Poly(hexamethylene adipamide) (nylon 66)	50	122	265	509
Poly(ω-aminoundecanoic acid) (nylon 11)	46	114.8	194	381
Poly(methyl methacrylate)—isotactic	45	113	160	320
Poly(vinyl acetate)	29	84.2	35–85	95–185
Poly(n-butyl methacrylate)	22	71.6	—	—
Poly(methyl acrylate)	9	48.2	—	—
Poly(vinylidene chloride)	−17	1.4	190–198	374–388
Polypropylene	−19	−2.2	176	348.8
Poly(ethyl acrylate)	−22	−7.6	—	—
Poly(butyl acrylate)	−55	−67	—	—
Polyoxymethylene	−68	−90.4	180–200	356
Polyisoprene (cis)	−73	−99.4	28–36	82–97
Polyethylene (branched)	−80	−112	105–125	221–257
Polyethylene (linear)	−80	−112	137	278.6
Polybutadiene (syndiotactic)	−85	−121	154	309.2
Poly(dimethyl siloxane)	−123	−189.4	—	—
Polystyrene	—	—	70–115	158–239
Polybutene-1	—	—	125–135	257–275
Polyoxymethylene	—	—	165–185	329–365
Polypropylene	—	—	160–170	320–338

unit area of a surface to the negative of the temperature gradient in the direction of flow:

$$K = \frac{dQ/dt}{dT/dx} \tag{4}$$

Methods for measuring K require the establishment of a known temperature distribution and measurement of the heat flux. The measurements can be made at the steady state or under transient conditions.

For crystalline polymers, the contributions of the crystalline and amorphous regions are assumed to be additive; the thermal conductivity of a crystalline polymer is greater than that of an amorphous material, but its dependence on temperature is the same. K is found to be proportional to the square root of $\overline{M}_w$. It increases with cross-linking and increases in the direction of orientation but decreases in the perpendicular direction as the result of molecular alignment. The effect of diluents is equivalent to that of lowering the molecular weight.

MEASUREMENT TECHNIQUES

The glass transition temperature can be measured by several methods. Its important to note, however, that not all methods are in agreement. The reason for this is the kinetic, rather than thermodynamic, nature of the transition. That is, T_g depends on the heating rate of the experiment plus the thermal history of the specimen.

In the first-order thermodynamic transition, such as a crystalline melting point, there are discontinuities in such properties as heat content and specific volume (first derivatives of the Gibbs free energy), associated with the heat of fusion and the volume change on fusion. In a second-order transition, such changes do not exist. The second derivatives of the free energy—heat capacity and volume expansion coefficient—however, do change abruptly. These changes are characteristic of the glass transition and are often used to determine T_g.

Rate effects are manifested in the dependence of T_g on the preceding thermal history of the specimen. The polymer is not in thermodynamic equilibrium at T_g when cooled at any finite rate. On subsequent heating after rapid cooling, the volume-temperature curve may lose its simple two-straight-line form and become nonlinear. It is usually possible, however, to cool the specimen slowly enough (say, around 0.3°C/min) that a further decrease in cooling rate changes T_g only a negligible amount. Experiments of this sort are easily carried out in a simple glass dilatometer (which is illustrated in Figure 3.3).

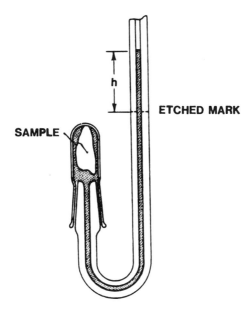

Figure 3.3 Volume dilatometer with mercury as confining fluid.

In addition to measurement of heat capacity, specific volume, or density as a function of temperature, the glass transition can be located by a variety of different tests. Among the more common techniques are thermal analysis, the temperature dependence of refractive index or dielectric constant or loss, dynamic mechanical properties, impact modulus, infrared or nuclear magnetic absorption spectra, and numerous industrial methods based on softening point, hardening point, heat deflection temperature, hardness, elastic modulus, or viscosity. Some of these methods are dynamic in the sense that temperature is varied at a constant rate until a change in property, such as softening or deflection, occurs. These tests may yield end-point temperature as much as 20°C higher than T_g values derived from other methods. Before describing some of the measurement techniques, it is important to have an understanding of the theories of glass transition.

The combination of some characteristics of thermodynamic transitions with obvious rate effects has led to much confusion regarding the origins of the glass transition. There are three opposing but not mutually contradictory views. The first of these is the so-called equilibrium theory of Gibbs (Gibbs, 1956; Gibbs and DiMarzio, 1958). This theory concludes that the observed glass transition is the result of kinetic manifestations of the approach to a true equilibrium thermodynamic transition. At infinitely long times, a second-order thermodynamic transition could be achieved under equilibrium conditions. The approach to the transition is viewed as the change with temperature of the configurational entropy of the material. As the temperature decreases, the number of states available to the polymer decreases and the rate of approach to equilibrium also decreases. At equilibrium, the configurational entropy becomes zero at T_g.

The "hole theory" of Hirai and Eyring (1958, 1959) regards the vitrification process as a reaction involving the passage of kinetic units (e.g., chain segments in which all the atoms move cooperatively) from one energy state to another. This requires a "hole" or empty space for the unit to move into, and the creation of this hole requires both a hole energy, needed to overcome the cohesive forces of the surrounding molecules, and an activation energy, to overcome the potential barrier associated with rearrangement via an activated state. All holes are characterized by a single mean volume. This theory can be interpreted physically when a rubbery liquid is cooled; the glass transition temperature, defined as the temperature of half-freezing of the hole equilibrium, depends only on the rate of cooling since there is ample molecular motion in the liquid for equilibrium to be achieved. In contrast, as a glass is heated, T_g depends not only on the heating rate but also on the thermal history of the sample since equilibrium cannot be achieved below T_g. Glasses that have been cooled differently represent different starting materials, which have different enthalpies and different time-dependent heat capacities in the glass transition range.

The "free-volume theory" of Fox and Flory (1950) is based on a free-volume model. The glass transition occurs when the free or unoccupied volume in the material reaches a constant value and does not decrease further as the material is cooled below T_g. The remaining or critical free volume is supposed to remain frozen in the glassy state; the residual small volume-expansion coefficient observed below T_g is believed to have the same origin as the thermal-expansion coefficient of a crystalline solid.

The fraction of free volume f can be written as follows:

$$f = \frac{V_f}{V_f + V_o} = 0.025 + 0.00048(T - T_g) \tag{5}$$

where V_f is the free volume and V_o is the occupied volume. The critical free volume is evaluated as 0.025 and is roughly a constant for many polymers. The relation of this increment in volume to the volume-expansion coefficients above and below T_g and the location of the pseudoequilibrium value of T_g at infinite time, T_∞, are shown in Figure 3.4. The figure also illustrates an alternative approach in which the critical free volume is measured by extrapolating the liquid line to the absolute zero of temperature and is about 0.113 at T_g. We may now discuss relevant experimental techniques for the measurement of T_g and other thermal behavior properties.

DIFFERENTIAL THERMAL ANALYSIS

Differential thermal analysis (DTA) is a method used for detecting the thermal effects accompanying physical or chemical changes in a sample as its temperature is varied through a region of transition or reaction. This is accomplished over a programmed heating or cooling range. Figure 3.5 illustrates a schematic of the basic instrumentation used.

Principal features include sample S, a reference material R, and a chamber contained in a heating block with electrical heating element H. Thermocouples placed in the centers of the sample and reference measure their temperature and also the temperature difference between them. This temperature difference depends on the density, thermal conductivity,

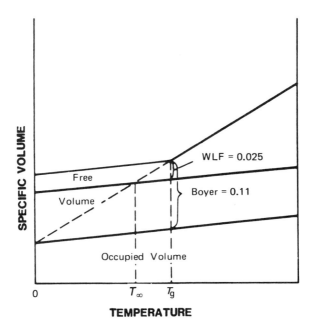

Figure 3.4 Specific volume around T_g based on free-volume theory.

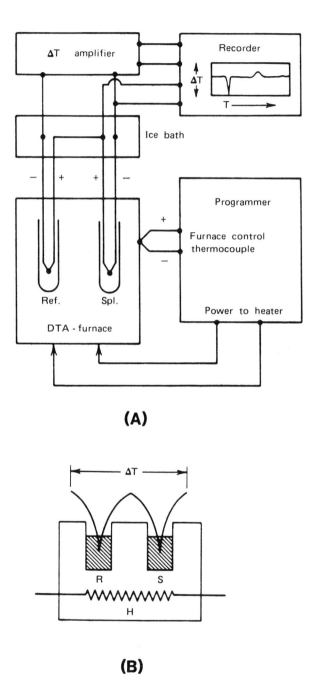

Figure 3.5 (A) Components of a DTA apparatus; (B) details of a DTA cell.

specific heat, and thermal diffusivity of sample and reference. It is often not possible to separate the effects of these individual factors from one another and from the influence of the geoemtry of the system.

In a typical experiment, the temperature of the heating block is programmed to increase linearly with time. The reference material is selected to be thermally inert except for slight changes in heat capacity with temperature. Materials such as fused quartz, porcelain, glass beads, or MgO are typically used as the reference. At the start of the experiment, the block, sample, and reference are all at the same temperature. As the block temperature rises, the temperatures registered by the thermocouples at the centers of sample and reference rise more slowly due to the finite heat capacities of these materials. A steady state is reached, however, in which the temperatures all rise uniformly with time. This is shown in Figure 3.6. At this point the temperature difference between sample and reference is zero, as indicated in Figure 3.6B.

As the temperature reaches the glass transition temperature of the sample, its heat capacity changes abruptly. The sample at this point absorbs more heat because of its higher heat capacity, and its temperature lag behind the reference sample is increased. Its temperature therefore changes more slowly than that of the reference, with the results indicated in Figure 3.6.

If the polymer sample is crystallizable but in the quenched amorphous state, crystallization will take place. The heat of fusion is evolved, the sample temperature rises, and the DTA trace shows an exothermic peak at T_c, the temperature at which the rate of crystallization is a maximum. On further heating the crystallites melt, heat is absorbed, and an endothermic peak occurs at T_m. This is defined as the temperature at which the rate of melting is highest (although this temperature is probably lower than T_m).

Other changes in the sample may occur. Oxidation or cross-linking (at T_{ox}) are exothermic, while decomposition (at T_d) is endothermic. Not all samples will exhibit all of the aforementioned phenomena.

To obtain accurate results by DTA, it is essential to achieve uniform temperature throughout sample and reference and to operate under steady-state conditions. Ideally, sample and surroundings should be in temperature equilibrium at all times. Some of the factors leading to abberations are mismatch in heat capacities between sample and reference, poor heat transfer, poor sample packing or improper particle size, lack of symmetry in the geometry, and effects arising from the presence of diluents.

The sample holder assembly or heating block must be made so that uniform heat transfer to both sample and reference takes place, without

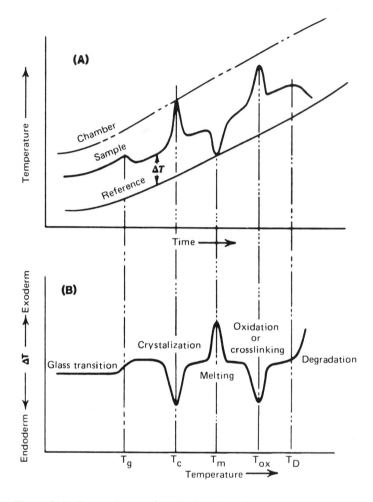

Figure 3.6 Dependence of DTA signal on time and temperature.

any fluctuations that might be mistaken for transitions in the sample. To obtain a reproducible heat flux that is relatively little affected by changes in the sample properties, a steep temperature gradient must be set up between the block and the sample. This is favored by the use of small samples and by keeping the sample and reference close together. Considerations of resolution and sensitivity cannot be neglected.

The selection of amount of sample and diluents is important. Diluents are inert materials (often identical to the reference) mixed with the sample to match its thermal conductivity and diffusivity to that of the reference. In general, small sample sizes and small amounts of diluent are preferred. A compromise must be made, however, since sensitivity increases with sample size but resolution decreases. In practice, samples of 25 to 50 mg are satisfactory when measuring large heats of transition but too small for the observation of glass transitions. Melting peaks can be sharpened by the use of smaller samples and lower heating rates. Peaks due to chemical reaction can be sharpened by increasing the heating rate. Maximum sensitivity and resolution are obtained by minimizing the sample size and increasing the efficiency of heat transfer from the sample to its thermocouple.

In addition to the amount of sample, the size and packing of the sample particles also influence the temperature measurement. Too small particle size increases the surface area and shifts the transition peaks to lower temperatures. In addition, small particles often undergo reorganization, causing thermocouple movement leading to artifacts in the temperature measured. Since thermal conductivity and diffusivity are influenced by sample density, tighter packing increases heat condition, reduces the chance of peak spreading, and improves reproducibility. Packing is especially important in cases where gaseous products are evolved or where samples are studied under a gaseous atmosphere.

The heating rate is an important parameter, is closely interrelated with sample size, thermocouple placement, and other variables. To avoid shifts in peak temperatures with heating rate, the sample thermocouple should provide the temperature base; otherwise, T_g tends to increase slightly with heating rate, and T_c and T_d significantly. T_m is little affected, but lower heating rates measure values of T_m closer to the equilibrium value as the result of annealing the sample during the DTA experiment. Unless the heating rate is linear, slight shifts in baseline can occur that may be mistaken for sample transitions.

DIFFERENTIAL SCANNING CALORIMETRY

Thermal analysis of polymers is effectively done using the technique of differential scanning calorimetry (DSC). In general terms, this test method provides a measurement of some material response to a programmed change of temperature. The parameter monitored can be mass, dimension, or an optical or acoustical property of the material. The DSC provides a measure of heat capacity (C_p). This property is important to polymers because the processing history, or any mechanical and/or thermal treatment, essentially confers its own "fingerprint" on the arrangement of polymer chains in the final product. This implies that the energy of the system is unique. Hence the change or integral of C_p with respect to a well-defined reference state such as the molten polymer characterizes the prior history of the sample.

Figure 3.7 schematically illustrates the operating principle of a DSC. The directly measured quantity can be either a differential power or temperature between two cells, where one cell contains the polymer sample and the other a thermally inert reference. In the case of a differential power instrument, each cell has its own heater and the differential quantity measured is the difference between the power input that is needed to maintain the same programmed temperature in each cell. If the sample is at its melting point, T_m, for example, the sample heater output must increase significantly as heat of fusion is supplied. When the instrument is based on a differential temperature measurement, both cells are regulated by a common power source. The temperature response signal, ΔT, arises from the differing thermal paths to, and properties of, the two cells. The sample temperature will remain constant at T_m until all the heat of fusion has been supplied. Since no such constraint applies to the reference cell, the quantity $\Delta T = T_s - T_r$ becomes quite negative in this region. Figure 3.8 illustrates the differential signal for a sample undergoing phase change. Both differential power and temperature instrument signals are similar, however, for the latter the sign is opposite. Note that the abscissa of the signal response plot is either time or temperature. For all practical purposes the two are essentially synonymous.

The signal from either type of DSC instrument is proportional to the difference in heat capacity between the sample and reference. Operation requires three experiments with the sample cell, which contains first an empty pan, second, pan + sample, and third, pan + calibrant. Through-

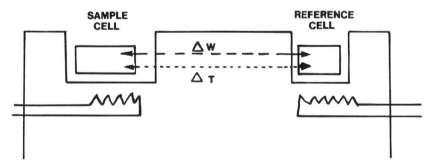

Figure 3.7 Operation of differential power (dashed line) and differential temperature (dotted line) DSC.

Thermal Analysis of Polymers

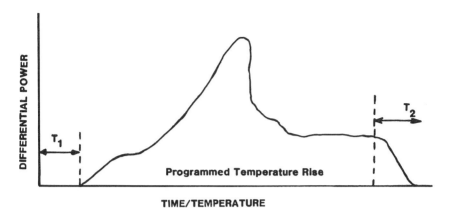

Figure 3.8 DSC signal (differential power) for sample undergoing a phase change.

out the experiment the reference cell is left undisturbed. It normally contains an empty pan, whose purpose is to balance roughly the effect of the pan in the sample cell. The following heat balances correspond to the three experiments:

Case 1—empty pan: $S_e = K\Delta Q$ (6a)

Case 2—pan + sample: $S_s = K(\Delta Q + m_s C_{Ps})$ (6b)

Case 3—pan + calibrant: $S = K(\Delta Q + m_c C_{Ps})$ (6c)

where ΔQ is the difference in heat capacity between the empty cells of case 1 and S is the signal. From the above, the governing expression for DSC is

$$C_{Ps} = \frac{(S_s - S_e) m_c C_{Ps}}{(S_c - S_e) m_s} \quad (7)$$

where m is the mass and C_P is the specific heat capacity. The calibration material varies, but is most often an alumina.

The accuracy of the instrument is quite good, with reproducibility typically to within $\pm 1\%$ by adiabatic calorimetry. Measurements can be made by either a heating or cooling cycle. It is generally good practice to check the agreement between the two modes.

Figure 3.9 illustrates the DSC record of the cure behavior of a typical EPDM. The record provides an indication of the various states or history

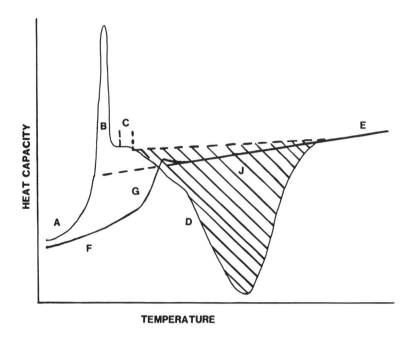

Figure 3.9 Heat capacity and curing exotherm of an EPDM.

of the sample. Reactants were compounded to a glassy material (A) that liquefies at the glass transition (B). Region (C) represents the purely liquid region, which is quite limited because the exothermic curing reaction (D) depresses the apparent heat capacity. At the termination of the reaction cycle, the heat capacity value reflects that of a rubbery cross-linked material (region E). Upon cooling to room temperature and reheating, a simple curve is generated. The cooling curve reveals a glass transition (G) where the low-temperature glass (F) transforms to a rubbery state (J). Note that the latter curve merges with the high-temperature region (E) from the previous experiment. The shaded region in Figure 3.9 represents the apparent heat of reaction.

An important property of state obtained by DSC is the T_g. The glass transition can be described as a kinetically dominating event in which the time scale for molecular motion is comparable with that of the experiment. Below T_g some molecular motion is frozen. The implication from Figure 3.9 is that T_g is not unique; in other words, when we examine the extremes of quenching and slow cooling (i.e., annealing), the latter results in a lower T_g value because relaxation takes place in a more stable state during cooling. This is illustrated best by examining specific volume on enthalpy curves, as shown in Figure 3.10. T_g values are essentially the point of intersection of curves for the glassy and liquid states. The calorimetric equivalent of specific volume is enthalpy. Recall that DSC gives C_P, which is the derivative of enthalpy. This has led to some confusion in the use of DSC, as it is far simpler to take some well-defined point from the C_P region as T_g (e.g., an extrapolated onset, the temperature of maximum slope, or other). However, when we examine quenched versus annealed samples, the expected order of values is reversed. In Figure 3.10, curve (q) represents the quenched case and (a) the annealed glass. The latter is obtained by slow cooling or by allowing the quenched material to relax at temperatures slightly below T_g. Upon reheating the "q" form superheats considerably less than the "a" form and the apparent glass-transition temperatures, as measured from the discontinuity in the DSC curve, are reversed. The problem of kinetic source can be eliminated at very slow heating rates; however, the advantage of a rapid test method no longer exists. Richardson (1984) notes that it is possible to obtain T_g values characteristic of the low-temperature glass that are independent of the instruments heating history by following some simple guidelines. The peaks of Figure 3.9 reveal the tendency for annealed glasses to superheat at typical DSC scanning rates (also illustrated in Figure 3.4, where the DSC peak represents the enthalpy increment required to catch up with the equilibrium liquid line). Selecting T_1 and T_2 as two temperatures which are un-

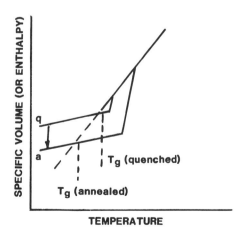

Figure 3.10 Details of specific volume curves.

ambiguous in the glassy and liquid regions, one can integrate the DSC derived C_P-T curve to obtain $H_1(T_2) - H_g(T_1)$, where H is enthalpy. The heat capacities below T_1 and above T_2 can be approximated by linear relations and integrated to give

$$H_g(T) = aT + 0.56T^2 + c \tag{8}$$
$$H_1(T) = AT + 0.5BT^2 + C \tag{9}$$

The integration constants, c and C, can be obtained by substituting T_1 in equation (8) and T_2 in equation (9). From C-c values, T_g is computed as a solution of the expression $H_g(T_g) = H_1(T_g)$.

One of the first uses of DSC, still prevalent today, is the study of crystalline polymers. The area under the melting peak will increase as the degree of crystallinity increases. Traditional, less accurate analyses have been aimed at measuring this area using a planimeter or simple area counting techniques and based on a known control sample, estimating the level of crystallinity of synonymously, the heat of fusion. Richardson (1984) notes, however, that there are several subtleties in the interpretation of this area, which can lead to inaccurate speculation. To interpret the DSC thermogram properly, one must view the data not in terms of C_P, but rather in terms of its integral (i.e., an enthalpy change). It is important to note that $H_1(T_r) - H_x(T)$ (where x refers to the degree of crystallinity) may be obtained as a function of T over any temperature range from T to T_R, where T_R is a reference temperature in the molten state. The analysis requires construction of a proper baseline that can be derived through the definition of $\Delta H(T)$:

$$\Delta H(T) = H_1(T) - H_x(T) = [H_1(T_R) - H_x(T)] - [H_1(T_R) - H_1(T)] \tag{10}$$

This represents the behavior of the supercooled liquid as denoted by the second term in brackets in equation (10). This corresponds to the low-temperature extrapolation of the liquid C_P on the DSC curve. This is a straightforward procedure because C_{p1} is usually a linear function of temperature. For high-melting polymers it is more complex to extrapolate to room temperature.

An obvious question that arises is: Can ΔH be related to room-temperature crystallinity, to which many other measurements refer? When a crystalline polymer is cooled, the degree of crystallinity, x, increases to some low-temperature limit s_c, as illustrated in Figure 3.11. This is a function of the conditions of the crystallization process. Consider two

Thermal Analysis of Polymers

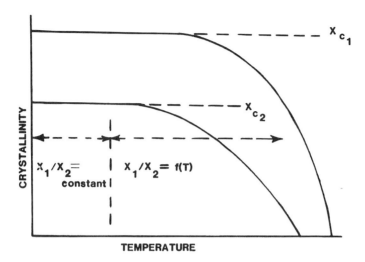

Figure 3.11 Variation of crystallinity with temperature.

samples with different crystallization histories. The ratio x_1/x_2 will vary with temperatures until both are in their x_c regions, in which case $x_{c1}/x_{c2} \neq f(T)$. The definition of enthalpic crystalinities is

$$x(T) = \frac{\Delta H(T)}{\Delta H_\infty(T)} \tag{11}$$

where H_∞ is the enthalpy of a perfectly crystalline polymer.

The onset of melting in one of the samples can be found by observing at what temperature the ratio $\Delta H_1(T)/\Delta H_2(T)$ is no longer constant. This enables one to relate high-temperature DSC measurements to more traditional methods which are usually based on room-temperature conditions. DSC runs can also be initiated well below room temperature for direct measurement of ΔH at this point. Relative crystallinites can be obtained from DSC results with a high degree of accuracy by relating measurements to some internal standard such as material crystallized in the DSC. Absolute values of calorimetric crystallinities require measurement of $H_\infty(T)$. This can be obtained from a measurement of x using x-rays.

An important aspect to consider is the existence of multiple melting peaks. This can be attributed to several factors. First, crystallization is very sensitive to the thermal history that the polymer is subjected to during processing. If the material is maintained at different temperatures, crystal growth will be promoted in those regions. As a result, on heating, a minimum can be detected at the annealing temperature, since those crystals formed on annealing will melt above the annealing temperature and those crystals formed on cooling from the annealing temperature will melt below that temperature. An example of processing pretreatment is illustrated in Figure 3.12. The three EPDMs are chemically and structurally identical, but they were processed and annealed under different conditions. Thermograms of this type offer the possibility of controlling processing conditions. Also, the influence of thermal history can be distinguished from structural differences by introducing a constant thermal history to the samples under study, such as cooling at 20°C/min from above the melt. If the rescans between samples are then identical, the initial differences may be attributed to thermal history. On the other hand, if the rescans continue to show differences, the samples are likely to be structurally different.

Multiple melting peaks can be observed in blends and mixtures. When two polymers are mixed together, they may retain their individual thermal characteristics. Figure 3.13 illustrates the thermogram of a blend of a high- and a low-density polyethylene. Branching tends to impede crystallite perfection; hence the low-density constituent melts at the lower tempreature, being almost completely resolved from melting of the crystallites formed from the linear molecules. Scans of materials such as waxes are comprised of mixtures of chemically similar and structurally different molecules. Melting thermograms of these materials can show multiple peaks as well. Wax thermograms in general are highly characteristic and can be used as a means of identification.

Since all materials will supercool, the rate of crystallization may be faster than the rate of melting. In this case, the crystallization exotherm may be taller than the melting endotherm. If complete crystallization has

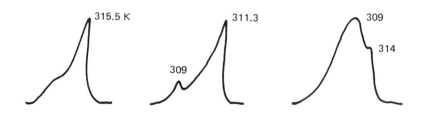

Figure 3.12 Effect of pretreatment on an EPDM's melting behavior.

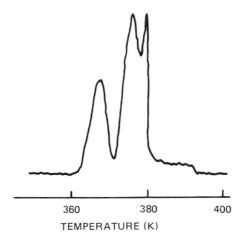

Figure 3.13 Example of multiple melting peaks from a physical blend of a branched and linear polymer.

occurred, the peak areas (i.e., energies) will, however, be the same. A problem in crystallization studies is, however, that a number of materials do not crystallize completely, or even partially, on cooling. For example, organic constitutents that have been crystallized from solution often tend not to crystallize from the melt. With polymers, the process of crystallization is not instantaneous but, instead, involves the processes of nucleation and propagation. As a result, the extent of crystallization also depends on the rate of cooling.

In cold crystallization, polymers crystallize very slowly so that they can be cooled rapidly to a completely amorphous glassy state. When they are heated through the glassy region, molecular mobility increases and the substance crystallizes. This behavior is referred to as cold crystallization since it occurs at temperatures far below the melting point.

As noted earlier, polymers tend to melt over a broad temperature range because of the surface free energy of the smaller, less perfectly formed crystals. Once these crystals are melted, however, the free energy of the liquid can be higher than that of the more perfectly formed crystals, which are stable in the temperature range. Consequently, it is within thermodynamic principles for the liquid to crystallize as shown in Figure 3.14. Controlled cooling of the copolymer results in partial crystallization, as witnessed by the absence of the cold-crystallization exotherm observed for the amorphous sample. On the other hand, the morphologies of the crystals formed give rise to recrystallization, as evidenced by the exother-

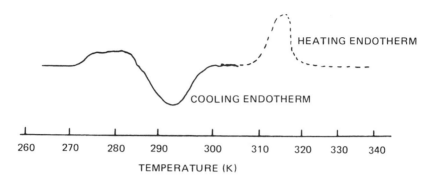

Figure 3.14 Thermograms of a copolymer.

mic peak at about 315 K, during the melting cycle of the experiment. These crystallization effects on heating influence determination of the true crystallinity of the sample. Simply considering the area under the fusion peak as a measure of crystallinity is likely to result in large errors.

A final note on crystallization concerns tests that are conducted under isothermal conditions. It is important to realize that isothermal operation can affect the degree of crystallization as well as the size distribution of the melting curve. It is important to standardize measurements to account for the degree of supercooling on the rate of crystallization.

The DSC is used routinely in polymer characterization studies. Specific thermodynamic properties measured include specific heats and heats of fusion. The technique can be used for distinguishing folded chain from extended chain morphologies, for evaluating the effects of comonomers or chain substitutes on morphology and general thermal behavior, for crystallinity determinations, and for determining thermal stability and crystallization rates. In addition, the DSC technique can be used for evaluating the effects and effectiveness of additives such as plasticizers, antioxidants, and accelerators, as well as catalyst and nucleating effects on the polymer's thermodynamic properties. Although DSC methods are not generally thought of in terms of processability testing, the technique enables one to establish useful correlations between DSC derived data and finished product quality. Many polymer processing operations are critically dependent on the control of thermal treatments; examples are curing, heat setting, annealing, shock cooling, and hot drawing. Each of these techniques can be studied and controlled using DSC techniques. A simple example of this is given in Figure 3.15, which shows the thermograms of two polymers made with the same catalyst system

Thermal Analysis of Polymers

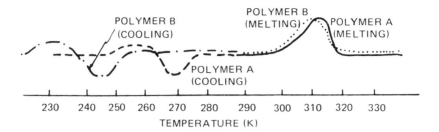

Figure 3.15 Cold crystallization detected by DSC. Differences between polymers can be related to mold shrinkage.

(solution polymerization) but at greatly different reactor mixing conditions. Note that both polymers show the same heats of fusion and therefore have the same level of crystallinity. In addition, they both show nearly identical melting ranges and peaks, but upon cold crystallization, polymer B (well mixed in reactor) crystallizes more that 20° below polymer A. We could conclude that polymer B is a more homogeneous polymer in terms of its ethylene distribution. In an injection molding operation we may expect polymer A to display the problem of mold shrinkage. In simple mixing/milling tests both polymers were compounded identically, rolled into sheets, and allowed to set up for 8 hr. It was observed that the polymer A showed nearly twice as much shrinkage as polymer B. Hence we may conclude that DSC can be employed routinely for processability testing in a manner that involves empirically relating thermodynamic properties to product processing performance.

The basic differences between DTA and DSC lie in the design of the heating system and the mode of operation of the instrument. In DSC, the sample and reference are heated separately by individually controlled elements. The power to these heaters is adjusted continuously in response to any thermal effects in the sample, so as to maintain sample and reference at identical temperatures. The differential power required to achieve this condition is recorded as the ordinate on an X-Y recorder, with the (programmed) temperature of the system as the abscissa.

In classical DTA, the sample forms a major part of the thermal conduction path. Since its thermal conductivity changes in a way that is generally unknown during a transition, the proportionality between temperature differences and energy changes is also unknown. This makes the conversion of peak areas to energies uncertain and relegates DTA to an essentially qualitative type of test.

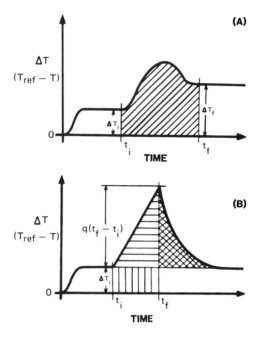

Figure 3.16 Typical DSC traces of (A) slow and (B) rapid transitions involving an enthalpy change.

In contrast, as noted above, in DSC the peak area is a true measure of the electrical energy input required to maintain the sample and reference temperatures equal, independent of the instrument's thermal constants or any changes in the thermal behavior of the sample. The calibration constant relating DSC peak areas to calories is known and constant, permitting a quantitative analysis.

The calculation of enthalpy from DSC differs depending on whether the transition is slow enough that a steady-state condition is maintained (Figure 3.16A) or so rapid that the change of the sample temperature is halted until sufficient heat is supplied (Figure 3.16B). Considering the former, assuming that there is no difference in heating rates for sample and reference, so that the DSC trace returns to the original baseline after the transition, then

$$\Delta H = \int_{T_i}^{T_f} C_P \, dT = \int_{T_i}^{T_f} \frac{K \Delta T}{q} \, dT \tag{12}$$

Thermal Analysis of Polymers

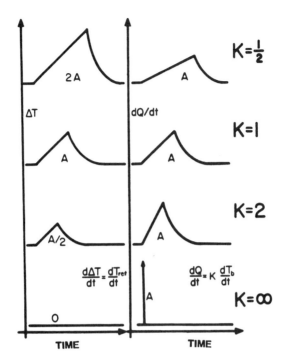

Figure 3.17 Effect of a change in thermal conductivity on DTA and DSC traces.

where the subscripts i and f refer to the initial and final temperatures of the transition. The effect of a change in K on peak area and shape for a sharp transition is illustrated for DTA and DSC in Figure 3.17. In DSC, peak area is independent of K since this constant enters only into the time lag between the temperature of the sample and its holder. Resolution improves as K increases. In DTA the latter is also true, but peak area decreases as K increases.

MEASUREMENT OF THERMAL CONDUCTIVITY

Several steady-state methods can be used. Figure 3.18 illustrates one approach where the thermal conductivity of a sample is determined by comparing the time required to vaporize a given amount of a known liquid when heated by conduction through the sample and through a reference material. In the apparatus a pure liquid is boiled in A, heating a silver plate S_1, and returned after condensation to A. An upper vessel, B, contains

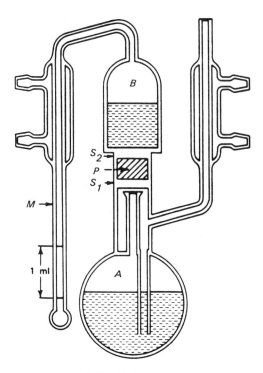

Figure 3.18 Colora apparatus.

a liquid boiling 10 to 20°C lower than that of the liquid in A. A second silver plate S_2 forms the bottom of B, and the sample is fitted between the two plates at P. Vapor from liquid boiling in B is condensed into a graduated vessel M.

Since the plates S_1 and S_2 are maintained at the constant-boiling points of the liquids in A and B, a fixed temperature gradient $\Delta T = T_A - T_B$ is imposed on the sample of thickness l and area A. When the steady state is reached, the time t necessary to distill a fixed amount (1 ml) of the liquid in B into M is measured. If the heat of vaporization of this amount of liquid is ΔH_v,

$$K = \frac{l \, \Delta H_v}{A t \, \Delta T} \tag{13}$$

where l and A are the thickness and area of the sample, respectively. In practice, one takes as a calibration parameter the thermal resistance $R' = t\,\Delta T/\Delta H_v$ in s · °C/cal,

$$K = \frac{l}{R'A} \qquad (14)$$

R' is first determined for a standard material of known thermal conductivity, for a range of sample thicknesses yielding times encompassing those anticipated for the polymer sample. A plot of R' versus t is constructed and values read off for the observed distillation times for polymer samples, again at several values of l. Substitution into equation (14) completes the calculation.

THERMOGRAVIMETRIC ANALYSIS

Thermogravimetric analysis (TGA) is a dynamic technique in which the weight loss of a sample is measured continuously during a state where its temperature is increased at some constant rate. An alternative approach is to measure weight loss as a function of time at constant temperature. The main use of TGA in application to polymers has been in studies of thermal decomposition, auto-oxidation, and stability. A DTA trace provides information of the temperatures at which a thermal event occurs. TGA provides data on whether the event is accompanied by a weight loss. The use of gas chromatography or mass spectrometry to analyze the effluent gases provides positive identification. A typical thermogram is illustrated in Figure 3.19, comprised of a plot of change in weight versus temperature. A small initial weight loss, such as that from w to w_0, generally results from desorption of solvent. If it occurs near 100°C, this is usually assumed to be loss of water. In the figure, extensive decomposition starts at T_1, with a weight loss $w_0 - w_1$. Between T_2 and T_3 another stable phase exists; then further decomposition occurs.

In many thermograms, phemomena occur so closely spaced that it is difficult to assign appropriate temperatures. This is more easily done with reference to the differential curve of rate of weight change versus temperature. The area under the curve provides the change in weight.

Increasing the heating rate in TGA increases the apparent decomposition temperature, and too high a rate may cause a two-step process to be seen as one because equilibrium conditions are not achieved. The pres-

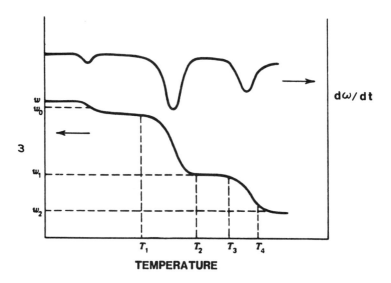

Figure 3.19 Typical TGA thermogram.

ence of a self-generated gaseous atmosphere can obscure details of the process, and it is often advantageous to supply a small flow of inert gas through the furnace chamber. The shape of the sample container may be important during volatilization or when a large volume of gas is evolved. The reason for this is that these processes depend on the surface area of the sample and the way it changes with time. Also, large sample particles, or thick sections, may lead to inefficient heat transfer and temperature gradients, either from the applied heat or from heats of reaction. Problems with the diffusion of volatiles may also arise. In general, decreasing the particle size lowers the temperature of both onset and completion of thermal decomposition. The effect of solubility of gases in the sample is pronounced and difficult to measure or eliminate. The shape of the TGA curve is a function of the kinetics of the decomposition reaction, specifically the order of the reaction and the activation energy and frequency factor in the Arrhenius equation. Table 3.2 provides a summary of the various test methods used for thermal behavior analysis, together with factors that may influence measurements.

NOTATION

A area
C_p heat capacity

Table 3.2 Summary of Test Methods for Thermal Analysis

Method	Property varying	Principle behind measurement
Dilatometry (linear or volume)	Volume, V, or dV/dT	Equilibrium thermodynamics
Refractometry	Refractive index, β-ray absorption	Equilibrium thermodynamics
Dilatometry (volume)	Compressibility, $(dV/dP)_T$	Equilibrium thermodynamics
Calorimetry (DSC)	Specific heat	Equilibrium thermodynamics
Infrared spectroscopy	Vibrational energy levels	Transport theories
Nuclear magnetic resonance	Proton environment	Transport theories
Stress birefringence	Refractive index	Transport theories
Dielectric constant and loss	Dipole moment	Transport theories
Creep and stress relaxation	Viscosity and elastic modulus	Transport theories
Dynamic mechanical properties	Mechanical energy absorption	Transport theories
Impact resistance	Failure mode	End-use properties
Softening point	Viscosity	End-use properties
Heat deflection	Viscosity	End-use properties
Hardness	Viscosity	End-use properties

f fraction of free volume
H enthalpy
H_v heat of vaporization
H_∞ enthalpy of perfectly crystalline polymer
K constant; also thermal conductivity
l thickness
$\overline{M}_w$ weight-average molecular weight
m mass
Q, q heat flow
R' thermal resistance
S signal

T absolute temperature
T_g glass-transition temperature
$T_{g\infty}$ glass-transition temperature at infinite molecular weight
T_m melt temperature
t time
V_f free volume
V_o occupied volume
w weight fraction

Greek Symbol

α thermal expansion coefficient

REFERENCES AND SUGGESTED READINGS

Brown, R. P., ed., *Handbook of Plastics Test Methods,* George Godwin, London, 1971.

Fox, T. G., and P. J. Flory, *J. Appl. Phys., 21,* 581–591 (1950).

Gibbs, J. H., *J. Chem. Phys., 25,* 185–186 (1956).

Gibbs, J. H., and E. A. DiMarzio, *J. Chem. Phys., 28,* 373–383 (1958).

Hirai, N., and H. Eyring, *J. Appl. Phys., 29,* 810–816 (1958).

Hirai, N., and H. Eyring, *J. Polym. Sci., 37,* 51–70 (1959).

Legge, N. R., G. Holden, and H. E. Schroeder, eds., *Thermoplastic Elastomers,* Hanser, Munich, 1987.

Richardson, M. J., pp. 101–116 in *Measurement Techniques for Polymeric Solids,* R. P. Brown and B. E. Read, eds., Elsevier, New York, 1984.

Walsh, D. J., J. S. Higgins, and A. Maconnachie, eds., *Polymer Blends and Mixtures,* NATO ASI Series, Martinus Nijhoff, Boston, 1985.

4
Measurement of Stress-Strain Properties

INTRODUCTION

In this chapter we review the basic concepts and test methods applied to the mechanical testing of rubbers and plastics. Mechanical testing of materials are performed for a variety of reasons, the principal ones being:

For comparing materials and establishing a basis for material selection and in product development
As a criterion in quality control
To provide data for design purposes
To establish a basis for predictions of service performance
As a starting basis for the formulation of theories in material science

The mechanical testing of plastics and rubbers has evolved from a collection of simple concepts to a subject area of high sophistication, interwoven with literally hundreds of standard tests. Indeed, there are extensive volumes devoted entirely to this subject; key references are given at the end of this chapter. The purpose of this chapter is simply to highlight the important concepts and tests often applied in largely product development and quality control programs.

HARDNESS

The term *hardness* is generally accepted as an end-performance property that can have slightly different meanings, depending on the application for which the plastic or elastomer is intended. It is generally accepted as an indication of the resistance to indentation, scratch resistance, and/or rebound resilience. Resistance to indentation is an interpreted measurement prior to fracturing. It is a function of the sample's rigidity or modulus. The measurement itself is a function of temperature, time, and the particular test method employed, for which there are several. Specific standards will not be reviewed or compared here, but rather, only the general test measurement concept is discussed. It is important to emphasize that hardness values obtained from one method in general cannot be compared with those derived from another, although data can be empirically correlated.

ISO Standards report three methods for measuring hardness: a Shore hardness, a ball indentation method, and Rockwell hardness. The Shore A method utilizes a truncated cone as the identor, with a minimum specimin thickness of 5 mm required. This is generally applied to elastomeric compounds and softer plastics. The Shore D method is used to characterize harder plastics and utilizes a sharp cone having a slightly rounded end. Shore D is applied to thinner speciments (usually down to 3 mm in thickness). The critical dimensions of the indentors are shown in Figure 4.1.

The viscoelastic properties of polymers necessitates that measurements be taken after a prespecified or standard time period. This period of

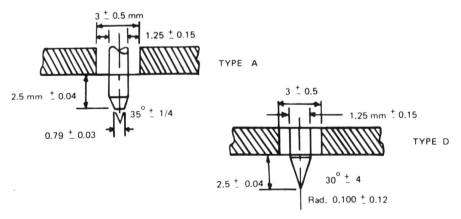

Figure 4.1 Key features of Shore durometer indentors.

Measurement of Stress-Strain Properties

time is usually 15 ° 1 sec; however, in some cases an instantaneous hardness may be meaningful after 1 sec of application. The instrument used to obtain the measurement is called a durometer.

In the ball indentation method, a 5-mm-diameter hardened steel ball is pressed into the specimen under a specified load. The indentation is between the limits 0.07 to 0.10 mm for method A and 0.15 to 0.35 mm for method B. The load is applied for a specified period of time (30 sec) before the depth reading is taken, and the test sample is 4 mm thick. The ball indentation hardness value is computed for the ratio of the applied load to the surface area of the impression left by the ball.

The Rockwell hardness (discussed later again) involves the determination of Rockwell and differs from the Standard Rockwell scales. Rockwell α correlates with the hardness as determined by the ball indentation method, using the following correlation:

$$h = \left(\frac{441.4}{150 - R_\alpha} \right)^{1.23} \tag{1}$$

where h is the hardness by ball indentation and, R_α is the Rockwell α hardness number.

In *British Standards,* a "softness number" is used for a measure of the hardness of plastics. This is essentially the same test as the ISO ball indentation method. The softness number is defined as the depth of indentation of the ball in 0.01-mm units caused by an increase in force (from 0.3 to 5.7 N) after it has been applied for 30 sec. Specimen dimensions specified are 8 × 10 mm thick for the standard test; however, other thicknesses are permissible provided that specimen dimensions are reported along with softness numbers.

In *U.S. Standards,* Shore A and D durometers that use the normal Rockwell scales are employed, as well as a test based on the Barcol impressor. Shore hardness measurements are nearly identical to the ISO test methods. A Rockwell hardness tester will also measure Brinell hardness. The Rockwell scales are summarized in Table 4.1. There are two ASTM test methods (procedure A and procedure B). In the first method, the minor load (Table 4.1) is applied for 10 sec and then the major load for 15 sec. The hardness reading is obtained from the scale 15 sec after the major load has been removed, but with the minor load still applied. This method is limited to scale R only. In procedure B, indentation is recorded 15 sec after application of the major load, but with the latter still applied. Test values are reported as Rockwell α hardness numbers and are computed by subtracting the indentation reading from ISO.

Table 4.1 Partial Listing of Rockwell Scales Given in ASTM

Rockwell hardness scale	Minor load (kg)	Major load (kg)	Indentor diameter (mm)
R	10	60	12.700 ± 0.0025
L	10	60	6.350 ± 0.0025
M	10	100	6.350 ± 0.0025
E	10	100	3.175 ± 0.0025
K	10	150	3.175 ± 0.0025

The Barcol impressor is illustrated in Figure 4.2. The device consists of a hardened steel truncated cone. The indicating arrangement has 100 divisions, each denoting 0.0076 mm of penetration. The higher the reading, the harder the material. The housing of the indentor is manually applied by hand, and the highest dial reading is recorded.

German Standards are essentially the same as those of ISO, with the exception that the major load must be selected from 49, 132, 358, and 961 N, with a minor load of 9.81 N in all procedures. The basic formula used to calculate the hardness number is

$$h = \frac{0.0535F}{d - 0.04} \tag{2}$$

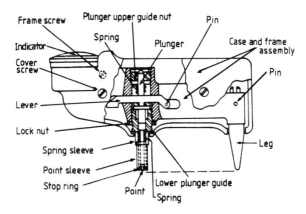

Figure 4.2 Barcol impressor.

Measurement of Stress-Strain Properties

where d is the penetration (mm) and F is the applied major load.

A variety of miscellaneous hardness scales have been developed throughout the rubber and plastics industry which are occasionally used for in-house quality control testing. The Brinell method is one such technique, which employs a spherical indentor. The hardness number is computed from the following formula:

$$h_B = \frac{2F}{\pi D^2 \{1 - [1 - (d_i/D)^2]^{1/2}\}} \quad (3)$$

where F is the load (kg), D the indentor diameter (mm), and d_i the diameter of impression (mm).

The impression is created again over a time period of 15 sec. Indentors employed are 1, 2, 5, or 10 mm in diameter, with loads selected to give F/D^2 values of 30, 10, 5, or 1. Values of h_B are reported along with F/D^2 ratios employed. The Brinell scale and method was originally developed for metals, and hence the F/D^2 values of 5 or 1 are usually needed for plastics.

Other hardness tests occasionally employed are the Vickers diamond pyramid test, the Wallace microhardness tester, the Knoop hardness test, the Sward hardness test, and the TNO hardness test. A brief description of each of these tests is given by Brown (1981) and Turner (1983). Table 4.2 provides some general comments on each of these methods.

STRESS-STRAIN PROPERTIES AND TESTING

We will consider how stress-strain information provides an understanding of the mechanical behavior of polymers, and examine the influence of molecular structure on the ultimate properties of polymers. Stress-strain behavior depends greatly on temperature, the thermal history of the sample, strain rate, magnitude of the strain, and other test conditions. As such, it is necessary to conduct experiments at a variety of conditions in order to obtain a full spectrum of information.

The most frequently applied stress-strain measurement is made in tension (i.e., stretching the specimen), as shown in Figure 4.3. A tensile stress can thus be defined by

$$\sigma_1 = \frac{F_1}{A_0} \quad (4)$$

where σ_1 is the tensile stress, F_1 the tensile force, and A_0 the cross-sectional area of specimen. If the specimen is stretched by the tensile stress to a length l_1, the tensile strain is

Table 4.2 Miscellaneous Hardness Tests

Test name	Origin	Application	Description/comments
(Vickers) diamond pyramid test	Metals testing	Hard plastics	Employs a right diamond pyramid on a square base, with an apex angle of 136° between opposite facets; object is to simulate one indentor in the Brinell range Principal equation: $h_v = \dfrac{2F \sin(\theta/2)}{d_w^2}$ where F is the applied load (kg), d_w the mean diagonal width (mm), θ the apex angle of pyramid (136°), and $F = 5$ kg load normally
Wallace microhardness	Adapted from diamond pyramid test	PVC, acrylics, and similar plastics; an adaptation is available for rubbers	Nondestructive test that relates to surface of specimen; a wedge system and dial gauge permit indentation to be measured to 0.00025 mm; the small size of specimens required allows hardness measurements of small O-rings and irregularly shaped articles; results are read directly in International Rubber Hardness Degress.
Knoop hardness test	Plastics	Soft to nonbrittle plastics	Instrument used is a commercial device known as the Tuken tester; consists of a diamond but with ratio of long and short diagonals of ~7:1 and depth of indentation of ~1/30 of its length
Sward hardness test	Paint films	Nondestructive testing of plastics	—

Measurement of Stress-Strain Properties

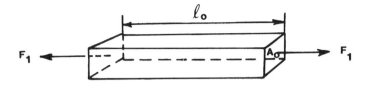

Figure 4.3 Specimen in tension.

$$\varepsilon_1 = \frac{l_1 - l_0}{l_0} = \frac{kt}{l_0} \tag{5}$$

where k is the rate of extension and, t is the time. Continuing the stressing test to the ultimate, that is, measuring the force at which the material breaks, results in a tensile strength known as the ultimate tensile stress:

$$\sigma = \frac{F}{A} \tag{6}$$

where F is the force at failure and A is the area of cross section at failure.

During the process of stretching, the specimen's dimensions orthogonal to the axis of the applied force decrease and so does the area of cross section. To normalize test results, however, tensile strengths are usually based on the original cross section (also, A_0 is most readily measured at the beginning of the experiment).

Elongation at break (called ultimate elongation) is usually expressed as a percentage of the original sample length:

$$\varepsilon = \frac{l - l_0}{l_0} \times 100 \quad \% \tag{7}$$

where l is the length at failure and l_0 is the original length.

Changes in stress-strain behavior within a single amorphous polymer as a function of temperature are illustrated in Figure 4.4. A modulus-temperature curve is also shown to illustrate the regions to which the stress-strain curves apply. Below T_g, polymers exhibit a brittle stress-strain curve, which gradually changes through a ductile to an elastomeric type. As the strain rate (i.e., speed of testing) increases for an elastic material, the trend is reversed. Ultimately, the response time of the polymer exceeds the time scale of the experiment, and the material behaves like a brittle

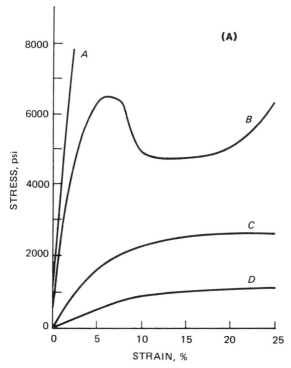

Figure 4.4 (A) Stress-strain curves at different temperatures; (B) corresponding modulus temperature curve.

solid. Stress-strain measurements on amorphous polymers can be treated by time-temperature equivalence to produce a master curve that is useful in predicting long-term data from short-term experiments in the rubbery region.

Practical information on polymer properties can be obtained from consideration of the material's failure envelope (Figure 4.5). The failure envelope is obtained as a plot of stress versus strain at failure over a range of strain rates. As temperature is increased or the strain rate decreased, the point of failure moves clockwise, the stress at break decreasing continuously while the strain at failure first increases during the transition from brittle to ductile behavior, then decreases again as the material enters the rubbery region.

At temperatures $\ll T_g$, the molecular weight of the polymer and branching or cross-linking have little influence on stress-strain behavior. Near T_g, however, and even somewhat below T_g for the more ductile

Measurement of Stress-Strain Properties

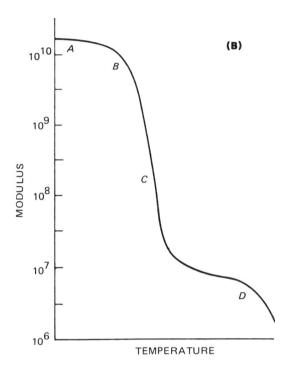

materials that can cold draw, the elongation at break is greater for samples of high molecular weight compared to those at low molecular weight. In the rubbery state, elongation at break generally increases with molecular weight. With broader polymers, the tensile strength decreases while the elongation increases. Cross-linking tends to decrease the elongation at break. Tensile strength increases with cross-linking initially, but goes through a maximum and decreases at high degress of cross-linking.

Crystallinity adversely affects the strength of polymers below T_g because the crystallites can act as stress concentrators. However, relatively high degrees of crystallinity (30 to 85%) can convert a rubbery material into a tough rigid polymer (examples are polyethylene and polypropylene). The type of morphology has a strong influence on the stress-strain relationship. An increase in crystallinity tends to raise the modulus but decreases the elongation at break. The interpretation of stress-strain curves of crystalline or semicrystalline polymers is difficult without a knowledge of bulk rheological and morphological characteristics of the material.

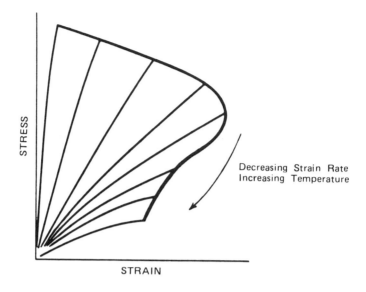

Figure 4.5 Failure envelope.

Under proper conditions, most polymers will cold draw. The resulting orientation leads to significant changes in material properties. Cold drawing is very sensitive to temperature and strain rate.

In addition to thermoplastics, polymeric materials of practical interest include polyblends, fiber (and filler)-reinforced plastics, elastomers, and thermoset materials. The stress-strain properties of these materials depend not only by the nature of the polymers but also by the adhesive strength of the interface.

The most widely used instrument for stress-strain measurement is the Instron Tensile Tester. This instrument is essentially a device in which a sample is clamped between grips or jaws which are pulled apart at constant strain rates varying from 0.5 to 500 mm/min. The stress on the sample is followed with load cells whose capacity ranges from 2 g to 500 kg or more. The elongation of the sample can be measured separate from the motion of the jaws to improve precision and avoid errors arising from slippage. There are a variety of jaws which can hold different samples, from filaments or fibers to large bars. Jaw design and specimen shape and preparation are selected so as to minimize the introduction of extraneous stresses or strains. The Instron can be preset to strain a sample at a given rate to a given extension; to return the crosshead automatically to the initial point after a given maximum extension of the sample; to cycle the crosshead between two adjustable points of extension (for hysteresis measurement); to stop the extension at a preselected point for stress-relaxation studies; or to stop the return.

The tensile modulus of elasticity, E, is defined as the ratio of stress to strain, and is determined from the initial slope of the stress-strain curve:

$$E = \frac{\sigma}{\dot{\gamma}} = \frac{Fl_0}{A(l - l_0)} \qquad (8)$$

where $\dot{\gamma}$ is the strain rate.

The methods described in this section fall under the category of short-term mechanical properties. In general, these provide only limited information on the properties and responses of materials. As such, the methods and concepts are applied principally in quality control, rather than fundamental property characterization.

Tests can typically be conducted in one of two modes: using a constant rate of loading or a constant rate of traverse. The latter method is the most extensively employed in polymer testing. Among the constant-rate-of-loading tests, the most often used apparatus is based on the so-called steelyard principle illustrated in Figure 4.6. In this arrangement, the poise weight is driven at a steady rate from left to right, from a position of equilibrium about the fulcrum, creating a situation where there is no force on the test speciment. The moment of the poise weight about the fulcum is increased steadily, resulting in the force on the specimen to increase at a steady rate. The poise weight is operated by an engaging screw running parallel to the steelyard. This type of device is most often used in adhesives testing.

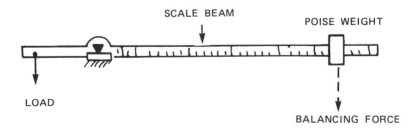

Figure 4.6 Constant rate loading steelyard-type test machine.

The constant rate of traverse machine has already been described; however, a few more notes on this subject are needed. As described earlier, instruments such as the Instron tester operate by straining the specimen and measuring the resulting force. It is important to note that, in fact, very few instruments ever actually accomplish a constant rate of extension of the specimen. Tensile specimen rarely extend uniformly over their entire length. As such, even if a constant rate of separation of the jaws holding the ends of the specimen is achieved, extension will only be increased at a constant rate on the average over the entire length of the sample. Furthermore, the separation rate of the driven jaws is the sum of the deformation of the specimen plus the movement of the jaws themselves. This means that the rate of jaw separation, whether constant or not, is related to the specimen deformation by some system variable. The deformation process includes, of course, the actions of stretching, bending, and/or compression. A true constant rate of extension can only be achieved by obtaining a gauge length of the specimen, over which straining properties are uniform, and by controlling the separation rate of the jaws with reference to the extension of the gauge length. It must be concluded that a constant-rate-of-separation experiment, which involves one crosshead holding one jaw from the other crosshead, which holds the other jaw, is simply an experiment conducted under the condition of *constant rate of traverse*. Many instruments thought to operate on this basis may in fact not. For example, the dynamometer, which is a force-measuring device, is found to "yield" significantly under the load it is measuring. In other words, it contributes to the apparent extension of the tensile specimen. Although current instrumentation has these shortcomings, within the practical property limitations of most plastics, and certainly extended elastomers, the deviation from constancy of rate of traverse is not serious, provided that the load is increased in the absence of surges and the time to break is accomplished over a well-defined time frame.

Sample Preparation and Test Procedures. Different tests are performed to charcterize different properties and as such they require different test specimens. Tensile tests on different materials therefore employ specimens of different sizes. To conduct a stretching or tensile test, an elongated specimen capable of being gripped at both ends is required. A rectangular bar would not be suitable because even if its thickness is small, fracture can occur anywhere along its length. In such a case, a failure load would be obtained which is lower than the characteristic value for the material in the specimen cross section. Furthermore, the failure can occur outside the region of the test sample being studied for elongation (i.e., the gauge length). To select a portion in which failure is to occur and thus obtain a breaking load on a cross-sectional area that is unaffected by the gripping mechanisms, a "dumbbell" specimen is employed. The basic types of dumbbell configurations and dimensions recommended by ISO are illustrated in Figure 4.7A–E. The American specifications differ only in number of dimensional details and are based essentially on Imperial units. With rigid plastics, the specimen can be machined on a lathe, molded, or simply cut out from thin, flat sheets.

The narrow-waisted dumbbell shown in Figure 4.7B is generally preferred in testing plastics with a high elongation at break and is also used for testing rubbers. The broad-waisted dumbbell (Figure 4.7C) is used for materials displaying a low-to-moderate elongation at break. Dog-bone test pieces (Figure 4.7D) are used for thermosetting molding powders only and do not allow an elongation measurement to be made. The parallel-strip (Figure 4.7E) is suitable for reinforced thermoplastic and thermosetting sheets. The end pieces of the grips are normally bonded onto the test piece in order to obviate the problem of sample fracturing near the grips. The parallel strip configuration can also be used for testing plastic films. The reader can refer to the specific standards listed at the end of the chapter for details on test methods.

We now direct attention to the measurement of strain itself. As noted earlier, it is inaccurate to measure strain or elongation simply by noting the rate of separation of the grips. Problems with grip slippage, deformation of the grips and crosshead of the test apparatus, and deformation of the force-measuring device contribute to the apparent stretch of the specimen. An additional complication is that dumbbell-shaped test pieces simply do not stretch uniformly over their entire length. This means that gauge marks must be used to follow the elongation of the test piece. One must measure the distance between marks, preferably in a continuous fashion. In standard instrumentation, this is accomplished by ensuring that the electrical output from the instrument is proportional to the elongation. The electrical signal is then fed to the x-axis of a chart recorder while the tensile force is simultaneously fed to the y-axis. Several types of devices are used to perform this function. For materials such as reinforced plastics with very low elongations at break, strain gauges can be bonded directly to the test piece. For strains in the range 1 to 50%, a clip-on type of strain gauge extensometer is often used. For strains greater tahn 50%, optical extensometers or long-travel clip-on extensometers are used.

Elongation of plastic films is most often measured by grip separation for several reasons. First, tensile forces are relatively small and hence errors due to the compliance of the apparatus are usually small. Also, parallel strip specimens are preferred because they extend uniformly over the entire length, unlike dumbbell test pieces. Finally, a contacting type of

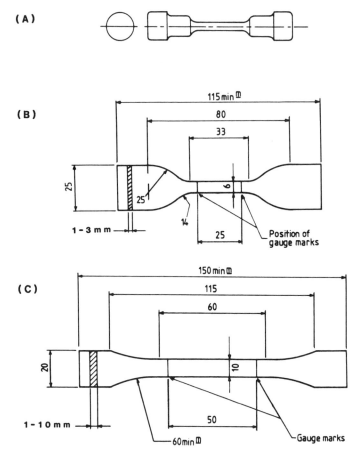

Figure 4.7 (A) Dumbbell test specimen; (B) narrow-waisted dumbbell (ISO/DIS 527 type A); (C) broad-waisted dumbbell test (ISO/DIS 527 type B); (D) dog-bone dumbbell (ISO/DIS 527 type C); (E) parallel-strip tensile test piece (ISO/DIS 527 type D).

extensometer cannot be used since a film is unable to support it without becoming distorted.

According to Hooke's law, for an ideal elastic material, stress is proportional to strain, and the constant of proportionality is called the *modulus*. For any given stress, the lower the modulus, the greater the strain or deformation of the specimen. The modulus of a material is therefore a measure of the specimen's ability to resist deformation. Unfortunately,

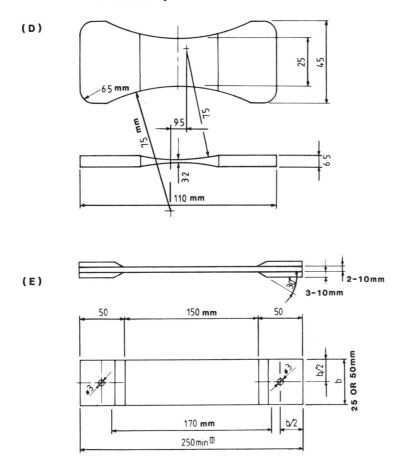

few materials display ideal behavior, and in fact most plastics and elastomers depart from Hookean behavior at very low strains in a tensile test. Despite this, Hookean theory is applied in standard test methods to include the measurement of tensile modulus. Tensile modulus is also referred to as the elastic or Young's modulus. The measurement is obtained by constructing a tangent to the initial, nearly linear portion of the stress-strain curve and computing the slope. In rubber technology, the terms 100% modulus, 200% modulus, and so on, are referred to when discussing tensile data. These in fact are not modulus values but rather, values of the tensile stress at the stated elongation.

Compression Stress-Strain, Shear Stress-Strain, and Flexural Stress-Strain Concepts

The antithesis of tensile stress-strain is compression stress-strain. Deformation in both cases is accomplished by application of two equal and opposite collinear forces acting in the reverse direction to each other. In the compression mode, instead of stretching the test piece, the material is crushed. This, of course, eliminates some of the contributing effects of gripping, and in general, the deformation of the test piece can be more accurately related to the movement of the apparatus crosshead than in tensile testing. In this type of test, however, it becomes most critical to supply a truly axial force. An inherent problem is that the test piece has a tendency to bend or buckle, and friction between the sample and end plates prohibits lateral expansion of the piece during compression, resulting in barrel distortion. Figure 4.8 illustrates a compression cage generally used on a tensile testing machine. The purpose of the cage is to reverse the test direction on the test piece while maintaining a tensile force on the load cell that measures the force. It is important to note that such cages will generate frictional forces, and as such, compressive forces to be measured must be significantly greater in order to maintain test accuracy.

Compressive forces are usually significantly greater than the tensile forces generated in a tensile test, mostly because larger test pieces are required in the former. Most universal tensile testing machines capable of performing both tests generally have greater capacity than machines designed only for tensile stretching tests.

Terminology applied to compression stress-strain tests are analogous to that used in tensile testing, with the exception of the term *slenderness ratio*. The slenderness ratio is defined as the ratio of the length of a solid or uniform cross section to its least radius of gyration, where the least radius of gyration, ξ_i, is referenced to a set of standard configurations. Table 4.3 summarizes ξ_i definitions for three geometries.

Both tensile and compressive stress-strain testing are based on the application of forces normal to the plane on which they act. Let's now consider the application of forces parallel to the plane as illustrated in Figure 4.9. Shear stress is defined as

$$\tau = \frac{F}{a^2} = \frac{F}{A} \tag{9}$$

and shear strain is

$$\gamma = \frac{\delta a}{a} \tag{10}$$

Measurement of Stress-Strain Properties

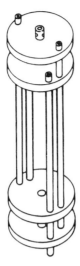

Figure 4.8 Compression cage for tensile machine.

Table 4.3 Least Radius of Gyration for Commonly Used Configurations

Geometry	Formula for calculating ξ_i[a]
Right rectangular prism	$\dfrac{a}{3.46}$
Right circular rod	$\dfrac{b}{3.46}$
Right circular tube	$\dfrac{\sqrt{D^2 + d_i^2}}{4}$

[a] a, Length of the side of the square prism; b, length of the shorter side of the rectangular prism; d, diameter of the cylinder; D, outer diameter of the tube; d_i, inner diameter of the tube.

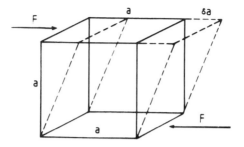

Figure 4.9 Specimen in shear.

From Hooke's law, the shear modulus is

$$G = \frac{\tau}{\gamma a} \tag{11}$$

where G is referred to as the modulus of rigidity.

To give rise to shear forces, parallel but opposite forces must act through the centroids of sections that are at some inifinitesimal distance apart. In the ideal case, this produces pure shear, but in practice this is extremely difficult, if not impossible to achieve. Instead, approximations must be made. One such simulation comes from the punch shear test, which uses a punch that bears down on a flat sheet of test specimen testing on a die. The nearer the internal diameter of the die and the external diameter of the punch are, the closer is the approximation to pure shear. This type of test has found wide acceptance among plastics characterization. A typical punching tool assembly documented in British Standards is illustrated in Figure 4.10. This type of arrangement is most frequently applied to molding materials and sheet materials. The experiment is typically conducted with the tool mounted in a testing machine operated in the compression mode over a specified time frame (British Standards cell for the punching to be completed between 15 and 45 sec from first application of the load). The shear strength is computed from as follows:

Method A-BS: test pieces are molded disks 25.3 ± 0.1 mm in diameter $\times\ 1.6 \pm 0.1$ mm thick:

$$\text{shear strength} = \frac{F}{\pi DT} \tag{12}$$

Measurement of Stress-Strain Properties

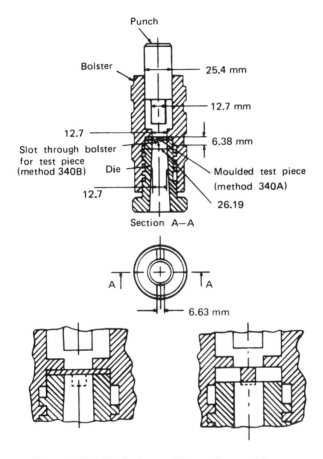

Figure 4.10 Typical punching tool assembly.

Method B-BS: test pieces are rectangular bars 6.4 ± 0.2 mm wide × 32 mm long × equal thickness to that of sheet under test:

$$\text{shear strength} = \frac{F}{2.096BT} \tag{13}$$

where F = force at fracture
D = punch diameter
B = mean width of test piece
T = mean thickness of test piece

A similar test is given by ASTM (see the reference section).

The last stress-strain type of test to consider is that of flexural or bending. Consider what happens when a rectangular beam is bent. A continuous change takes place from maximum tensile stress on one surface through the thickness to maximum compressive stress on the other. Figure 4.11A illustrates a homogeneous isotropic material undergoing pure bending. The top surface of the specimen is in tension while the bottom surface is in an equal compression. If the tensile and compressive moduli are equivalent, the stress at the midpoint of thickness is zero. At this point tension diminishes to zero before compression begins building up. The dashed line in Figure 4.11A denotes the neutral axis. In reality, the combination of only tensile and compressive forces is never achieved (there is always some transverse shear component).

Bending formulas for small deformations summarized below are based on the assumption of pure bending, assisted by having a span (length of loaded section) of beam that is large compared with the thickness. In this case, bending approaches that of a true arc of a circle.

Figure 4.11B illustrates the case of three-point bending for a rectangular beam with midpoint support. The flexural stress for this situation is

$$\sigma_F = \frac{3FL}{2bh^2} \qquad (14a)$$

Force F is that force at the midpoint. Refer to Figure 4.11B for other symbol definitions. At the point of fracture, the flexural strength is computed from the maximum froce, F_m, recorded:

$$\sigma_s = \frac{3F_m L}{2bh^2} \qquad (14b)$$

Some textbooks refer to σ_s as the "cross-breaking" strength. An alternative expression that further accounts for the horizontal component of the flexural moment is

$$\sigma_F = \frac{3FL}{2bh^2}\left(1 + \frac{4d^2}{L^2}\right) \qquad (15)$$

The apparent modulus of elasticity in flexure is

$$E_b = \frac{L^3}{4bh^3} n \qquad (16)$$

Measurement of Stress-Strain Properties

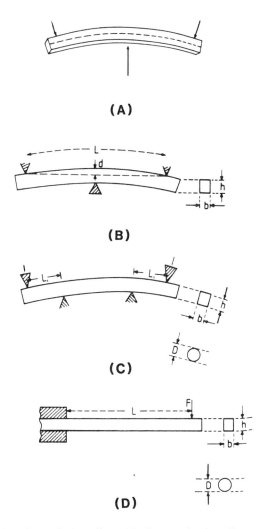

Figure 4.11 (A) Specimen in bending; (B) three-point bending; (C) four-point bending; (D) cantilever bending.

where n is the slope of the initial linear force-deflection curve. E_b is often assumed equal to Young's modulus as a first approximation. The case of four-point bending for a rectangular rod is shown in Figure 4.11C, and its flexural strength is calculated from the following expression:

$$\sigma_F = \frac{6F_x L_1}{bh^2} \tag{17}$$

where F_x is the force on each bearing point (i.e., half total force).

Finally, Figure 4.11D illustrates cantilever bending, for which the flexural strength of a rectangular beam is

$$\sigma_F = \frac{6FL}{bh^2} \tag{18}$$

Azbel and Cheremisinoff (1983) discuss the formula for bending, treating both circular and rectangular rods in detail.

ISO refers to the terminology "flexural stress at the conventional deflections." This parameter is defined as the flexural stress at a deflection of 1.5 times the thickness of the test piece, and is particularly meaningful for materials that do not fracture under test. Bending tests can be performed on either tensile or compression test machines. Examples of the test mode are illustrated in Figure 4.12.

Variables specified in a standard test are specimen size and shape, test speed, temperature, and the radius of curvature of the bearing rods. Cross-breaking strength tests are often used since they are straightforward and deflection data are readily obtained. With ideal materials subjected to pure bending, cross-breaking strength and bending modulus are the same as tensile strength and Young's modulus, respectively. In practice, however, materials are quite often not homogeneous and isotropic, and these equalities must be considered approximations only. Standard code lis-

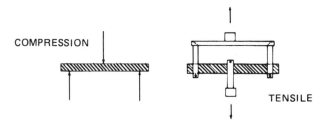

Figure 4.12 Bending tests in compression and tensile apparatus.

tings at the end of the chapter should be consulted for specific test methods.

Tear and Impact Testing

In plastics applications, tear tests are applied to films, whereas for rubbers they are applied to standard test sheets of 2-mm nominal thickness. Although these are perhaps among the simplest types of physical properties tests, they often create the most frustration since reproducibility is generally poor. Properties along and across the machine direction can vary, and as such, in plastic film characterization in particular, both should be determined in order to obtain proper characterization.

The two major types of tear are the trouser tear and the Elmendorf tear. Both techniques suffer from the fact that the tear occasionally propagates in a direction other than along the length of the test piece. Codes usually specify the extent of allowable deviation in the tear propagation for results to be acceptable. In the case of the trouser tear, an additional problem is that a steady tearing force is not always observed. The manner in which the force-extension curve should be analyzed to extract the tearing force properly is often not rigorously applied, leading to subjective rather than quantitative results.

The Elmendorf equipment is illustrated in Figure 4.13. The apparatus consists of a stationary jaw and a movable jaw supported on a pendulum. The pendulum is allowed to swing about on an ultralow-friction bearing. The test piece is typically of the trouser tear type and is clamped in the two jaws. Upon operating the sector release mechanism, the potential energy of the pendulum causes the cut in the film to propagate along the specimen's length. The energy absorbed by the process is recorded directly on a calibrated scale on the pendulum. The scale reading can be converted to a tear strength from information on the length of the tear created, the sample thickness, and an apparatus conversion factor.

In contrast, the trouser tear method employs a standard tensile testing machine. The test piece normally consists of a strip of film that is approximately 200 mm long × 50 mm wide. A clean cut is made in one end of the specimen to a distance of 75 mm (the cut is normally made central to the width of the film specimen). The two trouser legs are fixed in the jaws of the testing machine and a grip separation rate (250 mm/min) is maintained. The force-extension curve typically shows an initiated rise, followed by a plateau region, as the tear propagates. The tear strength is defined as the force in this plateau area divided by the film thickness.

We now direct our attention to the property of toughness. In the general sense, toughness is the work done in breaking a specimen, molding an article, and so on. It is derived from the load-extension graph (refer

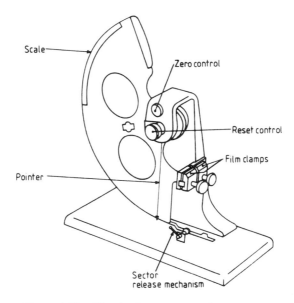

Figure 4.13 Elendorf-type tear testing apparatus.

back to Figure 4.4A). The area under the stress-strain curve is an integration of the force-distance relation (i.e., the extension over which force is exerted), thus providing total work or toughness. By application of high-speed tests, a material's behavior can be characterized in terms of its plane face properties. This is of obvious engineering importance to many consumer-type articles manufactured from plastics and cured elastomers. A serious limitation, however, is that the area derived from a conventional tensile or bending test curve is limited mainly because what is of greatest interest is the toughness under conditions of rapid deformation: for example, when an article is dropped or has something impacting it. This has led to the evolution of impact tests, or shock-resistant testing. This area has been studied rather extensively, and suggested references at the end of this chapter should be consulted for both theoretical and practical considerations.

Standard impact tests can be divided into two general categories. First, there are those tests that employ apparatus where a pendulum of known energy strikes a specimen of defined size and geometry. The other types of tests are those where weights or other impactors are allowed to fall freely through known heights onto the specimen, from whence the impact strength is computed from the minimum combination of height and weight required to produce fracture. Tests of the first type are subcategorized into cantelever (Izod) and supported beam (known as Charpy) variants. These employ test pieces comprised of flat plane faces or molded or machined notches to assess the sensitivity of weakening by notches. Tensile impact can also be assessed using dumbbell-shaped test pieces. Test data from the first type of test, particularly Izod, are most frequently employed.

The premise behind the Izod test is to allow a pendulum of known mass to fall through a specified height and strike a standard specimen at the lowest point of its swing. One must record the height to which the pendulum continues to swing. When the striking edge of the pendulum is sited to conicide with the center of percussion of the pendulum, the bearings of the pendulum are frictionless, and hence there is no loss of energy to windage. It is believed that the product of the mass of the pendulum and the difference between the fall distance and the height it reaches after impact is the impact strength of the test specimen. Results can be expressed in units of joules. Test specimens can be plain rectangular bars and are usually carefully notched either by a molding process or machining into the face to be struck. The purpose of the notch is to help the test simulate the resistance of a material to shock where notches or "stress raisers" are often present. The ratio of the impact strength's unnotched to notched value is regarded as a measure of the material's notch sensitivity.

There are two types of notch that may be cut or molded in the test piece. Both types form an angle of 45°, but the radius at the tip of the notch differs depending on the ISO or ASTM method used (each code specifies three separate methods). The notch is cut into the y dimension and is of depth $y/5$, so that the x dimension becomes the notch length. The designation of x and y dimensions is important, particularly when tests are conducted on laminated test specimens. These test pieces are then tested in edgewise and/or flatwise directions (Figure 4.14A). The height to which the pendulum rises after impacting the test piece is recorded by an idler pointer on a scale, providing direct units of energy absorbed. ASTM standards reports values in units of energy per unit length of notch.

The Charpy test ressembles the Izod impact test in that both are performed by a pendulum striking a bar-shaped specimen. In general, the Charpy test has had wider acceptance in Europe for plastics applications. Briefly, a rectangular bar, which may be notched or unnotched, is supported at both ends in such a manner that the pendulum, upon releasing strikes the specimen in the center and directly behind the notch, is included. The energy absorbed in the impact is recorded by means of an

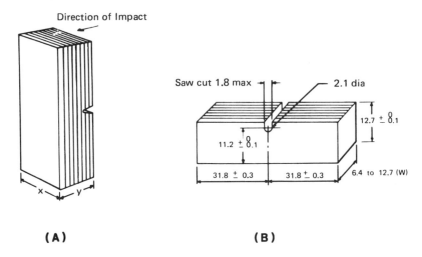

Figure 4.14 (A) Izod impact testing on laminated specimen; (B) details of Izod impact resistance test piece.

idler pointer from a scale that is calibrated to account for frictional and windage contributions. In general, the Izod and Charpy tests have significant differences that prohibit direct comparisons between the data derived from each test.

Another class of experiments for impact testing is based on a falling weight experiment. In this method the resistance to shock of a thin flat disk of material is assessed by determining how resistant the test piece is to fracturing when impacted by a ball that falls centrally on its surface. The height through which the ball falls, as well as its mass, can be varied, so that the minimum value of the product of full height and mass resulting in fracture provides the material's impact strength. An inherent problem with this test is that when the same test piece is subjected to successive impacts, there is no way of assessing how much damage short of fracture has actually occurred. In other words, there is no way of assessing how much progressive damage has occurred over the test cycle, and therefore the derived impact strength will be misleading. For this reason each test specimen must be subjected to impact only once, and the impact resistance computed from the spectrum of behavior of go/no-go fracture versus regularly varied impact values. This means, of course, that a large number of samples must be employed, even for a single material evaluation.

The apparatus described in British Standards (a similar device is described by ASTM) is sketched in Figure 4.15. The specimen supported is a

Measurement of Stress-Strain Properties

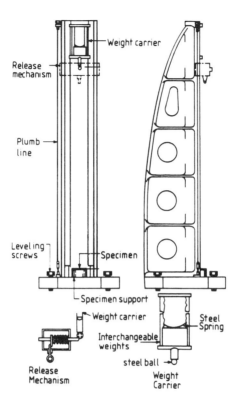

Figure 4.15 Details of falling weight apparatus.

hollow cylinder, whereby its axis is aligned to coincide with the line of fall of the striker. A soft shock-absorbing disk is positioned inside the cylinder to rest on its base. The test piece is usually clamped onto the support. The striker itself has a hardened hemispherical striking surface. It is fitted with a carriage that accommodates weights. In this manner a specified series of increments of energy may be obtained from one of two different heights above the upper surface of the test piece. The striker can be supported electromagnetically and released by turning off the current.

A control sample is normally run at the beginning of a test. In this pre-experiment the striker is loaded in such a fashion that the product of weight and the fall height is equal to the expected impact strength. If the specimen does not fracture, the result is recorded as unbroken. The result is recorded as "broken" if the piece circles or tears through from one surface to the other. In the broken sample case, a second test piece is tested with an impact energy less than the first by a specified amount. If the second piece breaks, a third specimen is tested with an impact energy less than that applied to the second. The process is repeated until a specimen does not break. If, on the other hand, the first result was "unbroken," a second specimen is tested with an impact energy greater than the sample before it. The procedure is repeated until a specimen that breaks is found. The full test is now conducted where the energy of impact applied to any specimen is a Y-value more or less than that on the previous sample. The Y-value is specified by the particular code followed. In the British Standards a total of 18 samples are specified plus at least two additional pieces for the trail run. The impact strength is calculated from the following formula:

$$\sigma_s = \frac{1}{21 - m} \sum_{i=1}^{21} \gamma_{m+i} \tag{19}$$

where m is the number of blows in the trial run and γ_{m+i} is the impact energy of the ith blow of the testing run. Note that γ_{21} is the impact energy of the twentieth blow, decreased or increased by Y according to whether the specimen broke or did not break.

To this point, tests described have employed a flexural mode of deformation. In actual service, however, a part can also be subjected to shock loading in a tensile configuration. ASTM codes define a tensile impact strength derived from a pendulum impact machine. One end of a dumbbell-shaped test piece is retained in the pendulum head, with the other end held in a crosshead clamp. When the pendulum is released, the head is able to pass between two halves of a fixed anvil while the crosshead is not. Figure 4.16 illustrates the setup. The test specimen is fractured in tension and the absorbed energy of impact is obtained from the subsequent swing of the pendulum head, just as in the Izod and Charpy tests.

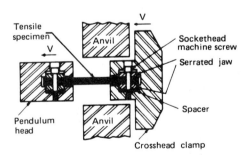

Figure 4.16 Tensile impact apparatus.

General Comments on Short-Term Stress-Strain Properties

All the test methods described thus far in this chapter are aimed at characterizing short-term stress-strain properties. A basic shortcoming of the data generated from such tests is that they provide property information that are largely indicative or comparative. In other words, information derived is not necessarily functional and quantitative, and is therefore of little direct value to a design engineer. This is particularly true for engineers whose experiences lie largely with metals and concrete, and must deal with the properties of rubbers. In metallurgy and deflections are small, and component behavior can be predicted from a number of relatively few material constants, such as Young's modulus and bulk modulus. Although elastomeric components is engineering and consumer-oriented applications can be designed using a fairly small number of design parameters, some significantly different defintions and concepts exist between metals and polymeric materials.

As noted in Chapter 1, rubber and plastic components are employed to perform a variety of engineering tasks, ranging from essentially static applications as in bridge bearings, to vibration or shock isolation, particularly in automotive components. In spring-type applications, the stiffness characteristics of the article are major parameters that establish the material's suitability for a particular application. Of course, there are other important considerations, such as strength, aging, weathering, and chemical resistance; however, these are often considered later in the development stage.

In the next section we review concepts important to the initial design of articles. In particular, polymer properties such as static and dynamic moduli, as well as principles related to compounding effects, are considered.

STATIC AND DYNAMIC PROPERTY BEHAVIOR

Discussions here will largely pertain to rubbers. The static stiffness or compliance of an elastomeric spring is one property of considerable design interest. The load-deflection analyses involving the Young's modulus of the material are given in many standard texts. A typical example is compression of a rubber disk bonded between steel plates. A thoretical solution for the load-deflection relationship is

$$F = Y \frac{A(1 + 2S^2)x}{t} \tag{20}$$

where S is a shape factor, defined in the case of a disk as $D/4t$ (D = the diameter and t = thickness), Y is the "Young's modulus" of the rubber, A is the bonded area, and x is the deflection. In this instance, the quantity Y is a property of vital importance to the designer.

Rubbers exhibit nonlinear elasticity, so that the stress/strain ratio varies markedly with the type and amplitude of the applied strain. Classical theories of elasticity were developed for large deformations of incompressible solids. They predict that the stress (σ_c) exerted by a rubber in homogeneous compression is

$$\sigma_c = G'_c(\lambda - \lambda^{-2}) \tag{21}$$

where λ is the extension ratio, $t/t_0 (= 1 + e)$. G'_c is a parameter that can vary not only with the strain but also with the type of strain (i.e., tensile or compressive). At infinitesimal strains, G'_c becomes equal to $E_0/3$ and equation (21) reduces to Hooke's law.

Similarly in shear, these theories predict that the shear stress (σ_s) is

$$\sigma_s = G'_s \gamma \tag{22}$$

where γ is the shear strain and G'_s is a parameter that may vary with strain. There is no a priori reason why G'_s would equal G'_c except at infinitesimal strain and the classical relationship that $E = 3G$ may not apply at finite strains.

From equation (21) one can conclude that the usual defintiion of modulus (i.e., stress-strain) is of limited applicability to rubbers. Even if G'_c is a constant, this modulus will alter with strain in both tension and compression. A more appropriate replacement for modulus is that proposed by Gregory (1984) whereby the ratio $\sigma/(\lambda - \lambda^{-2}) = H$. This removes the nonlinearity associated with incompressibility of the rubber. This ratio is an effective shear modulus, and in an ideal rubber where G'_c is strain independent, should equal the shear modulus $G'_s (= \sigma_s/\gamma)$. For unfilled rubbers, Gregory reports good correspondence between the shear modulus and the effective shear modulus obtained from tensile or compressive deformations.

The Malaysian Rubber Producers' Research Association publish a series of data sheets on natural rubber vulcanizates. These provide shear moduli, bulk moduli, and compression moduli of bonded units for a wide range of natural rubber formulations. Static shear moduli from 2 to 350% shear strain are provided at ambient temperature. This enables a designer either to select the modulus at a suitable strain or to express the shear modulus as a suitable function of strain. Moduli are provided for both initial deformation and deformation after nine cycles to 100% shear strain.

In view of the uncertainties associated with inhomogeneous strain situations, modulus data are also provided for bonded disks having various shape factors, in compression. Test pieces used in this analysis are shown in Figure 4.17. The cylindrical shear sample was chosen because of the ease of clamping in a rigid jig enables oscillation about zero strain and allows the same type of sample to be used for dynamic testing. Also, clamping of the end units avoids the contraction perpendicular to the direction of shear that occurs with quadruple shear tests. Finally, as samples are transfer molded, the small sample reduces the likelihood of anisotropy.

The compression test pieces are a compromise, being the largest samples that can be tested on a 10-ton test machine. Even so, allowances must be made for deflection of the steel end pieces under this load. Additional information on the static shear moduli is also provided in the form of low-frequency dynamic tests (0.1 Hz) over a wide range of temperatures. As the dynamic shear modulus at low frequencies is essentially the same as the static modulus, the designer is presented with shear moduli covering a wide range of both strains and temperatures. Extensive static testing is best conducted using computer control, where displacement values are stored for analysis. In this way, static moduli over a range of strains can be obtained with little more effort than required for determination of a point modulus. *Static shear moduli* are required for calculation of initial deflec-

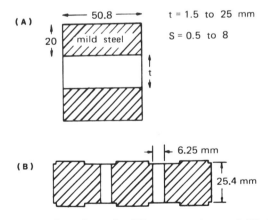

Figure 4.17 Test pieces for (A) compression and (B) shear.

tions of bearings. Most rubber bearings are used to accommodate motions of some description, in applications such as vibration or shock isolators. For these, the dynamic modulus of the rubber is required or, more precisely, the in-phase and out-of-phase components of the modulus.

For a simple linear viscoelastic solid, imposition of a sinusoidal strain cycle results in a sinusoidal stress cycle that is at an angle in advance of the strain. The following relationships can be used to characterize the response of the material:

$$\frac{\text{peak stress}}{\text{peak strain}} = G^* \tag{23}$$

$$\tan \delta = \frac{G''}{G'} \tag{24}$$

$$G^* = \sqrt{G'^2 + G''^2} \tag{25}$$

where G^* = complex modulus
G' = in-phase modulus
G'' = out-of-phase modulus
δ = loss angle

Two of these parameters must be known in order to characterize the specimen.

The measurement of complex modulus and loss angle is straightforward, provided that the displacement applied to the sample and the resulting load are monitored using suitable transducers. Peak loads and displacements can be monitored using voltmeters, and the phase shift measured using a phase meter. Alternatively, Fourier analysis of the output signals can be used to obtain these parameters.

Rubbers are generally nonlinear elsatic materials and the simple analysis above is not always valid. Input of a sinusoidal strain cycle can lead to a stress output in which higher harmonics contribute. These, however, are usually minor except where high-amplitude compressive strain cycles are involved.

The use of rubber springs as antivibration mounts relies on the frequency of the input vibration being at least one to five times the natural frequency of the spring-mass combination. The natural frequency of the system is

$$\omega_0 = \frac{1}{2\pi}\sqrt{\frac{K}{m}} \tag{26}$$

where K is the spring stiffness and m is the supported mass. The efficiency of isolation is determined by the ratio of the disturbing frequency to an effective natural frequency corresponding to the frequency and amplitude of the impact.

The effective natural frequency is

$$\omega'_0 = \frac{1}{2\pi}\sqrt{\frac{K_{\omega,e}}{m}} \qquad (27)$$

where $K_{\omega e}$ is the spring stiffness at the frequency ω and amplitude e of the disturbing cycle. The effects of frequency on shear modulus are well documented. In service a vibration isolator is subject to a static deformation by the load m, and a disturbing vibration is superimposed on this deformation. The stiffness toward the vibration is given by the change in force divided by the change in displacement. For an ideal nonhysteretic rubber, the stiffness in shear deformation is given by differentiation, as

$$K_\omega = \frac{GA}{t} \qquad (28)$$

while for compression

$$\begin{aligned}K &= \frac{A(1 + 2S^2)}{t}\frac{d\sigma}{de} \\ &= \frac{A(1 + 2S^2)}{t}(1 + 2\lambda^{-3})G\end{aligned} \qquad (29)$$

The dynamic stiffness can be obtained from the slope of a load-deflection curve obtained at the required frequency.

For many practical rubbers, these simple relationships are not universal. Imposition of a vibration on a deformed rubber results in partial retraction, and it is the stiffness in retraction rather than in extension that determines the stiffness toward a vibration. This is illustrated by the data of Gregory and is shown in Figure 4.18. The retraction stiffness must be higher than the extensional stiffness, and use of the tangent modulus to predict dynamic stiffness can result in underestimates of the natural frequency with consequent lower efficiency of isolation than predicted.

From equations (28) and (29) we can obtain an effective dynamic shear modulus (G_d) which is a function not only of frequency but also of the amplitudes of both static and dynamic strains. Figure 4.19 illustrates

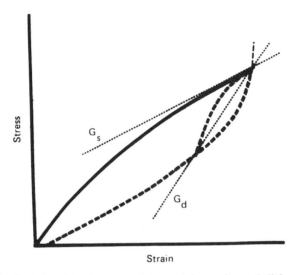

Figure 4.18 Relationships between static and dynamic moduli for a rubber subjected to vibration about a mean strain.

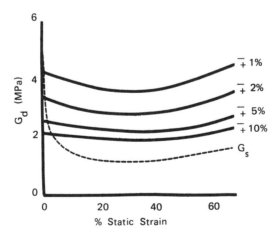

Figure 4.19 Vibration of dynamic modulus with static and vibration amplituded. Dashed curve represents tangent modulus from a static stress strain curve.

how this dynamic modulus is influenced by the amplitudes of both static and dynamic strains. The dynamic shear modulus is almost independent of the static strain imposed, but the dynamic strain amplitude has a marked effect on the dynamic modulus. Gregory's data in Figure 4.19 show that the dynamic shear modulus when a static strain is applied can be estimated from the conventional dynamic shear modulus measured in a symmetrical test (i.e., when cycled through zero strain). However, as the dynamic shear modulus is influenced by changes in strain rather than by the absolute strain, the appropriate conventional shear modulus is that at a zero-to-peak amplitude corresponding to the peak-to-peak amplitude of the dynamic strain applied to the deformed specimen.

Another material property of importance in vibration isolation is the loss angle. This parameter characterizes the peak transmissibility of the mounted system and also the transmissibility well above the natural frequency. Although antivibration mountings are designed to operate well above the natural frequency of the system, vibration at the natural frequency may be encountered during the starting up or shutting down of equipment. Also, low-amplitude vibrations at the natural frequency may be present in the vibration spectrum.

The loss angle of a rubber varies with temperature and frequency and if substantial amounts of fillers are used, with the amplitude of dynamic deformation. Information on the loss angles obtained with natural rubber is required over a range of amplitudes, frequencies, and temperatures in order that a designer can estimate the peak transmissibility in a given application.

The combination of static and dynamic moduli, including loss angle, here is necessary information for the initial design of rubber springs. These properties cannot be obtained from single-point measurements; instead, a range of measurements is required to describe the mechanical behavior of these polymers. Provided that the non-Hookean behavior of rubbers is accepted, characterization of the material properties introduces no major problems except for the increased number of measurements required.

Dynamic mechanical properties of unfilled elastomers at small dynamic strain are a function of temperature and frequency. When carbon black filler is introduced, the additional dependency on dynamic strain amplitude results. This means that a large data base is needed in order to specify the behavior of a carbon-filled rubber for a range of dynamic loading conditions. This information is required to enable the accurate prediction of component performance in various applications.

As described earlier, when a sinusoidal displacement is applied to a linear viscoelastic material, the response is a sinusoidal stress having the

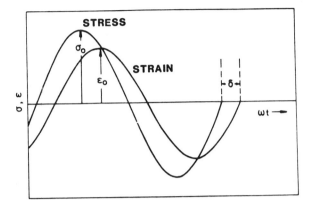

Figure 4.20 Out-of-phase oscillating stress (σ) and strain (ε) for linear viscoelastic materials.

same frequency f but leading the strain by a phase difference δ, called the phase angle. This is illustrated in Figure 4.20.

If the stress and strain cycles are taken to be represented by the real parts of

$$\sigma(t) = \sigma_0 \exp[i(\omega t + \delta)] \tag{30}$$

and

$$\varepsilon(t) = \varepsilon_0 \exp(i\omega t) \tag{31}$$

respectively, where $\omega = 2\pi f$, the stress/strain ratio at any instant in time defines a complex modulus

$$M^* = \frac{\sigma_0}{\varepsilon_0} \exp(i\delta) \tag{32}$$

This can be resolved into real (M') and imaginary (M'') components:

$$M^* = M' + iM'' \tag{33}$$

and

$$M' = \frac{\sigma_0}{\varepsilon_0} \cos\delta \qquad M'' = \frac{\sigma_0}{\varepsilon_0} \sin\delta \tag{34}$$

Measurement of Stress-Strain Properties

M' is the storage modulus and is proportional to the maximum energy stored per cycle, while M'' is termed the loss modulus and is proportional to the energy dissipated per cycle.

The loss factor (tan δ or d') is defined by the ratio

$$d' = \tan \delta = \frac{M''}{M'} \tag{35}$$

Dean et al. (1984) developed a dynamic mechanical apparatus based on the forced nonresonance technique, which is particularly suitable for the study of low-modulus, high-loss materials such as carbon-filled rubbers. The essential features are illustrated with the shear load stage in Figure 4.21. In this instrument samples can be accommodated in a variety of load stages, such as tensile, compression, shear, and combined compression and shear. An electromagnetic vibrator is driven from a signal generator (sinuosidal output) having a frequency range of 0.01 to 1000 Hz. The moving vibrator table is connected to one-half of the load stage, and a linear variable differential transformer (LVDT) transducer is used to measure its displacement. The other half of the load stage is bolted to a rigidly mounted force transducer. The transducers are connected to matched bridges,

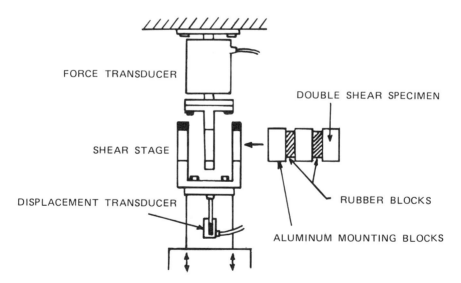

Figure 4.21 Details of dynamic mechanical apparatus showing shear load stage and double-shear specimen.

so that no significant extraneous phase angle is introduced, and the bridge outputs are fed to one of two processing devices. The first consists of a voltmeter and an accurate crystal-controlled timer that measures the maxima and minima of the force and displacement waveforms and the time interval between the centers (zero volts if no dc offset) of each waveform.

The second method of signal processing involves the use of a transient recorder. This equipment simultaneously digitizes the output from the stress and strain bridges at discrete time intervals, typically recording two cycles of each waveform. The resultant accuracy from these data is $\pm 1\%$ for moduli values, with a resolution of $0.1°$ for phase measurements.

Dean et al. report that the influence of frequency, temperature, and dynamic strain amplitude on the dynamic properties of rubber is most conveniently studied under a simple shear deformation. The complex shear modulus, G^*, is

$$G^* = \frac{\tau(t)}{\gamma(t)} \tag{36}$$

where $\tau(t)$ is the dynamic shear stress applied force/sample cross-sectional area, and $\gamma(t)$ is the dynamic shear strain = displacement/thickness.

The specimen is comprised of a double-shear arrangement and consists of two identical samples (rectangular, square, or circular cross section) bonded to aluminum mounting blocks. Materials tested were conditioned prior to the experiments by subjecting the shear sample to a higher strain than it would experience in the testing for a period of 1 min. The sample was then allowed to reach thermal equilibrium. This conditioning was carried out to ensure that reproducible results were obtained since the moduli of carbon-filled materials are highly dependent on recent strain history, particularly at low strain levels. This behavior is believed to be associated with the structural breakdown of the reinforcing carbon black at higher strain.

A nonlinear behavior was observed for materials. The stress response to an applied sinusoidal strain is a distorted sinusoid. Although a time lag still exists between each waveform, its value depends on the nature of the distortion and varies with time throughout a cycle. There is some concern, therefore, whether a measurement of phase angle will allow a valid determination of loss properties. It is for this reason that an alternative method of phase angle measurement using a transient recorder was developed. Calculations are based on the stress-strain loop for one cycle of deformation. The loop is derived from a cross-lot of instantaneous stress and strain values for one cycle.

Measurement of Stress-Strain Properties

Dean et al. found that the loss factor has only a small frequency dependence, the value of d at 100 Hz being between 10 and 20% higher than that of 0.2 Hz for all values of temperature and strain. This should be contrasted with the significant frequency dependence displayed by the shear modulus G'.

Figure 4.22 shows typical behavior of the loss factor with varying dynamic shear strain amplitude, there being a peak at approximately 2% strain. This is the case for the two temperatures plotted, and was found to be so over the range 0 to 100°C. A trend of decreasing loss factor and broadening of the peak at 2.0% strain was found with increasing temperature. [See Dean et al. (1989) for details.] The shear loss factor of the filled rubber considered showed only slight dependence on frequency. A description is therefore only necessary for the temperature and strain-amplitude dependence of loss factor, $d(\tau,\gamma)$.

For many applications, such as engine mountings and other antivibration devices, rubber components are subjected to compressive or combined compressive and shear deformations (Figure 4.23). For design requirements it is necessary to have a knowledge of the dynamic stiffness of

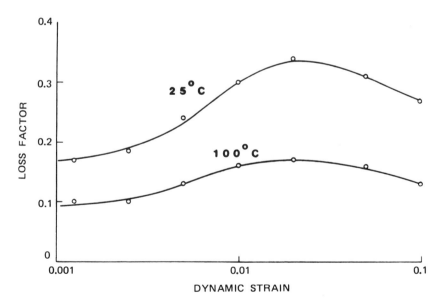

Figure 4.22 Vibration of loss factor with strain amplitude at two temperatures for one frequency.

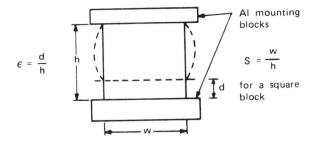

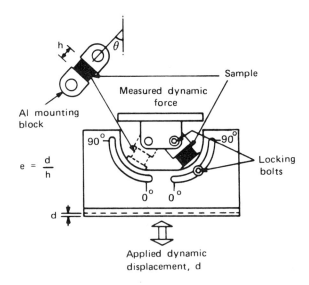

Figure 4.23 Load stages.

a component under such modes of deformation and its variation with temperature, frequency, and stain amplitude. The chosen geometry of a sample will also affect its dynamic stiffness.

When a rubber block is subjected to a load at some angle θ to its surface, it experiences both shear and compressive deformations. It is possible to relate the dynamic stiffness of the block $K^* = K' + iK''$ to compressive and shear properties. The procedure developed by Dean et al.

enables the contribution from compression to be related to shear so that the dynamic stiffness may be calculated from shear data at the appropriate strain and frequency. This procedure yields the following equation (valid for square blocks):

$$K'(f,e,\theta) = \frac{\omega^2}{h} G'(f,\gamma)[\sin^2\theta + 3\cos^2\theta(1 + ks)] \qquad (37)$$

where e is the dynamic applied strain given by the ratio of the applied displacement to the thickness of the rubber block. For compression loading, the appropriate shear modulus value is that measured at an equivalent shear strain γ, given by

$$\gamma = e(\sin^2\theta + 3\cos^2\theta)^{1/2} \qquad (38)$$

Figure 4.24 shows a comparison of the calculated modulus $M' = K'(h/w^2)$ (solid curve) of the sample under combined loading obtained from equivalent shear modulus data using equation (38) with measured apparent modulus values (data points) over a range of loading angle θ and at different dynamic strain amplitudes. The close agreement between the experimental data and the theoretical predictions show that it is possible to determine the combined loading behavior of V-blocks from just shear data using equations (37) and (38).

FATIGUE TESTING

Conventional fatigue data are presented in the form of stress versus (log) cycles to failure (S-N) curves. These curves are derived from a large number of tests on a closely specified material. Fatigue data are subject to considerable scatter. These tests are normally extended to include lives in excess of 10^7 cycles. Principally because of time constraints, these results are obtained by high-frequency tests, typically 50 Hz (approximate 2 days for 10^7 cycles). It is essential that the loading of the specimen be kept below a level that will avoid excessive heating. With plastics the viscoelastic behavior and low thermal conductivity combine to give considerable heating at quite low loading if the frequency is on the order of the 50 Hz. While the shortcoming is the primary reason for the lack of fatigue data in plastics, there are other factors that have precluded the collection of standard fatigue data. Although these limiting factors are closely linked, they will be separated into parameters that are related to the physical characteristics of the material and the working environment. The operational en-

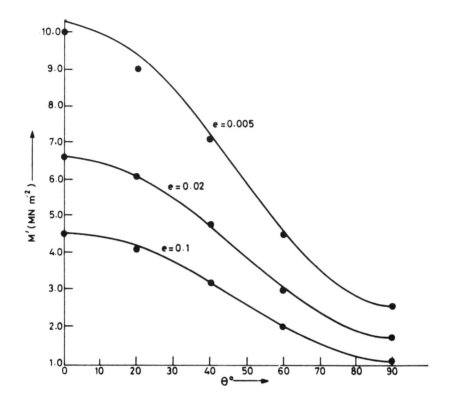

Figure 4.24 Vibration of apparent M' with loading angle for different dynamic strain amplitudes, e.

vironment can only be considered in detail with respect to the particular set of operating conditions, such as temperature and loading pattern, but with the limited operational temperature of most commercially available plastics, temperature and severity of loading are of extreme importance.

There are a variety of factors that have a dramatic effect on the mechanical properties and performance of polymers; among the more significant ones are molecular weight and distribution, degree of cross-linking, composition, material orientation, and the thermal history of the material.

Molecular weight and distribution are factors contrived during the manufacturing processes; however, under production conditions it is not easy to ensure constancy. Some studies have shown that a few percent of a low-molecular-weight material can halve the fatigue life of a specimen.

The degree of cross-linking has a marked effect on the modulus and other physical properties, but the degree of cross-linking may change with time, usually in the direction of further cross-linking. These progressive changes may be a natural tendency within the material, but cross-linking can be affected by external factors such as work hardening, work softening, and the environment (i.e., absorption or ultraviolet radiation).

Commercial plastics and elastomers contain additives such as plasticisers, extenders, pigments, fillers, and terminators. These can greatly modify the performance of the material. Some inclusions actually form the seat of a fracture, while others act to reinforce the material or to blunt cracks.

The manner in which the material is procesed will not only affect the distribution of the material but will also create distinctive flow lines and stress patterns. Further, the materials may be copolymers or composites in which case the distribution can be even more dramatic.

Thermal history extends from early in the manufacture of the basic material through to the finished component. Such factors as flow, shrinkage, internal stresses, and surface condition have to be considered at each stage and some interstage thermal treatment may be necessary. Temperature is another of the vital factors found in the working environment.

There are two generalized areas that must receive particular attention in any design: the test/working environment and the frequency, amplitude, and nature of the applied load.

A principal factor in the choice of a material is its corrosion resistance, in which case plastics have a major advantage. Plastics are, however, vulnerable to some environments, such as solvents, certain detergents, acids, alkalis, ultraviolet radiation, mineral oils (natural rubbers), and even air and water in certain instances. All of these can have a marked effect on the performance of the polymer particularly under fatigue loading.

Viscoelastic behavior of the material is perhaps the most crucial factor in restricting fatigue studies. Even at relatively low loads there is significant hysteresis in the load cycle, and since this represents work (which appears as heat), there is a problem of heat removal. This is exacerbated by the low thermal conductivity of most plastics. The results are that thermal gradients occur within the material which may be increased if the heat is forcibly removed.

The frequency and rest periods of a test are quite crucial to creep or fatigue life. Viscoelasticity is traditionally modeled as a spring-dashpot system, which defines the material as being time dependent. Thus there will be a delay in the response, an inherent error, and a marked difference between the applied stress and the resultant strain. There is a general acceptance that this difference in response is reduced significantly only if that particular state of stress or strain is maintained for a considerable period of time. Thus in dynamic testing there is always the problem of knowing the precise nature of the applied load. There is also a further complication of deciding the appropriate rate of load application. In general, most materials respond unfavorably to the sudden imposition of load. The severity of this load regime depends on the frequency, extent, and profile of the load. The profile of the signal may be random (white noise), but often it is sinusoidal, sawtooth, or square waves. Square waves are the most severe for any given frequency or amplitude. In practice, therefore, there is no absolute control over the precise stress-strain conditions within the material, especially if there is strain softening or strain hardening.

Whatever the loading pattern employed for the test, the response of the machine does not match the condition of the actual load in any direct way, and the loading axes have to be modified to obtain that match. In addition, some attempt must be made to accelerate the test program. However, it is not sufficient to accept so-called "stable failure" (i.e., any test where the specimen does not fail uncontrollably in a thermally dominated way) since this may be totally unrepresentative of the actual service conditions.

There are basically two general types of tests used to characterize the process of component failure: cyclic tests and static stress rupture tests. Fatigue life or endurance is defined as the number of cycles to failure at some specified stress level. In general, the fatigue life of polymeric components is reduced at higher temperatures. The fatigue limit (also referred to as the endurance limit) is the stress condition below which the material may endure an infinite number of stress cycles without failing. As a rule of thumb, the fatigue limit of many common polymers is about 25 to 30% of the static tensile strength.

The type of deformation imposed can be either tensile, compressive, flexural, or shear, or even a combination of any of these. The simplest deformation to produce is from a sinusoidally vary stress. Standard tests are conducted with constant amplitude of cycle stress or constant amplitude of cycle strain. Other approaches involve the superposition of oscillating stress (or strain) onto a static stress (or strain) or to increase the amplitude of the cycle stress (or strain) with time.

Figure 4.25 illustrates the basic components in a rotational fatigue apparatus. In this device, a cylindrical test piece is rotated about its axis of symmetry while the ends are constrained by bearings that are misaligned symmetrically. The test piece is subjected to a stress that varies sinusoidally between similar maxima of tension and compression.

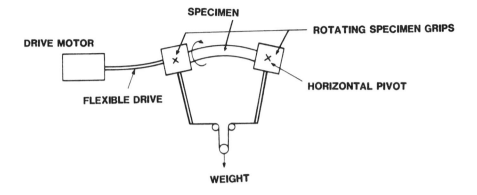

Figure 4.25 Rotational fatigue apparatus.

Flexural tests (i.e., without rotation) can also be performed on rectangular bars. These include, but are not limited to, cantilever bending and three- and four-point bending, described earlier. These types of flexural tests are relatively simple to perform in comparison to tensile-compression tests; also, problems of axiality are less critical. Data generated from such experiments are, however, considerably more limited, and more often do not have any direct correlation with actual end-use service of the material.

For flexible materials such as rubbers or plastic-coated fabrics, repeated bending of thick sheet specimens is often applied. Methods are outlined in ASTM. ASTM Standards basically recommend the constant-amplitude-of-force approach. Fluctuating stresses and strains need not be of constant amplitude or frequency. Two test methods are described, selected according to the specimen thickness and the stress range over which measurements are to be made. Test pieces have a triangular configuration with a rectangular cross section. This arrangement allows for a uniform stress distribution of respective test spans. It is important to note that there are some inherent limitations with the code tests, one major one being that only a single frequency, which may be too high for some plastics, is recommended. Plastics are relatively high-damping, low-thermal-conductivity materials. This means, of course, that when a sample is subjected to repeated straining, it experiences a temperature rise within and throughout its mass. If the stress-strain cycle is sufficiently rapid, significant heat will be generated, possibly casuing transition across the glass transition boundary and thus inducing thermal failure. This in itself can be an important parameter to study; however, this is a frequency-dependent parameter and its effects must be carefully separated from mechanical fatigue. Where thermal effects are to be minimized and cooling of the specimen is inadequate (the heat generated is from within the test piece body and not on its surface, where heat is removed by the test apparatus), much lower frequencies than recommended by the code must be used.

The advantages of performing such tests at reduced frequencies (i.e., a few hertz) are:

It still may allow frequencies to be used which are considerably above those in the loading regime, thereby gaining time.
The temperature will have a chance to stabilize to a level similar to that expected in the actual application, and thermal gradients within the specimen may be less severe.
The stress-strain relationship will be more closely related and observable.
Creep effects can be observed in terms of cumulative damage and the converse of allowing creep recovery to operate. Any creep recovery that is possible will reduce the "latching" effect of progressively hysteresis loading.

The environment can also be crucial to a component in service, and the effects of such factors as ultraviolet light, water, solvents, and temperature would have to be assessed in relation to the specific conditions. An ideal instrument would therefore be capable of controlling the test environment.

There are many multiaxial fatigue machines reported in the literature, a large proportion of which are biaxial (cruciform) systems. Very few are designed for universal application and most require specimens that have highly specialized geometry for ease of mathematical analysis or to force the failure into a prescribed zone. Much has been gained from these tests, but cross-correlation between one system and another is difficult since each test yields data unique to that configuration. To obviate some of this limitation, the requirements of a more generalized system are discussed by Lawrence (1984).

Lawrence notes that a basic requirement of any facility is for two surfaces that move toward and away from each other (push-pull), and that those surfaces should rotate with respect to each other (right- and left-hand torsion). A practical means of obtaining a triaxial state is to provide a high-pressure fluid to a hollow component (hoop stress), as shown in Figure 4.26. For adequate simulation these systems should operate in phase or out of phase with each other, either stress or strain dependent. They should have common specimen geometry and chucking and either prescribed waveforms or a facility to accept an external random (non-

Measurement of Stress-Strain Properties

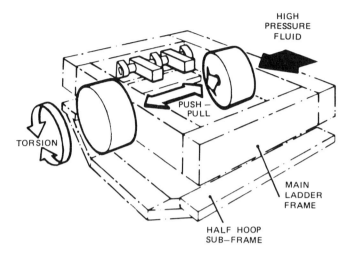

Figure 4.26 Arrangement for triaxial fatigue loading.

specific load) signal. While a multiaxial machine has to be designed as a complete entity, for clarity the basic elements of the machine will be considered separately as far as is possible. Since the potential loads can be applied in any combination, the basic design must be such that the structure and bearings will resist any combination of load without any significant deflection.

The second class of test is the stress rupture experiment. These tests are of a form similar to long-term continuous stress tests (see the references). They are essentially creep tests. Test pieces are subjected to a known stress (usually, some fraction of their ultimate short-term failing stress). The time that elapses between application of load and ultimate failure is recorded. Stress rupture tests are influenced by temperature, vibration, and so on, just as are creep and stress relaxation tests.

These can be extensive tests, depending on the material application, that can last from a few seconds to perhaps months. The most commonly applied test is that described in British Standards relevant to the integrity of plastic pipe service. These are essentially long-term hydraulic tests, in which a specimen of pipe is maintained at a constant internal pressure, and the time to burst (between 1 and 10 hours) is recorded. A second specimen is subjected to a similar experiment, except that the applied pressure is chosen to induce bursting after 100 to 1000 hours. The minimum wall thickness and mean outside diameter of each test piece are established prior to each experiment. The circumferential stress of the pipe is determined from the following formula:

$$P = \frac{2\delta_0 t'}{D - t'} \tag{39}$$

where

P = applied pressure, bar
δ_0 = circumferential stress, bar
t' = minimum wall thickness, mm
D = mean outside diameter, mm

Apparatus and methods employed for creep tests are generally applicable to stress rupture tests. In general, static-stress rupture tests are used as a measure of resistance to environmental stress cracking.

NOTATION

A_0	cross-sectional area of specimen
a	length of side of specimen
B	width of test specimen
b	sample width
D	diameter
d	penetration depth
d_i	diameter of impression
d'	loss factor
E	tensile modulus of elasticity
E_b	apparent modulus of elsaticity in flexure
e	amplitude
F	load of force
F_x	force at bearing point
G	Hooke's law shear modulus
G'_c, G'_s	strain parameters
G_d	effective dynamic shear modulus
G^*	ratio of peak stress to peak strain (complex modulus)
G'	in-phase modulus
G''	out-of-phase modulus
h	sample height; hardness number
K	spring constant
K^*	dynamic stiffness
k	rate constant for extension
L, l	length

M'	storage modulus
M''	loss modulus
M^*	complex modulus
m	mass; number of blows in impact trial
n	slope of initial linear force-deflection curve
P	pressure
R_α	Rockwell α hardness number
S	Shape factor
T	mean thickness of test piece
T_g	glass-transition temperature
t	time; thickness (where noted)
t'	minimum wall thickness

Greek Symbols

γ	shear strain
$\dot{\gamma}$	strain rate
γ_{m+i}	impact energy of ith blow in impact test
$\gamma(t)$	dynamic shear strain
δ	loss tangent; length factor
δ_0	circumferential stress
ε_1	tensile strain
θ	angle
λ	extension ratio
ξ_i	least radius of gyration
σ_1	tensile stress
σ_F	flexural stress
σ_s	cross-breaking strength
τ	shear stress
$\tau(t)$	dynamic shear stress force per unit area
ω_0	natural frequency

REFERENCES AND SUGGESTED READINGS

Azbel, D., and N. P. Cheremisinoff, *Chemical and Process Equipment Design,* Ann Arbor Science, Ann Arbor, Mich., 1983.

Braun, D., *Simple Methods for Identification of Plastics,* 2nd ed., Hanser, Munich, 1986.

Brown, R. P., ed., *Handbook of Plastics Test Methods,* George Godwin, London, 1981.

Brown, R. P., and B. E. Read, *Measurement Techniques for Polymeric Solids,* Elsevier, New York, 1984.

Coxon, L. D., and J. R. White, *Polym. Eng. Sci., 20,* 230 (1980).

Dean, G. D., J. C. Duncan, and A. F. Johnson, in *Measurement Techniques for Polymeric Solids,* R. P. Brown and B. E. Read, eds., Elsevier, New York, 1984.
Ferry, J. D., *Viscoelastic Properties of Polymers,* Wiley, New York, 1961.
Gregory, M. J., in *Measurement Techniques for Polymeric Solids,* R. P. Brown and B. E. Read, eds., Elsevier, New York, 1984.
Haworth, B., and J. R. White, *J. Mater. Sci.,* 16, 3263 (1981).
Haworth, B., G. J. Sandilands, and J. R. White, *Plast. Rubber Tnt.,* 5, 109 (1980).
Haworth, B., C. S. Hindle, G. J. Sandilands, and J. R. White, *Plast. Rubber Proc. Appl.,* 2, 59 (1982).
Hindle, C. S., J. R. White, D. Dawson, W. J. Greenwood, and K. Thomas, *39th Antec, Boston,* 783 (1981).
Kent, R. J., K. E. Puttick, and J. G. Rider, *Plast. Rubber Proc. Appl.,* 1, 111 (1981a).
Kent, R. J., K. E. Puttick, and J. G. Rider, *Plast. Rubber Proc. Appl.,* 1, 55 (1981b).
Kubat, J., and M. Rigdahl, *Int. J. Polym. Mater.,* 3, 287, 1975.
Kubat, J., R. Selden, and M. Rigahl, *J. Appl. Polym. Sci.,* 22 (1978).
Lawrence, C. C., in *Measurement Techniques for Polymeric Solids,* R. P. Brown and B. E. Read, eds., Elsevier, New York, 1984.
Legge, N. R., G. Holden, and H. E. Schroeder, *Thermoplstic Elastomers,* Hanser, Munich, 1987.
McCrum, N. G., B. E. Read, and G. Williams, *Anelastic and Dielectric Effects in Polymeric Solids,* Wiley, New York, 1967.
Mills, N. J., *J. Mater. Sci.,* 17, 558 (1982).
Nolle, A. W., *J. Polym. Sci.,* 5, 1, (1950).
Pethrick, R. A., *Polymer Yearbook,* Harwood Academic, New York, 1987.
Qayyum, M. M., and J. R. White, *Polymer,* 23, 129 (1982).
Qayyum, M. M., and J. R. White, *J. Appl. Polym. Sci.,* 28, 2033 (1983).
Saechtling, H., *International Plastics Handbook,* 2nd ed., Hanser, Munich, 1987.
Saffell, J. R., and A. H. Windle, *J. Appl. Polym. Sci.,* 25, 1118 (1980).
Sandilands, G. J., and J. R. White, *Polymer,* 21, 338 (1980).
Thomas, K., D. Dawson, W. J. Greenwood, J. R. White, C. S. Hindle, and M. Thompson, reported at 5th Int. Conf. on Deformation Yield and Fracture of Polymers, Cambridge, Apr. 1982.
Turner, S., *Mechanical Testing of Plastics,* 2nd ed., Longman, New York, 1983.
White, J. R., *Mater. Sci. Eng.,* 45, 35 (1980).
White, J. R., *J. Mater. Sci.,* 16, 3249 (1981a).
White, J. R., *Rheol. Acta,* 20, 23 (1981b).
Zener, C., *Elasticity and Anelasticity of Metals,* Chicago University Press, Chicago, 1948.

5
Processability Testing

INTRODUCTION

Characterization of the flow behavior of molten polymers is of tantamount importance to the fabricator since one of the first issues to assess is the feasibility of any intended operation/process involving metling. The ideal test or tests will enable extrapolation of a number of key properties such as thermal stability, relative ease of processing, production capacity (particularly for extrusion operations), and the quality of fabricated articles. The rubber and plastics manufacturer also has keen interest in this subject from the standpoint of providing adequate technical service to customers. In addition, there is considerable degree of control over processability and in designing new products by modifying parameters such as average molecular weight, molecular weight distribution, composition such as crystallinity, establishing the type and degree of branching, and by incorporating various types and amounts of plasticizers, lubricants, and stabilizers.

The traditional approach to product development involves the synthesis and bench-scale evaluation of polymer prototypes. From the standpoint of processing studies, many lab-scale equipment and methods practiced allow at best, only qualitative comparisons between prototypes, and provides almost no data suitable for projecting quantitative performance properties in commercial-scale operations. As such, there are very often a series of intermediate-scale tests, such as semicommercial trials on either near size or sometimes even short-term duration trials on full-scale equipment prior to actual customer trials of a new polymer. The systematic approach to product development outlined in Chapter 1 is often not followed because of time and money constraints and even the impatience of management. In the processing property area more so than others, the philosophy and approach to testing is the most critical because it is one of the most frequent causes for failure of a polymer, particularly in rather late stages of development. Many of the processability instrumentation, such as the Mooney viscometer, capillary rheometer, melt flow index tester, and others described in this chapter, are most often applied as quality control techniques because they provide a link between molecular weight and the viscoelastic properties of the neat polymer. What is often glossed over, and even ignored by the manufacturer sometimes, is the link between molecular properties and compound rheology and processability. The reasons for this stem largely from inadequate studies aimed at providing scale-up criteria between small scale-testing to factory fabrication, and the restriction imposed in manufacturing to implement a rapid, reproducible test that will ensure product consistency, although again this consistency is often achieved without full understanding of the customer's processing needs.

In this chapter we review the basic equipment and methods applied to processability testing of polymers. The standard practices are reviewed, shortcomings with each technique discussed, and general recommendations for enhancing testing for each method are made.

VISCOELASTIC PROPERTIES OF POLYMERS

Rheology is a subject that addresses the deformation of a fluid and fluidlike materials. When surface forces, or stresses, are applied to a fluid body, that body deforms or flows. The reactive behavior can be mathematically described by means of a set of *constitutive equations*. A constitutive equation defines the cause-and-effect relationships in terms of the properties or characteristics of the material. To develop such relationships, *balance equations* must be derived. A balance equation is a mathematical statement of the universal laws of conservation of mass, energy, and momentum that are specific to the system of interest. By way of a simple example found in many fluid dynamic textbooks [see Cheremisinoff (1982) and Azbel and Cheremisinoff (1983)], consider a shearing force applied to a rectangular body of incompressible fluid as illustrated in Figure 5.1. For steady-state, isothermal flow conditions, we may write the follow-

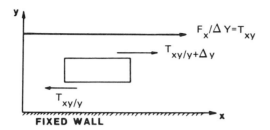

Figure 5.1 Fluid body subjected to constant force.

ing mass balance (continuity equation) and momentum balance (equation of motion):

Mass balance:

$$\frac{d}{dx} V_x = 0 = \frac{dV_x}{dx} \tag{1a}$$

and if the body undergoes deformation,

$$V_x \neq f(x) \tag{1b}$$

Momentum balance:

$$-\frac{d\tau_{xy}}{dy} = 0 \tag{2}$$

and $\tau_{xy} = C$ for any body where C is the applied stress. A limiting case is that of a *Hookean elastic solid*. The constitutive equation for this system is

$$\tau_{xy} = G\dot{\gamma}_{xy} \tag{3}$$

where $\dot{\gamma} = C/G$ and G is the shear modulus. The coordinate system for this model is defined in Figure 5.2.

The examples above are representations of a fluid's behavior when it is subjected to a force. A material's ability to undergo changes in shape or movement can be characterized by measuring its resistance to undergoing

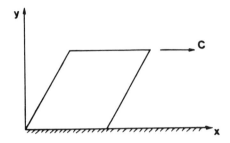

Figure 5.2 Body undergoing deformation.

deformation. The quantity that provides a measure of deformation is the shear strain $\dot\gamma_{xy}$. When $G = G(\dot\gamma_{xy})$, a solid body is non-Hookean and is described as being *elastic*. The shear modulus G is a rheological constant characterizing the material. Similarly, $G(\dot\gamma_{xy})$ is a material rheological function.

For a Newtonian fluid, the constitutive equation is Newton's law of viscosity:

$$\tau_{xy} = -\mu \frac{dV_x}{dy} = -\mu \frac{d\gamma_{xy}}{dt} = -\mu \dot\gamma_{xy} \tag{4}$$

The system is defined as shown in Figure 5.3, where

$$\frac{dV_x}{dy} = -\frac{C}{\mu} \tag{5}$$

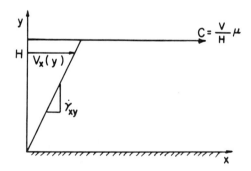

Figure 5.3 Coordinate system for developing constitutive equation for generalized Newtonian fluid.

Integrating this expression gives

$$V_x(y) = -\frac{C}{\mu}y + C_1 \tag{6a}$$

where $C_1 = 0$, and hence

$$V_x(y) = -\frac{C}{\mu}y \tag{6b}$$

or

$$V = -\frac{C}{\mu}H \tag{6c}$$

and C is

$$C = -\frac{V}{H}\mu = \tau_{xy} = -\mu\dot{\gamma} \tag{7}$$

The variables in the relationships above and Figure 5.3 are

μ = Newtonian viscosity (i.e., resistance to flow)
$\dot{\gamma}_{xy}$ = shear rate (i.e., measure of flow)

Constitutive equations can be generalized into two categories: linear and nonlinear. For a Newtonian material, equation (4) is the governing expression and it is noted that $\mu \neq \mu(\dot{\gamma})$. In this case, the resistance to deformation (viscosity) depends only on those property characteristics of the material that affect the intermolecular forces of the fluid. These property characteristics include the chemical structure of the fluid, temperature, and pressure. From thermodynamic principles, the viscosity of a Newtonian fluid can be defined approximately by

$$\mu(T,P) = \mu_0 \exp\left[\frac{\Delta E}{R}\left(\frac{T-T_0}{T_0 T}\right)\right] \exp\left[\beta(P-P_0)\right] \tag{8}$$

where

T, P = absolute temperature and pressure
ΔE = flow activation energy
β = viscosity pressure coefficient
R = universal gas law constant

In contrast, a nonlinear form of the constitutive equation is

$$\tau = f(mf_i \cdots mf_{nj} \dot{\gamma}) \tag{9}$$

Non-Newtonian materials are described by nonlinear constitutive equations, where the most general statement is

$$\gamma = \eta(\dot{\gamma}) \cdot \dot{\gamma} \tag{10}$$

Equation (10) is referred to as the generalized Newtonian fluid equation, where η is a scalar quantity. Restating equation (4) for an incompressible fluid yields

$$\tau_{xy} = -\frac{\mu}{\rho} \frac{d(V_x \rho)}{dy} \tag{11}$$

The quantity $V_x\rho$ is a momentum concentration since it has units of momentum per unit volume. Since $d(V_x\rho)/dy$ is a gradient, we may regard this quantity as a flux of the x-directed momentum in the y-direction. This constitutive equation is analogous to Fourier's law of heat conduction and Fick's law of diffusion.

Fourier's law:

$$\frac{q}{A} = -k \frac{d\theta}{dy} \tag{12a}$$

$$\frac{q}{A} = \frac{-k}{C_p \rho} \frac{d(\theta C_p \rho)}{dy} = -\alpha \frac{d(\theta C_p \rho)}{dy} \tag{12b}$$

Fick's law:

$$\frac{N_A}{A} = -D_{AB} \frac{dC_A}{dy} \tag{13}$$

where

A = area
q = heat flow
θ = temperature
k = thermal conductivity
α = thermal diffusivity
C_p = specific heat

C_A = concentration of component A

D_{AB} = binary molecular diffusivity for the system $A + B$

Equation (12b) is Fourier's law for an incompressible fluid and the quantities $\theta C_p \rho$ and $d(\theta C_p \rho)/dy$ are the volumetric thermal concentration and thermal concentration gradient, respectively. Fick's law [equation (13)] is stated for steady-state equimolal counterdiffusion of components A and B. The analogies between equations (4), (12), and (13) are striking: μ/ρ, α, and D_{AB} all have the same primary units of length^2/temperature. The quantity μ/ρ can be thought of as momentum diffusivity. Standard texts on transport phenomenon [see Bird et al. (1960)] conventionally term μ/ρ the kinematic viscosity.

Equation (4) is a linear expression and a plot of τ_{xy} versus $\dot{\gamma}$ results in a straight line passing through the origin and having slope μ. Fluid materials that display nonlinear behavior through the origin at a given temperature and pressure are non-Newtonian (Figure 5.4A). The broad classification of non-Newtonian fluids is: time-independent fluids, time-dependent fluids, and viscoelastic fluids. The first category is comprised of fluids for which the shear rate at any point is a function of only the instantaneous shear stress. In contrast, time-dependent fluids are those for which shear rate depends on both the magnitude and duration of shear. Some fluids in this second class also show a relationship between shear rate and the time lapse between consecutive applications of shear stress. Less common viscoelastic fluids display the behavior of partial elastic recovery upon the removal of a deforming shear stress.

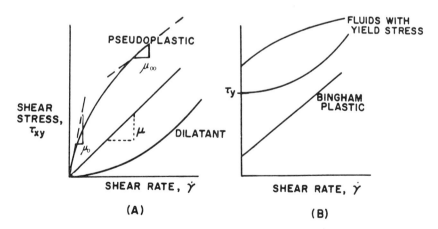

Figure 5.4 Flow curves for non-Newtonian, time-independent fluids.

Time-dependent non-Newtonian fluids can exhibit the property of a yield stress. The yield stress, τ_y, is a minimum stress value that must be exceeded in order for deformation to occur; that is, when $\tau_{xy} < \tau_y$, the fluid's internal structure remains intact, and when $\tau_{xy} > \tau_y$, shearing movement occurs. Flow curves for these types of materials are illustrated in Figure 5.4B. An idealistic fluid is the Bingham plastic, which models non-Newtonian behavior via a linear constitutive equation.

The rheological flow curve in Figure 5.4B for this fluid is

$$\mu_a = \eta + \frac{\tau_y}{\dot{\gamma}} \qquad (14)$$

where μ_a is the fluid's *apparent viscosity* and is analogous with the Newtonian apparent viscosity as restated in equation (4):

$$\mu_a = \frac{\tau_{xy}}{\dot{\gamma}} \qquad (15)$$

The Bingham plastic constitutive equation contains a yield stress τ_y and a term η referred to as the *plastic viscosity*. Equation (14) states that the apparent viscosity decreases with increasing shear rate. In practical terms this means that a value of the apparent viscosity can only be related as a flow property with a corresponding shear rate. Although the Bingham plastic fluid itself is idealistic, it can be applied to modeling a portion of a non-Newtonian's flow curve. Many materials show only small departure from exact Bingham plasticity and therefore can be described approximately by equation (14).

A large variety of industrial fluidlike materials may be described as being *pseudoplastic* in nature. The flow curve is illustrated in Figure 5.4A. Examining this curve, note that it is characterized by linearity at very low and very high shear rates. The slope of the linear region of the curve at the high shear rate range is referred to as the *viscosity at infinite shear* (μ_∞), whereas the slope in the linear portion near the origin is the *viscosity at zero shear rate* (μ_0). A logarithmic plot of τ_{xy} versus $\dot{\gamma}$ is found to be linear over a relatively wide shear rate range and hence may be described by a power law expression (known as the Ostwald–de Waele model):

$$\tau_{xy} = K\dot{\gamma}^n \qquad (16)$$

The slope n and intercept K are referred to as the flow behavior index (or pseudoplsaticity index) and consistency index, respectively. The power law exponent ranges from unity to zero with increasing plsticity (i.e., at $n = 1$, the expression reduces to the Newtonian constitutive equation). The value of the consistency index is obtained from the intercept on the τ_{xy} axis and hence represents the viscosity at unit shear rate. As shown later for a variety of polymers, K is very sensitive to temperature, whereas n is much less sensitive. By analogy to Newton's law, the apparent viscosity of a power law fluid is

$$\mu_a = K\dot{\gamma}^{n-1} \tag{17}$$

Since $n < 1$, the apparent viscosity of a pseudoplastic fluid decreases with increasing shear rate, and hence these materials are often referred to as *shear thinning*.

Another class of time-independent fluids are *dilatant* materials. Volumetric dilatancy refers to the phenomenon whereby an increase in the total fluid volume under application of shear occurs. Rheological dilatancy refers to an increase in apparent viscosity with increasing shear rate. The flow curve is illustrated in Figure 5.4A. As in the case of a pseudoplastic fluid, a dilatant material is usually characterized by zero yield stress. Hence the power law model may also be used to describe this fluid behavior but with n-values greater than unity.

Time-dependent non-Newtonian fluids are classified as thixotropic and rheopectic. Thixotropic fluids display reversible decrease in shear stress with time at a constant shear rate and temperature. The shear stress of such a material usually approaches some limiting value. Both thixotropic and rheopectic fluids show a characteristic hysteresis as illustrated by the flow curves in Figure 5.5. The flow curves are constructed from data generated by a single experiment in which the shear rate is steadily increased from zero to some maximum value and then immediately decreased toward zero. The arrows on the curves denote the chronological progress of the experiment. An interesting complexity of these materials is that the hysteresis is time-history dependent. In other words, by changing the rate at which $\dot{\gamma}$ is increased or decreased in the experiment alters the hysteresis loop. For this reason, generalized approaches to defining an index of thixotropy have not met with much success.

Rheopectic fluids are sometimes referred to as antithixotropic fluids because they exhibit a reversible increase in shear stress over time at a constant rate of shear under isothermal conditions. The location of the hysteresis loop shown in Figure 5.5B is also dependent on the material's time history, including the rate at which $\dot{\gamma}$ is changed.

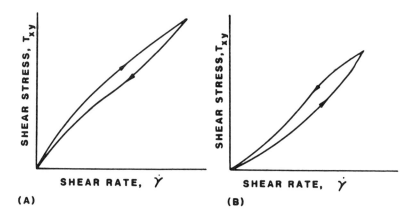

Figure 5.5 Hysteresis loops for time-dependent fluids. (A) Thixotropic hysteresis. (B) Rheopectic hysteresis.

An additional classification of non-Newtonians is that of viscoelastic fluids, which are materials exhibiting both viscous and elastic properties. For purely Hookean elastic solids the stress corresponding to a given strain is time independent, but with viscoelastic materials the stress dissipates over time. Viscoelastic fluids undergo deformation when subjected to stress; however, part of their deformation is gradually recovered when the stress is removed. These materials are sometimes described as fluids that have memory. Viscoelasticity is frequently observed in the processing of various polymers and plastics. For example, in the production of synthetic fibers such as nylon for ultrafine cable, material is extruded through a die consisting of fine perforations. In these examples the cross section of the fiber or cable may be considerably larger than that of the perforation through which it was extruded. This behavior is a result of the partial elastic recovery of the material.

In this chapter, an overview of techniques aimed at relating rheological characteristics observed from laboratory studies to commercial processing of polymers is presented. The materials discussed are primarily polymer melts and solutions. These materials show the properties of pseudoplasticity, exhibit normal forces in excess of hydrostatic pressure in simple shear flow, and often display viscoelastic behavior (i.e., stress relaxation, creep recovery, and stress overshoot). It should be noted that many polymers show all three characteristics. Engineering practices have not yet advanced to the point where we may develop a single generalized constitutive equation to describe the entire range of rheological features. Clearly, the more features that are incorporated into a constitutive equation, the more complex it becomes and hence the more difficult it is to apply its use in a balance expression needed to address a problem in fluid mechanics.

The use and interpretation of rheological flow curves are heavily relied upon to understand and project the processing characteristics of polymers and other non-Newtonian materials. It should be noted that the flow curves for many types of polymers display a Newtonian plateau or region, then a transition to non-Newtonian, and finally a power law region. The upper limit of shear rate of the Newtonian plateau depends on the polymer's molecular weight and temperature. This upper limit for many polymers is typically 10^{-2} sec^{-1} (nylon and PET are exceptions). This limit decreases with molecular weight and molecular weight distribution, as well as decreasing temperature. The onset of the transition region between the Newtonian plateau and power law can be related through the dimensionless Debora number:

$$D_b = \frac{\text{characteristic material time}}{\text{characteristic system time}} = \lambda \dot{\gamma} \tag{18}$$

where λ is the material's relaxation time.

The transition region is usually quite distinct for monodispersed polymeric melts, and broad for polydisperse melts. The magnitude of viscosity for each of the regions depends on both material and state variables. Another characteristic of pseudoplastic flow curves to keep in mind is that the power law index n in fact is not an absolute constant for a specific material. For many materials, however, it often reaches a constant value for a reasonably wide range of conditions.

The constitutive equation for a power law fluid was stated in equation (17), where apparent viscosity depends on two parameters, K and n. The consistency index K has units of $(N \cdot s^n/m^2)$. K is temperature sensitive and is found to follow the Arrhenius relationship:

$$K = K_0 \exp\left[\frac{\Delta E}{R_G}\left(\frac{1}{T} - \frac{1}{T_0}\right)\right] \tag{19}$$

where

ΔE = activation energy of flow per mole
R_G = gas constant per mole
T, T_0 = absolute and reference temperatures
K_0 = constant, characteristic of the material

For relatively small temperature ranges, equation (19) can be approximated by

$$K = K_0 \exp[-a(T - T_0)] \tag{20}$$

It should be noted that from thermodynamic consideration, K also depends exponentially on pressure.

The apparent viscosity of a power law fluid is a function of all the velocity gradients in nonsimple shearing flows. However, from a practical standpoint most laboratory viscometric techniques measure viscosity in only one velocity gradient. An important point to realize is that the power law expression is an empirical model and should therefore not be used for extrapolation of viscosity data. It does not level off to a Newtonian plateau, but instead keeps on increasing.

The Carreau model provides a constitutive expression which in fact does level off to a limiting viscosity η_0, which is either measured or estimated. It is, however, a more complex equation having additional model parameters:

$$\frac{\eta(\dot{\gamma}) - \eta_\infty}{\eta_0 - \eta_\infty} = [1 + (\lambda\dot{\gamma})^2]^{(n-1)/2} \tag{21}$$

where η_∞ is the solvent viscosity for solutions. For polymer melts it is common practice for $\eta_\infty = 0$. Note that we have changed the notation for apparent viscosity to the more conventional symbol η.

Another model that also incorporates the Newtonian plateau viscosity (η_0) is the Ellis model,

$$\frac{\eta}{\eta(\tau)} = 1 + \left(\frac{\tau}{\tau_{1/2}}\right)^{\alpha - 1} \tag{22}$$

where

$$\eta|_{\tau=\tau_{1/2}} = \frac{\eta_0}{2} \tag{23}$$

The exponent $\alpha - 1$ is the slope obtained from a plot of $(\eta_0/\eta - 1)$ versus $\log(\tau/\tau_{1/2})$.

The Ellis model can also be expressed as

$$\tau_{xy} = \frac{1}{A + B\tau_{xy}^{\alpha-1}} \dot{\gamma} \tag{24}$$

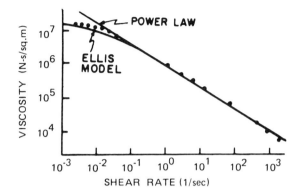

Figure 5.6 Comparison of power law and Ellis model for ABS polymer melt.

where the coefficients A and B have units of $L^2\tau^{-1}F^{-1}$ and $L^{2\alpha}\tau^{-1}F^{-\alpha}$, respectively. Term α is dimensionless. Figure 5.6 provides a comparison of the power law and Ellis models for ABS at two temperatures. Note that the power law model provides an adequate fit of the data for a large portion of the shear rate range, but departs from the measured values at the low shear rates.

Later in this chapter the principles used in obtaining viscosity measurements are described. The instruments used are viscometers, which are devices that enable determination of the relationships between τ_{xy} and $\dot{\gamma}$. Viscoelastic materials require measurement of both normal and shearing stresses as functions of shear rate, and consequently require more sophisticated measurement techniques. We must first consider a general problem of fluid mechanics that will help us to understand the operational principles of standard viscometers.

Consider the classic problem of fluid rod climbing in Couette flow as illustrated in Figure 5.7. The mass balance for this system is

$$\frac{\partial V_\theta}{\partial \theta} = 0 \tag{25}$$

and for the θ-momentum:

$$-\frac{1}{r^2}\frac{\partial}{\partial r}(r^2 \tau_{r\theta}) = 0 \tag{26}$$

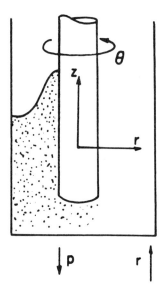

Figure 5.7 Couette flow system.

For a generalized Newtonian fluid we may write

$$\tau_{r\theta} = -\eta(\dot{\gamma}_{r\theta})\dot{\gamma}_{r\theta}$$
$$\dot{\gamma}_{r\theta} = r\frac{d}{dr}\left(\frac{V_\theta}{r}\right) \tag{27}$$

Combining the expressions above gives

$$\frac{1}{r^2}\frac{d}{dr}\left[r^3\eta\frac{d}{dr}\left(\frac{V_\theta}{r}\right)\right] = 0 \tag{28}$$

This expression can be solved by specifying $\dot{\gamma}_{r\theta}$. Additionally, a torque balance on the rotating rod is:

$$\Upsilon = R_i H 2\pi R_i \tau_{r\theta}|_{r=R_i} \tag{29}$$

The foregoing set of constitutive equations describes a generalized Newtonian fluid, showing that only a torque Υ is needed to sustain Couette-type flow. In contrast, a non-Newtonian such as a polymer melt flowing in

Couette flow shows $p = p(r)$. That is, a nonzero dp/dr term exists. Consequently, the analysis above does not adequately relate to non-Newtonian behavior.

To try to understand what happens with a polymer melt or similar material in this type of flow, we must examine the r-component momentum equation and keep track of all the stress terms that can be nonzero. This results in

$$-\rho \frac{V_\theta^2}{r} = -\frac{dP}{dr} - \frac{1}{r}\frac{d}{dr}(r\tau_{rr}) + \frac{\tau_{\theta\theta}}{r} \tag{30}$$

In creeping Couette flows, it is observed that

$$\rho \frac{V_\theta^2}{r} \simeq (10^{-1} \text{ to } 10^{-2}) \frac{dP}{dr}$$

This condition would result in a depression of the liquid level nearest the rod and not a climbing effect. We can therefore neglect this term (which is the centrifugal force). Hence the model simplifies to

$$\frac{dP}{dr} = -\frac{1}{r}\left(r\frac{d}{dr}\tau_{rr} + \tau_{rr}\right) + \frac{\tau_{\theta\theta}}{r}$$

$$\frac{d}{dr}(P + \tau_{rr}) = \frac{\tau_{\theta\theta} - \tau_{rr}}{r}$$

Solving yields

$$(P + \tau_{rr})|_{R_o} - (P + \tau_{rr})|_{R_i} = \int_{R_i}^{R_o} \frac{\tau_{\theta\theta} - \tau_{rr}}{r} dr \tag{31}$$

Experimentally, it is observed that the integral term is less than zero, and hence

$$\tau_{\theta\theta} - \tau_{rr} < 0 \tag{32}$$

The analysis reduces to attempting to define which stress component $\tau_{r\theta}$ and $(\tau_{\theta\theta} - \tau_{rr})$ contributes to shearing Couette flow. An appropriate rheological constitutive equation must be introduced in order to make this assessment. One approach is to use the Criminale-Ericksen-Filby equation (Criminals et al., 1956), which is general for steady shearing flows with one gradient.

$$\tau = -\eta(\dot{\gamma}) \cdot \dot{\gamma} - \left[\frac{1}{2}\Psi_1(\dot{\gamma}) + \Psi_2(\dot{\gamma})\right]\dot{\gamma} \cdot \dot{\gamma} + \frac{1}{2}\Psi_1 \frac{\mathscr{D}\dot{\gamma}}{\mathscr{D}t} \quad (33)$$

where $\mathscr{D}\dot{\gamma}/\mathscr{D}t$ is the corotational time derivative, defined as

$$\frac{\mathscr{D}\dot{\gamma}}{\mathscr{D}t} = \frac{D\dot{\gamma}}{Dt} + \frac{1}{2}[\omega \cdot \dot{\gamma} - \dot{\gamma} \cdot \omega] \quad (34)$$

where

ω = vorticity tensor
$\eta(\dot{\gamma})$ = viscosity function
$\Psi_1(\dot{\gamma})$ = first normal stress difference coefficient function
$\Psi_2(\dot{\gamma})$ = second normal stress difference coefficient function

Applying the condition of Couette flow,

$$\dot{\gamma} = \begin{bmatrix} 0 & \dot{\gamma}_{\theta r} & 0 \\ \dot{\gamma}_{r\theta} & 0 & 0 \\ 0 & 0 & 0 \end{bmatrix}$$

equation (33) can be solved to give

$$\begin{aligned} \tau_{r\theta} &= -\eta(\dot{\gamma}) \cdot \dot{\gamma}_{\theta r} \\ \tau_{\theta\theta} - \tau_{rr} &= -\Psi_1(\dot{\gamma}) - \dot{\gamma}_{\theta r}^2 \\ \tau_{rr} - \tau_{zz} &= -\Psi_2(\dot{\gamma}) \cdot \dot{\gamma}_{\theta r}^2 \end{aligned} \quad (35)$$

The fluid modeled in this approach exhibits shear as well as normal stresses over and above P_{hyd} in Couette flow. Furthermore, we may note that

$$\dot{\gamma} = \begin{bmatrix} 0 & \dot{\gamma}_{12} & 0 \\ \dot{\gamma}_{12} & 0 & 0 \\ 0 & 0 & 0 \end{bmatrix}$$

and hence

$$\begin{aligned} \tau_{12} &= -\eta \cdot \dot{\gamma}_{12} \\ \tau_{11} - \tau_{12} &= -\Psi_1 \cdot \dot{\gamma}_{12}^2 \\ \tau_{22} - \tau_{33} &= -\Psi_2 \cdot \dot{\gamma}_{12}^2 \end{aligned} \quad (36)$$

It can be concluded that the proper application of rheometry must be directed at evaluating all three functions: η, Ψ_1, and Ψ_2. From the standpoint of expediency, as is often the case in industrial environments, the tools of rheology (i.e., viscometers) do not always measure all the appropriate functions. Consequently, many viscometers are applied in a very limited fashion as crude quality control devices and generally do not provide suitable scientific or engineering data applicable to projecting end-use material processability. In later discussions some of these shortcomings are noted. For now, Figure 5.8 provides a summary of the types of shear flow geometries to which non-Newtonian fluids can be subjected to in order to evaluate the appropriate rheological functions.

DESCRIPTION OF CONVENTIONAL VISCOMETERS

There are a variety of viscometers used for characterizing non-Newtonian materials. These devices fall into two general categories: scientific instruments designed to make basic measurement of rheological properties, and instruments that are employed primarily for industrial quality control purposes. The most widely used techniques and instruments used in industrial labs are described in this section for polymer characterization as well as other materials.

The *capillary tube viscometer* is one of the simplest and most widely used instrument for rheological characterization. Before describing its operation we first consider the problem of capillary flow. Consider the system in Figure 5.9, where a fluid is forced at a steady rate from a large reservoir into a small-diameter capillary tube of length L. Assume steady isothermal flow of an incompressible fluid and $L/R \gg 1$. The coordinate system is defined along the r, θ, z axes. Ignoring entrance and exit losses in the capillary tube gives

$$V_\theta = 0 \quad V_r = 0$$
$$V_z \neq 0 \quad V_z \neq f(\theta) \quad \text{(axisymmetry)} \tag{37}$$

Then we write the continuity equation,

$$\frac{\partial P}{\partial t} + \frac{1}{r}\frac{\partial}{\partial r}(\rho r V_r) + \frac{1}{r}\frac{\partial}{\partial \theta}(\rho V_\theta) + \frac{\partial}{\partial z}(\rho V_z) = 0 \tag{38}$$

where density ρ is assumed to be constant. At steady state $\partial P/\partial t = 0$, and

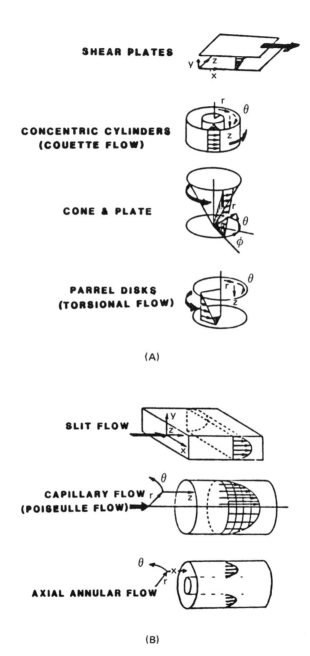

Figure 5.8 Common shear flow geometries. (A) Drag flows. (B) Pressure flows.

Processability Testing

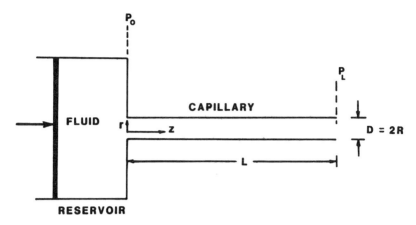

Figure 5.9 Capillary tube flow.

since $V_r = 0$, $V_\theta = 0$, the second and third terms on the left-hand side of Eqn. (38) drop out as well, leaving

$$\frac{dV_z}{dz} = 0 \tag{39}$$

The momentum equation for the z-component is

$$\rho\left[\frac{\partial V_z}{\partial t} + V_r\frac{\partial V_z}{\partial r} + \frac{V_\theta}{r}\frac{\partial V_z}{\partial \theta} + V_z\frac{\partial V_z}{\partial z}\right]$$
$$= -\frac{\partial P}{\partial z} - \left(\frac{1}{r}\frac{\partial}{\partial r}(r\tau_{rz}) + \frac{1}{r}\frac{\partial \tau_{\theta z}}{\partial \theta} + \frac{\partial \tau_{zz}}{\partial z}\right) + \rho g_z \tag{40}$$

From the constraints of the system imposed, the z-component momentum equation reduces to

$$\frac{dP}{dz} = -\frac{1}{r}\frac{d}{dr}(r\tau_{rz}) \tag{41}$$

Both sides of this expression are a function of r [i.e., $f(r)$] and both sides are constant. We can therefore assume that $dP/dz = \Delta P/L$ (note that $dP/dz < 0$) and can therefore integrate equation (41)

$$r\tau_{rz} = -\frac{\Delta P}{L} + C_1$$

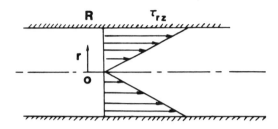

Figure 5.10 Shear stress profile in tube flow.

$$\tau_{rz} = -\frac{\Delta P}{L}r + \frac{C_1}{r} \tag{42}$$

At $r = 0$, $\tau_{zr} \neq \infty$; therefore, $C_1 = 0$, and hence

$$\tau_{zr} = -\frac{\Delta P}{2L}r \tag{43}$$

Equation (43) describes the shear stress profile for any incompressible fluid, and the profile across the capillary tube will have the form shown in Figure 5.10.

Now, if the fluid's rheological properties follow that of a power law constitutive equation (i.e., $\tau = -K|\dot{\gamma}|^{n-1}\dot{\gamma}$, where we note that $\tau = \tau_{zr}$ and $\dot{\gamma} = dV_z/dr$ and K and n are constants), then

$$-K\left|\frac{dV_z}{dr}\right|^{n-1}\frac{dV_z}{dr} = -\left(\frac{\Delta P}{2L}\right)r \tag{44}$$

Note that $|dV_z/dr| = -dV_z/dr$ since $dV_z/dr < 0$.

Integrating this expression and invoking the condition of no slip at the wall (i.e., $V_z(r = R) = 0$) gives

$$V_z(r) = \frac{nR}{n+1}\left(\frac{\Delta PR}{2KL}\right)^{1/n}\left[1 - \left(\frac{r}{R}\right)^{(n+1)/n}\right] \tag{45}$$

Equation (45) is the velocity profile of a power law fluid for a tube flow. For a Newtonian fluid (i.e., $n = 1$, $K = \mu$) the velocity profile of equation (45) reduces to the parabolic velocity profile expression

Processability Testing

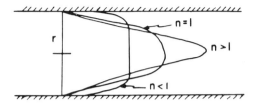

Figure 5.11 Velocity profiles: $n = 1$, Newtonian fluid; $n > 1$, dilatant fluid; $n < 1$, pseudoplastic fluid.

$$V_z(r) = \frac{\Delta P R^2}{4\mu L}\left[1 - \left(\frac{r}{R}\right)^2\right] \quad (46)$$

The theoretical velocity profiles computed from equation (45) have the shapes shown in Figure 5.11: for $n = 1$, the parabolic shape [Newtonian, i.e., equation (46)]; for $n = 1$, a dilatant fluid; and for $n < 1$, a pseudoplastic fluid. For polymer melts and solutions, $n < 1$, and hence velocity distribution is "pluglike." This means that there are very high velocity gradients in the vicinity of the tube wall, as shown in Figure 5.11.

The volumetric flow rate of the fluid through the tube is

$$Q = 2\pi \int_0^R V_z(r) r \, dr \quad (47)$$

Substituting in the velocity profile expression [equation (45)] and integrating yields

$$Q = \frac{\pi n R^3}{3n + 1}\left(\frac{R\Delta P}{2KL}\right)^{1/2} \quad (48)$$

Equation (48) can be written in a linear form as

$$\log Q = \log \frac{n\pi R^3}{3n + 1} + \frac{1}{n}\log \frac{R}{2KL} + \frac{1}{n}\log \Delta P \quad (49)$$

Note that the first two set of expressions on the right-hand side are constant. Hence $Q = f(\Delta P)$, and a logarithmic plot of Q versus P results in a straight line. Flow data for real pseudoplastic materials tend to be nonlinear, as shown in Figure 5.12.

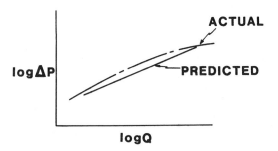

Figure 5.12 Comparison of actual to theoretical flow curve prediction.

Orienting the capillary tube so that the flow is vertically downward, a force balance gives

$$\frac{\pi D^2}{4} \Delta P = \pi D L \tau_w$$

or

$$(\tau_{rx})_{r=R} = \tau_w = \frac{D \Delta P}{4L} \tag{50}$$

Consider the flow in a slice of fluid to constitute an annulus between r and $r + dr$:

$$dQ = V 2\pi r \, dr$$

where V is the linear velocity at r. Hence

$$Q = \pi \int_0^R V 2r \, dr = \pi \int_0^{R^2} V \, d(r^2)$$

Using no-slip ($V = 0, r = R$), the expression becomes

$$\frac{Q}{\pi R^3} = \frac{8Q}{\pi D^3} = \frac{1}{\tau_w^3} \int_0^{\tau_w} \tau_{rx}^2 f(\tau_{rx}) \, d\tau_{rx} \tag{51}$$

where $f(\tau_{rx}) = -dV/dr$. By appropriate manipulation and replacing τ_w with $D \Delta P/4L$, the Rabinowitsch-Mooney equation of shear rate at the tube wall for steady, laminar flow of a time-independent fluid results:

Processability Testing

$$3\left(\frac{8Q}{\pi D^3}\right) + \frac{D\,\Delta P}{4L}\frac{d(8Q/\pi D^3)}{d(D\,\Delta P/4L)} = \left(-\frac{dV}{dr}\right)_w \quad (52)$$

Since $Q = VA = (1/4)\pi D^2 V$, equation (52) can be rearranged as

$$3\left(\frac{8Q}{\pi D^3}\right) = \left(-\frac{dV}{dr}\right)_w = \frac{3}{4}\left(\frac{8V}{D}\right) + \frac{8V}{D}\frac{d[(\tfrac{1}{4})(8V/D)]/(8V/D)}{d(D\,\Delta P/4L)/(D\,\Delta P/4L)}$$

or

$$-\dot{\gamma}_w = \left(-\frac{dV}{dr}\right)_w = \frac{8V}{D}\left(\frac{3}{4} + \frac{1}{4n}\right) = \frac{3n+1}{4n}\frac{8V}{D} \quad (53a)$$

where $n = d\ln(D\,\Delta P/L)/d\ln(8V/D)$. We may also state this expression as

$$-\dot{\gamma}_w = \frac{3}{4}\Gamma + \frac{\tau_w}{4}\frac{d\Gamma}{d\tau_w} \quad (53b)$$

where $\Gamma = 4Q/\pi R^3$ is the shear rate for a Newtonian fluid.

The Poiseuille equation for a Newtonian fluid [see Bird et al. (1960)] is

$$\Delta P = \frac{8\mu LV}{R^2} = \frac{32\mu LV}{D^2}$$

which can also be written as

$$\ln\frac{D\,\Delta P}{4L} = \ln\frac{8V}{D} + \ln(\mu)$$

Since μ is constant, then

$$\frac{d\ln(D\,\Delta P/4L)}{d\ln(8V/D)} = n = 1.0$$

Substituing this expression into equation (53) shows that $(-dV/dr)_w = 8V/D$. [See Skelland (1967) for details.] This is true only for a Newtonian fluid since viscosity is a constant. In the case of a non-Newtonian material, n varies along the log-log plot of $D\,\Delta P/4L$ versus $8V/D$. To apply a capillary tube viscometer to develop a non-Newtonian fluid's flow curve, data must be converted into a logarithmic plot of $DP/4L$ versus $8V/D$, where n is determined as the slope of the curve at a particular value of the wall shear stress (i.e., $\tau_w = D\,\Delta P/4L$). The shear rate corresponding to this wall shear stress value is determined from equation (53). It is important to note that equation (53) is derived for laminar flow, and hence one must ensure that capillary tube flow is in this regime by checking the Reynolds number. Metzner and Reed (1955) have derived the generalized Reynolds number for a power law fluid as

$$\text{Re}_{gen} = \frac{D^n V^{2-n} \rho}{\dot{\gamma}} \tag{54}$$

or

$$\text{Re}_{gen} = \frac{D^n V^{2-n} \rho}{K 8^{n-1}} \tag{55}$$

This expression applies to all fluids that are not time dependent. Metzner and Reed have shown that such non-Newtonian fluids obey the conventional Newtonian friction factor versus Reynolds number. Therefore, laminar flow prevails for $\text{Re}_{gen} < 2100$.

Data generated in any capillary tube viscometer will require corrections for four contributions to frictional losses: (1) the head of fluid above the tube exit, (2) kinetic energy effects, (3) tube entrance effects, and (4) effective slip near the tube wall. The last effect is not always present and can generally be detected as explained later.

The first three corrections can be made through application of the total mechanical energy balance. Skelland (1967) outlines a correction approach and provides an illustrative example. An alternative approach which generally gives good results is the Bagley (1954) shear stress correction method. This method was originally developed using a variety of polyethylenes. The resulting flow curve is independent of the capillary's L/D ratio; however, the procedure requires the assumptions of zero slip at the tube walls and time independency of the fluid material. The method uses a fictitious length of tubing $N_f R$ that is added to the actual length of tube L, such that the measured total ΔP across L is that which would be obtained in fully developed flow over the length $(L + N_f R)$ at the flow rate used in the experiment. The procedure is analogous to the application of an effective length in estimating kinetic resistances in pipe flow problems. The shear stress at the wall for fully developed flow over the length $(L + N_f R)$ is then

$$\tau_w = \frac{R \, \Delta P}{2(L + N_f R)} \qquad (56)$$

If the fluid is time independent and there is no slip at the tube wall, then τ_w is a unique function of $4Q/\pi R^3$ ($\simeq \Gamma$) in laminar flow (see Skelland for derivation). Hence

$$\tau_w = \frac{R}{2} \frac{\Delta P}{L + N_f R} = f\left(\frac{4Q}{\pi R^3}\right) \qquad (57)$$

from which

$$\frac{L}{R} = -N_f + \frac{\Delta P}{2f(4Q/\pi R^3)} \qquad (58)$$

The procedure involves obtaining a series of ΔP measurements made on several tubes of various L/R (or L/D) ratios, while maintaining $4Q/\pi R^3$ constant. From these data a plot of L/R versus ΔP, as shown in Figure 5.13 for different shear rates can be constructed. Equation (58) is a straight line, and hence the intercept on the L/R ordinate at $\Delta P = 0$ gives $-N_f$. A nonlinear plot as shown by the data for 3.6 and 10.8 sec^{-1} would suggest that one of our model assumptions is incorrect (i.e., the presence of time dependence or effective slip).

From a linear regression of each of the plots in Figure 5.13, the correc-

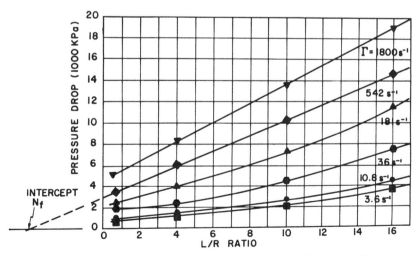

Figure 5.13 Plot of L/R versus ΔP for evaluation of Bagley correction term.

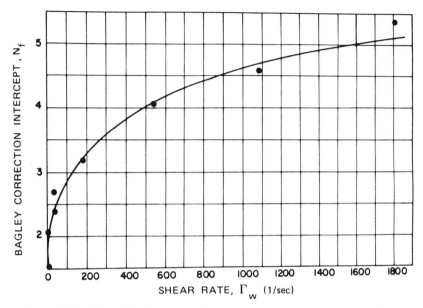

Figure 5.14 Plot of Bagley correction term versus Newtonian shear rate.

tion term N_f (intercept) can be derived. A plot of N_f versus Γ ($\equiv 4Q/\pi R^3$) is constructed in Figure 5.14 and further regressed to provide an analytical solution to equation (57). Substitution in the expression $\tau_w = R\,\Delta P/2(L + N_f R)$ allows calculation of τ_w free of end effects and should lead to a flow curve that is independent of the L/D ratio used in the capillary viscometer. Illustrations of the results are shown in Figures 5.15 and 5.16. Figure 5.15 shows a plot of τ_w versus Γ using the raw data obtained from the capillary tube.

Figure 5.16A shows τ_w data for L/D ratios of 2, 5, and 10 collapse to a single correlation. The data for $L/D = 0.25$ are in closer agreement for the high shear rate range ($\Gamma > 36$ sec^{-1}), but data points fall away from the average curve at the low shear range. Since the resulting flow curve is linear we may assume there is no time dependency for the material, and hence the lack of agreement in τ_w data obtained at $L/D = 0.25$ is probably due to slippage. The power law coefficients k and n can be readily regressed from the single line in Figure 5.16A ($\tau_w = K\Gamma^n$). The corrected shear rate at the wall can now be computed using equation (53). A plot of τ_w versus the effective shear rate $\dot{\gamma}$ computed from equation (53) is shown in Figure 5.16B.

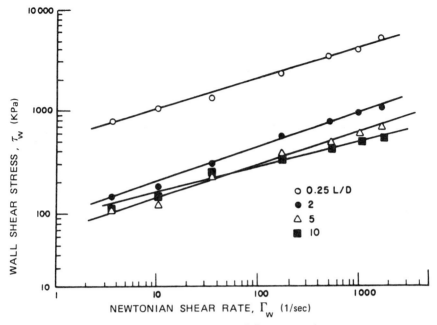

Figure 5.15 Plot of uncorrected flow curve data.

To minimize entrance losses and therefore minimize lengthy correction calculations, the L/D ratio should be made as large as is practically possible. Thomas (1963), for example, had to resort to an L/D ratio of 1000 in work on suspensions. At excessive L/D ratios, frictional corrections become negligible in viscometric data.

In the example above, the data resulted in a linear flow curve. If the data showed a nonlinear response, it would have been necessary to employ standard tangent-drawing procedures. By constructing tangents to the curve on Figure 5.16A, we would obtain corresponding values of n from the tangent slopes and K-values from the tangent intercept at $\Gamma = 1$. We may then recalculate $\dot{\gamma}_w$ from the Rabinowitsch-Mooney equation.

It is important to note that some non-Newtonian fluids display a peculiar orientation of their molecules in the vicinity of the tube walls. In the case of an aqueous suspension, the discrete phase may actually move away from the wall, leaving a thin layer of the continuum phase in the immediate proximity of the wall. In this situation there is a reduction in the apparent viscosity in the vicinity of the wall, known as the phenomenon of "effective slip" at the wall. For a capillary tube viscometer, an effective slip coefficient β can be evaluated as a function τ_w using the following

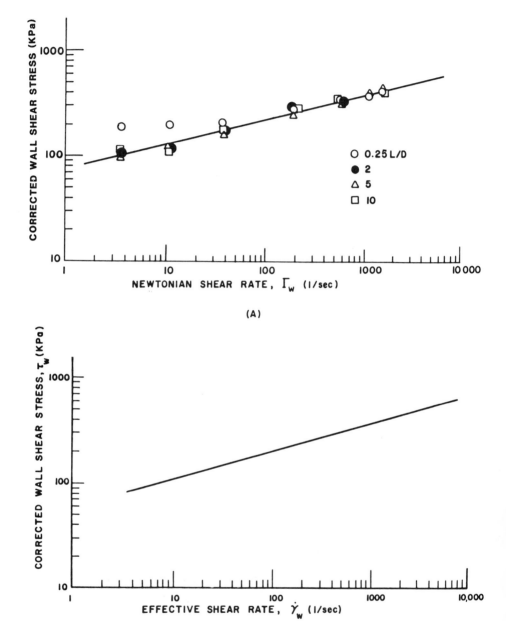

Figure 5.16 Plot of corrected flow curve data using (A) Newtonian shear rate and (B) effective shear rate data.

Processability Testing

relationship:

$$\frac{Q}{\pi R^3 \tau_w} = \frac{\beta}{R} + \frac{1}{\tau_w^4} \int_0^{\tau_w} \tau_{rx}^2 f(\tau_{rx}) \, d\tau_{rx} \qquad (59)$$

A series of capillary tube measurements are needed to evaluate β. For a range of tubes of various R but constant L, a plot of $Q/R^3\tau_w$ versus τ_w can be constructed. If no slip occurs with the material (i.e., $\beta = 0$) all the curves will coincide in accordance with equation (59). If, however, we obtain a family of curves, then β has some value. From the family of curves, select some fixed value of τ_w and obtain values of $Q/\pi R^3\tau_w$ for each corresponding R-value curve. Now prepare a plot of $Q/\pi R^3\tau_w$ versus $1/R$. The slope of this curve provides a β-value for the selected τ_w, according to equation (59). By repeating this procedure for different τ_w values a relationship between τ_w and β can be established. Figure 5.17 illustrates the method. Once a relationship between the effective slip coefficient and τ_w is established, the measured volumetric flow rate from the capillary tube can be corrected for slippage:

$$Q' = Q_{\text{measured}} - \beta \tau_w \pi R^2 \qquad (60)$$

where Q' is the volumetric flow rate corrected for slippage and $\tau_w = D \Delta P / 4L$ corresponding to Q measured.

In rubber processing, the capillary rheometer has been automated into so-called processability testers, the most notable of which is the Monsantor Precessability Tester (MPT) illustrated in Figure 5.18. The instrument is designed for automatic testing of viscoelastic materials such as unvulcanized rubber and thermoplastics. The MPT is essentially a controlled shear-rate capillary rheometer with a laser die-swell measuring device. Fully programmable test sequences allow measurement of viscosity and die swell at four different shear rates, plus a stress relaxtion response. Data are presented on digital panel meters, and also as a continuous plot of barrel pressure and die swell versus time on a two-pen recorder. Data representing average values for each shear-rate condition may also be collected by a printer or computer at the end of the test. Operation is flexible with a wide range of temperatures and shear rates which may be used to simulate conditions approximating production process. The test material is placed in a 19-mm-diameter heated barrel and the downward movement of a heated piston forces the material through a small capillary at the bottom of the barrel. The barrel pressure and die swell are measured simultaneously. A closed-loop hydraulic system drives the piston, while a microprocessor is programmed for precise servo con-

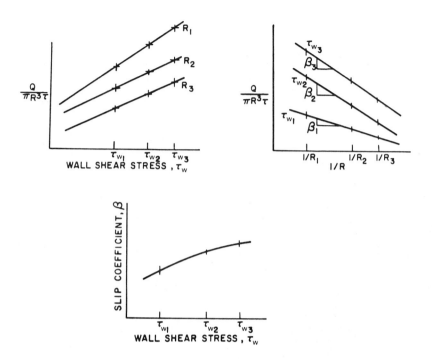

Figure 5.17 Procedure for correcting for effective wall slip.

trol of the piston movement at selected rates. By choosing a capillary with the desired diameter and appropriate piston rates, specific shear rates may be obtained. Shear rates from 1 to 10^4 sec^{-1} may be run using the standard capillaries supplied with the instrument.

The shear stresses developed within the sample at the test shear rate can be calculated from the pressure decrease along the length of the capillary. A pressure transducer located just above the entrance to the orifice in the barrel is used to measure the extrusion pressure, and the barrel pressure is displayed on both a panel meter and the strip-chart recorder. The shear stress in the capillary will depend on its length and diameter, as well as the barrel pressure.

After leaving the capillary, the extrudate passes a laser scanning device that measures its diameter and reports it as a percentage of the capillary size, or in actual dimensions. The extrusion chamber, where the laser die swell measurements are made, is heated so that the sample will reach a final die swell in a short time span. The microprocessor can be

Processability Testing

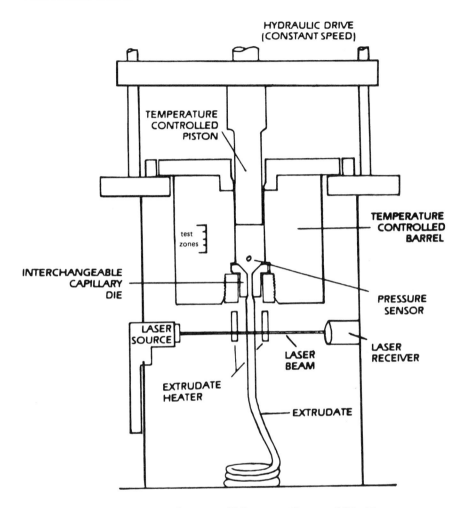

Figure 5.18 Key features of Monsanto Processability Tester.

programmed to stop the piston for a time to measure the relaxed die swell of the stationary extrudate. Die swell represents the stored elastic energy in the test material as it exits the capillary. A stress relaxation test may also be run by the MPT to evaluate the relaxation response of the sample.

A typical test of a rubber compound would result in a set of data describing the flow properties of the compound over a range of flow rates. A series of these tests (two or three) over a range of temperatures would result in data describing flow properties for the conditions experienced in processing. The flow rate, determined from the volume flow rate and the geometry of the die, can be expressed in terms independent of the overall flow geometry as a shear rate, in seconds^{-1}. The resistance to flow is measured by the MPT as the pressure drop across the length of the capillary. This can also be reduced to a term that is independent of geometry—namely shear stress, in units of pascal. The ratio of shear stress to shear rate is the viscosity of the test material. As noted earlier, apparent viscosity is a material property that varies with either shear stress or shear rate, but is characteristic for the material at the same temperature and flow conditions. If the flow conditions and temperatures of the MPT and a production process are similar, MPT flow data can be used to predict processing flow, as discussed later.

As described above the shear stress-temperature response for many polymers is independent of the shear stress/shear rate response. It is often possible to fit MPT data to a simple expression (i.e., power law) to describe apparent viscosity as a function of shear rate. A curve-fitting program has been incorporated into a computer program to monitor MPT test results. As a result, the flow characteristics of a rubber compound over a range of temperatures and shear rate can be measured automatically by running two or three tests with the MPT.

Elastic properties are more difficult to measure than flow properties, because they are time dependent. The MPT measures elastic properties by means of several tests. The "running" die swell is measured as the extrudate passes the laser during steady flow. If the flow rate is rapid, the running die swell is a short time elastic response to the stresses in the stock just before exiting the die. The "relaxed" die swell, measured after extrusion has been halted, is the long time elastic response to the same stresses (Figure 5.19). An extrudate heater helps to eliminate cooling stresses as a variable in the die swell measurement. The stress relaxation test involves a rapid downward movement of the piston at the end of the flow tests. The high pressure developed in the barrel is allowed to decay by relaxation of the stress in the stock. The time for the pressure to relax to a preset percentage of the peak pressure is reported as the stress relaxation time. The stress relaxation test often correlates with the die swell, as both measure the elastic memory of the stock.

Another way the MPT can measure elastic effects is in measuring the pressure developed as the stock is accelerated from the barrel into the capillary. This entrance pressure is a function of the shear rate and the die entrance geometry, but is independent of the die length. By running an MPT test with two dies having the same diameter but different lengths, the

Processability Testing

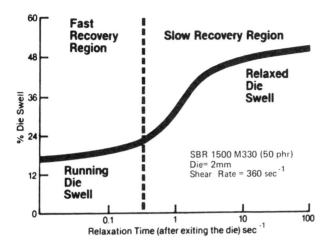

Figure 5.19 Effect of recovery time on die swell.

entrance pressure can be measured. This type of test is the Bagley correction test described earlier. The pressure drop across the die length leads to the calculation of the shear stress. The MPT measures the entrance pressure plus the die pressure drop, and a die length associated with the entrance pressure (entrance correction) can be calculated from the Bagley test. The entrance correction is another measure of the elastic property of the test stock.

In most rubber processes, an acceleration of the stock from a large geometry to a small geometry is commonly practiced. The MPT entrance correction may be used to estimate the stresses developed in these flow conditions. Stresses are released in shaping processes such as extrusion, calendering, and milling. The elastic memory measured by MPT die swell can be found to correlate with the final dimensions of a shaped rubber part, usually different by some equipment geometry constant. In making a comparison of MPT and processing die swell, the temperature and stress states should be the same.

The stress relaxation test is useful as a separate measure of elastic properties for control purposes. Direct application of the stress relaxation data to processing behavior is difficult. Materials that have the same stress relaxation response and the same viscosity should be similar in processing. This type of test can be very useful for testing raw materials such as polymers.

Concentric cylinder rotary and rotating cylinder viscometers are another common instrument, but most often applied to solution viscosity mea-

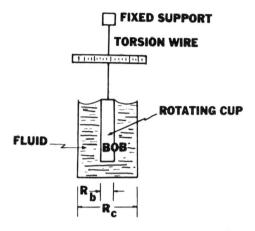

Figure 5.20 Basic elements of concentric cylinder rotary viscometer.

surements, and are usually limited to shear rates or less that 100 sec^{-1}. The basic elements of the concentric cylinder rotary viscometer are shown in Figure 5.20. The unit operates by applying shear to a fluid located in the annulus between the concentric cylinders. One cylinder (usually the cup) rotates while the other is fixed in space. From a series of measurements of the angular speed of the rotating cup and of the torque applied to the stationary cylinder, a flow curve for the fluid under shear can be derived.

The development of the constitutive equations for this type of visometer is as follows. A torque balance about the surface of the fixed bob while the cup revolves at a steady angular velocity is

$$\tau_b(2\pi R_b l)R_b = 2\pi R_b^2 l \tau_b = \Upsilon \tag{61}$$

where Υ is the torque, R_b the bob radius, and τ_b the shearing stress at the bob surface.

The balance equation neglects end effects at the base of the bob and assumes that shear occurs only at the cylindrical surface of the bob. In practice, however, end effects are important and can be accounted for by including an effective bob length l_{eff}, in equation (61). Solving equation (61) for the bob surface shear stress yields

$$\tau_b = \frac{\Upsilon}{2\pi R_b^2 l}$$

Processability Testing

The shear rate at the bob surface is a complex relationship, which for brevity is not included here. An approximate relationship proposed by Calderbank and Moo-Young (1959) based on the analytical derivation of Kreiger and Maron (1954) is as follows:

$$\left(\frac{dV}{dr}\right)_b = \frac{4\pi N}{1 - (R_c/R_b)^2} C_r \tag{62}$$

Equation (62) is applicable for cup-to-bob radii ratios < 1.75. The term C_r represents a series expansion of the shear rate/geometric relationship developed by Kreiger and Maron for time-independent power law fluids. Table 5.1 provides values of C_r that can be used in equation (62). The rheological flow curve can be developed by a logarithmic plot of τ_b versus $(dv/dr)_b$. By employing different bob or cup sizes, the different ranges of shear rates can be studied. In the extreme, for very small annular gaps (i.e., $R_c/R_b \to 1.0$), the shear rate approaches a constant value across the annulus and can be approximated simply by

$$\frac{dv}{dr} = \frac{2\pi RN}{R_c - R_b} \tag{63}$$

It should be noted that close clearance viscometers such as this are not suitable for use with suspensions where particle sizes are comparable to the annular gap.

End effects can be corrected for by substituting an effective length l_{eff} for the true bob length l in equation (61). The effective length can be determined by calibrating the viscometer using a Newtonian fluid of known viscosity μ and applying the following formula:

$$l_{\text{eff}} = \frac{\Upsilon}{2\pi R_b^2 \mu (dV/dr)_b} \tag{64}$$

To correct for effective slip in rotary viscometers, the correction method of Mooney (1931) can be used. The method involves obtaining measurements using three different R_c/R_b ratios. For R_1, R_2, and R_3 radii, such that $R_1 < R_2 < R_3$, three radii ratios can be defined ($S_a = R_2/R_1$, $S_b = R_3/R_2$, $S_c = R_3/R_1$). The rotational speed of the cup required to obtain the same torque for each of these cases can be measured as N_{12}, N_{23}, and N_{13}. The effective slip coefficient corresponding to the stress at the surface $2\pi R_2 l$ is

Table 5.1 Values of Calderbank/Moo-Young Shear-Rate Term (C_r)

n''	R_c/R_b Ratio[a]					
	1.070	1.150	1.166	1.250	1.400	1.746
0.050	2.722	4.999	5.435			
0.100	1.708	2.617	2.801	3.735	5.184	
0.200	1.287	1.622	1.689	2.031	2.593	3.615
0.300	1.162	1.342	1.377	1.554	1.843	2.382
0.400	1.102	1.213	1.234	1.340	1.511	1.826
0.500	1.068	1.139	1.153	1.220	1.327	1.522
0.600	1.045	1.091	1.100	1.144	1.212	1.335
0.700	1.029	1.058	1.064	1.091	1.134	1.209
0.800	1.017	1.034	1.037	1.053	1.077	1.119
0.900	1.007	1.015	1.016	1.023	1.034	1.052
1.000	1.000	1.000	1.000	1.000	1.000	1.000
1.250			0.971		0.941	0.910
1.500			0.952		0.903	0.853
1.750			0.939		0.877	0.814
2.000			0.929		0.857	0.785
2.250			0.921		0.842	0.763
2.500			0.915		0.830	0.746
2.750			0.910		0.820	0.732
3.000			0.906		0.812	0.720
3.250			0.902		0.806	0.711
3.500			0.899		0.800	0.702
3.750			0.897		0.795	0.695
4.000			0.894		0.791	0.689
Coefficient, A	0.903	0.861	1.144	0.898	1.169	1.155
Exponent, B	−0.304	−0.505	−0.320	−0.547	−0.418	−0.485
Coefficient of fit, r	0.946	0.963	0.891	0.974	0.928	0.957

[a] Each set of R_c/R_b data sets has been regressed using least squares to a power law expression. For approximate C_r-values, apply the model coefficient above: $C_r = A(n^*)B$.

$$\beta_2 = \frac{2\pi^2 R2^2}{\Upsilon}(N_{12} + N_{23} - N_{13}) \tag{65}$$

Repeating this procedure for different torque values provides a correlation between β_2 and the surface shear stress τ_2. To correct for slippage, for a given torque value, τ_b can be computed from equation (61). The corresponding expression for wall shear stress at the cup surface is

$$\tau_c = \frac{\Upsilon}{2\pi R_c^2 l} \tag{66}$$

The effective slip velocities at the bob and cup surfaces are $V_{s,b} = \beta_b \tau_b$ and $V_{s,c} = \beta_c \tau_c$, respectively. The effective velocity of the fluid at the rotating cup surface will be less than that of the cup by the amount $V_{s,c}$. The effective velocity of the fluid at the stationary bob surface is $V_{s,b}$. The true shear rate corresponding to τ_b can be calculated by multiplying either equation (62) or (63) by the ratio of the actual velocity difference (i.e., from cup to bob) to the velocity difference without slip [i.e., $(V_{sc} - V_{sb})/V_{\text{meas}}$].

A modification of the concentric cylinder viscometer is the *Brookfield viscometer*, in which the cup radius is effectively extended to infinity. Flow curves are generated from measurements of the torque needed to rotate a cylindrical rod at various speeds when immersed in the fluid of "infinite" volume. Basically, the fluid is contained in a cup whose radius is much larger than the rod. With this arrangement, the walls of the retaining cup do not influence the shearing movement of the fluid. Equation (61) defines the shearing stress at the rod surface (R_b = radius of the cylindrical rod, l = length of rod immersed in the fluid). As before, an effective rod immersion length, l_{eff}, can be developed to correct for end effects. The shear rate at the surface of the rotating rod for any time-independent fluid is given by

$$\left(-\frac{dV}{dr}\right)_b = \frac{4\pi N}{n''} \tag{67}$$

where n'' is the slope of a logarithmic plot of torque versus rotational speed N, evaluated at the particular N for which Υ was measured. A plot of τ_b versus $(-dV/dr)_b$ provides the flow curve. With either device we may also generate a plot of $D\,\Delta P/4L$ versus $8V/D$ (Γ is the Newtonian shear rate) for laminar tube flow. In general, both devices enable studies over a very limited shear rate range. Commercial units are usually equipped with an electrical heating jacket and appropriate heat exchange to remove heat generated by friction in the fluid.

The *cone-and-plate visometer* is a widely used instrument for shear flow rheological properties studies. The device can be used to measure both viscosity and normal stresses. Applications and analysis of this type of viscometer for studying polymer melts are covered by Lee and White (1974), Lobe and White (1979), Chapman and Lee (1970), Minagawa and White (1976), King (1966), Meissner (1969), and Weissenberg (1948). The principal features of this viscometer are illustrated in Figure 5.21, consisting essentially of a flat, horizontal plate and an inverted cone, the apex of which is in near contact with the plate. The angle between the plate and cone surface is small (typically, <2°). The fluid sample is placed in this small gap between the cone and plate. A flow curve is developed from measurements of the torque needed to rotate the cone at different speeds. For a constant speed N, the linear velocity at r is $2\pi rN$. The gap height at r is $r \tan \phi$. Hence the magnitude of the shear rate at r is

$$\frac{2\pi rN}{r \tan \phi} = \frac{2\pi N}{\tan \phi} \tag{68}$$

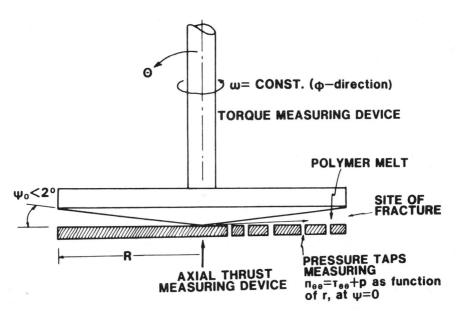

Figure 5.21 Schematic of cone-and-plate viscometer.

Processability Testing

The shear rate is constant between $0 > r > R$, which means that τ_{yx} is also constant over this range. For small ϕ, $\tan \phi \simeq \phi$, and hence the magnitude of the shear rate $\simeq 2\pi N/\phi$. The following expression defines the relationship between measured torque and shear stress:

$$\Upsilon = 2\pi \tau_{yx} \int_0^R r^2 \, dr = \tfrac{2}{3} \pi R^3 \tau_{yx}$$

or

$$\tau_{yx} = \frac{3\Upsilon}{2\pi R^3} \tag{69}$$

Equation (69) enables τ_{yx} to be obtained at different measured torque values. Shear rate can be obtained from equation (68), and hence τ_{yx} can be plotted against shear rate for the flow curve.

Commercial instruments that are capable of measuring normal stresses as well as the shear stress are the *Weissenberg Rheogoniometer* (manufactured by Sangamo Controls, Bognor Regis, U.K.) and the *Mechanical Spectrometer* (manufactured by Rheometrics Inc., Union, N.J.). Another type of shear flow instrument, used primarily for polymer melts, is the *sandwich viscometer* (also known as the *parallel-plate viscometer*). The measurement principle is illustrated in Figure 5.22. A viscous material such as an elastomer is sheared between two parallel steel plates or as in Figure 5.22, in a sandwich between three plates. The shear rate is

$$\dot{\gamma} = \frac{V}{H} \tag{70}$$

where H is the interplate distance.

The applied force can be determined as a function of time, which after normalization with area provides shear stress data as a function of time:

$$\tau = \frac{F}{2A} \tag{71}$$

It is possible to measure transient startup stresses of the material with this device. After a sufficiently long time and if the material is stable, the steady-state stress can be measured, at which point the viscosity can be determined as a function of shear rate. With many materials such as elas-

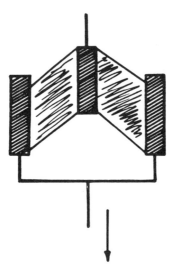

Figure 5.22 Schematic of sandwich viscometer.

tomers, both large times and equipment strains are needed. These conditions often result in slippage. To minimize slippage, contact surfaces are usually knurled. This type of viscometer is capable of making measurements at very low shear rates on materials that usually exhibit slippage in nonpressurized rotational devices (e.g., cone-and-plate viscometers). The instrument is described in detail by Zakharenko et al. (1962), Middleman (1969), Goldstein (1974), and Furuta et al. (1976).

A *parallel disk viscometer* (also known as a "shearing disk" viscometer) measures viscosity in torsional flow between a stationary disk and a rotating disk. Design variations are illustrated in Figure 5.23. The operation involves the use of a serrated disk that is rotated in a sample fixed in a pressurized cavity. This system was developed for rubber applications where both pressurization and the use of serrated surfaces are intended to avoid slippage. Properties are measured based on the shear rate at the outer radius of the disc.

$$Y_a = \frac{a\omega}{H} \tag{72}$$

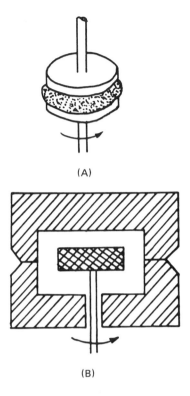

Figure 5.23 (A) Parallel disk viscometer; (B) Mooney shearing disk viscometer.

Shear stress and normal stress differences are given by the following relationships:

$$\tau|_a = \frac{3\Upsilon}{2\pi a^3}\left(1 + \frac{1}{3}\frac{d\ln\Upsilon}{d\ln\gamma_a}\right) \tag{73}$$

$$P_{11} - 2P_{22} + P_{33}|_a = \frac{2F'}{\pi a^2}\left(1 + \frac{1}{2}\frac{d\ln F'}{d\ln\gamma_a}\right) \tag{74}$$

where F' is the normal thrust. The parallel disk apparatus is limited to low shear rates.

The Mooney viscometer (Figure 5.23B) consists of a rotating disk in a cylindrical cavity. This instrument has become the standard quality control instrument of the rubber industry. The American Society for Testing and Materials (ASTM) specifies the diameter of the disk to be 19.05 mm and thickness 5.54 mm. A single-point viscosity measurement is made at a standardized rotor speed of 2 rpm, after a period of 4 min and 100°C. In the elastomers industry some manufacturers still report measurements at 8 min and 127°C; however, considerable effort is under way to standardize to the ASTM guidelines. Measurements are reported in terms of a calibrated torque measurement on a standard dial (referred to as ML-4 reading). The designation ML-4 means: M, *Mooney* viscometer; L, *large* disk size [a smaller diameter disk (S) is used for vulcanization scorch tests], and 4 min of test.

ML-4 readings can be converted to torque, shear stress, and viscosity values. Nakajima and Harrel (1979) provide the following correlations:

$$\Upsilon = 8.30 \times 10^{-2} D_R \tag{75a}$$

$$\tau_a = 0.382(D_R)f' \times 10^4 \tag{75b}$$

where D_R is the Mooney unit dial reading. Υ and τ_a have units of N/m and Pa, respectively. f' is a function defined by the following relationship for shear stress at the disk's outer radius:

$$\tau|_a = \frac{\Upsilon}{\pi a^3} f' \tag{76}$$

$$f' = \frac{n'+3}{4} \left(1 + \frac{(2/n')^{n'}(n'+3)hH^{n'}}{2a^{n'+1}\{1 - [a/(a+\delta)]^{2/n'}\}^{n'}} \right)^{-1} \tag{77}$$

where δ is the distance between the outer disk radius and cavity, h is the disk thickness, and n' the power law exponent ($\simeq d \ln \Upsilon / d \ln \dot{\gamma}_a$). Function f' typically has a value of around 0.55 for the standard measurement conditions.

A slit rheometer is an analogous instrument to the capillary viscometer. The primary difference between the devices is the orifice cross section. Han (1971, 1974, 1976); Wales (1969), and Wales and Janeschitz-Kriegel (1967) describe the use of this instrument in the study of polymer melts. The device makes use of a series of flush-mounted transducers located along the flow tube. These transducers measure the pressure gradients along the direction of flow, which can then be converted to wall shear stress values via

$$\tau_w = \delta_c \frac{dP}{dZ} \tag{78}$$

where δ_c is the half thickness of the channel.

Han (1976) gives the following expression for the shear rate at the channel wall:

$$\dot{\gamma}_w = \frac{3Q}{4ab^2}\left[\frac{2}{3} + \frac{1}{3}\frac{d(\ln(3Q/4ab^2))}{d(\ln \tau_w)}\right] \tag{79}$$

where a is the half-width. In general, the instrument is capable of operating over comparable shear rate ranges to the capillary viscometer.

The *melt indexer* (illustrated in Figure 5.24) is widely accepted as a standard quality control instrument in the plastics industry. The device,

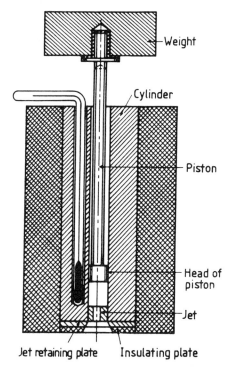

Figure 5.24 Details of melt flow index apparatus.

developed by DuPont, is traditionally used to specify polyethylenes, polypropylene, and polystyrene but is not restricted to these materials. It is essentially an extrusion rheometer that is in fact similar to a capillary rheometer. The instrument employs a single capillary (diameter of 0.655 mm and L/D ratio of 12) in which extrusion of material is accomplished due to pressure applied by a dead weight. The melt index is given as a measure of the weight in grams extruded for a 10-min test.

The ASTM specifies a series of test conditions involving temperature and applied dead-load settings. These conditions are designations A through K, and they cover the following ranges; for temperature: 125 to 275°C, for applied dead load: 325 to 21,600 g (or in terms of pressures, 0.5 to 29.5 atm). Specifications are selected so that melt indices lie between the values of 0.15 and 25 g per 10-min test. The most frequently used condition in the industry is specification E (190°C, 2.95 atm).

Although this is a widely used instrument, its value is that of a quality control tester which can only empirically relate to end-use processability of products. The melt index (MI) is related to the inverse of viscosity; however, no end-loss corrections have been developed, nor can the MI be easily related to the Weissenberg-Rabinowitsch shear-rate expression.

Referring to Figure 5.24, the cylinder is of hardened steel and is fitted with heaters, lagged and controlled for operation at the specified temperature (125 and 300°C at ± 0.5°C by ISO standards). The dimensions of the cylinder, especially the radius, are closely specified. The piston is steel and has a diameter considerably smaller than the internal diameter of the cylinder. The die has an internal diameter of 2.095 or 1.180 ± 0.05 mm, depending on the code method, and is made of hardened steel. All surfaces that must come in contact with the molten polymer are highly polished. Weights are employed so that the piston can be loaded to provide a total weight of between 325 and 21,600 g, depending on the procedure.

Prior to each test, the cylinder, piston, and jet must be thoroughly cleaned by treatment with hot solvents and all contaminants must be removed. The cylinder is allowed to equilibrate thermally by maintaining the unit at the test temperature for at least 15 min. It is then charged with between 4 and 8 g of sample and the unloaded piston is reinserted. After 4 to 6 min the weight is added and the molten polymer begins to extrude through the die. The rate of extension can be measured by simply cutting of the extrudate at suitable time intervals. Several of these cutoffs are obtained up to 30 min after charging the sample. These must be taken when the piston head is between 50 and 20 mm above the upper end of the die (marked by scribe lines on the piston). Usually, the first cutoff sample contains bubbles from an entrapment and must be discarded. Usually from three samples, an average mass is obtained (to the nearest 0.001 g). Maximum and minimum values of the individual weightings must be within specified values of the average; otherwise, results are discarded and a new test with fresh sample initiated. The melt flow rate (i.e., melt flow index) is then computed as the weight in a reference time as specified in the standard (usually 600 sec). Along with test data, the temperature of the test as well as the piston loading must be reported.

APPLICATION OF CAPILLARY RHEOMETRY TO PROCESSING PREDICTIONS

Sezna (1984) and, more recently, Cheremisinoff (1987) describe application of capillary rheometry to predicting the processing performance of rubbers. The following considers two common processing operations, mixing and extrusion.

The process conditions in an internal mixer are complex, since the rotors are continuously transferring, stretching, and shearing the material at different locations in the mixing chamber. The mixing process itself alters the characteristics of the rubber stock with time. Most of the energy consumed in mixing is concentrated in the shearing flow between the rotor and the wall of the mixer. The power consumed in shear mixing is related to the product of the frictional forces and the rotor speed. The frictional forces can be represented by the shear stress in the mixer, and the rotor speed can be reduced to a shear-rate value. The shear rates in a conventional batch mixer used for rubber processing such as Banbury mixer, typically ranges from 10 to 100 sec^{-1}. The energy consumed by the mixing process is related to the average power consumed over this range of shear rates:

$$\text{Energy} \simeq \text{Constant} \times \text{Average [(shear stress)} \times \text{(shear rate)]} \quad (80)$$

The shear stress/shear rate relationship can be measured using a capillary rheometer. By conducting a test in both the mixer and the capillary, the relationship between the model and the process can be established. The procedure presented by Sezna in implementing the mixing model is as follows:

1. Measure the temperature and shear rate relationships to shear stress for the polymer samples by testing in the capillary viscometer.
2. Estimate the effect of shear heating during mixing.
3. Calculate the average power consumed over the range of shear rates in the mixer:

$$\tilde{P} \simeq \frac{\Delta \tau \propto \dot{\gamma}}{\Delta \dot{\gamma}} \tag{81}$$

4. Calculate the correlation constant between the calculated average power term and mixing energy measured in an actual mixing process.
5. Compare predicted mixing energy with measured mixing energies to evaluate the model.

Typical capillary test results are the result of a regression analysis fitting the data to an equation of the power law form [equation (16)]. To account for temperature, this expression can be modified along with the Arrhenius equation to give

$$\tau = K_1 e^{k_3/T} \dot{\gamma}^n \tag{82}$$

The energy consumed in the mixer may be correlated with an average power term by combining equations (81) and (82):

$$\tilde{P} \simeq K(100^{n+1} - 10^{n+1}) \tag{83}$$

In Sezna's study, five different samples of SBR-1500 and five samples of high-Mooney-viscosity medium-ethylene-content EPDM polymers were selected for evaluation. Each of the samples was produced by a different manufacturer. Three batches of each SBR stock and four batches of each EPDM stock were processed. A conventional mix procedure was used for the SBR, mixing the polymer alone for 1 min before adding the fillers. The EPDM masterbatches were mixed "upside down," as is usual for a highly extended stock.

The masterbatches were all mixed in a laboratory-size inernal mier. Energy consumed during the mix was determined with a power integrator. This instrument subtracts the power consumed by the empty mixer from the power consumed by the loaded mixer and integrates the instantaneous power values over time. The electrical energy used to drive the mixer is displayed in watt-hours, and the instantaneous power curve is recorded on a chart. Both the raw polymer samples and the masterbatches were tested for flow characteristics and elastic response over the range of shear rates and temperatures typical in rubber mixing, using an MPT.

The shear heating of the polymers during mixing is related to the shear stress. Figure 5.25 provides a comparison of the shear stress values, as measured by the capillary devices to mixing dump temperatures for the EPDM and SBR compounds. The shear heating correlation may be used to account for differences in mixing temperature, which should affect the

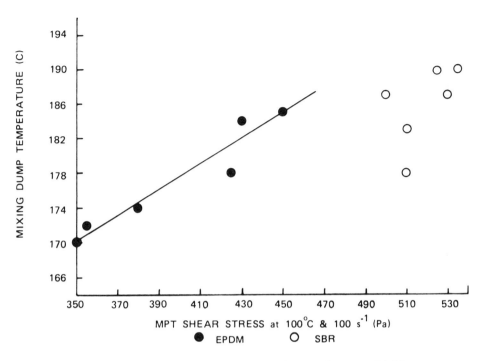

Figure 5.25 Shear heating in mixing. [Data of Sezna (1984).]

mixing energy levels. In this study the differences in mixing temperatures were not sufficient to affect correlations between mixing energy and capillary stresses.

The energy consumed in mixing is related to the flow properties of the material in the mix. This material undergoes a significant change in flow properties throughout the mix cycle, with the early stages of the mix corresponding to polymer flow, and the latter stages of the mix corresponding to the mixed compound properties.

In the case of SBR mixing, the polymer Mooney viscosity test results were found to have a high degree of correlation with the early mix energy, as did MPT shear stress values. The average power coefficient was also found to agree wtih these correlations. The variation in the flow index, slope n, was so small for the SBR samples that the shear stress levels, or power law constant K, were the dominant variables. The variation in flow index was also very small for the SBR masterbatches, and the masterbatch MPT stress values and Mooney viscosities were found to correlate closely with the end of cycle mix energy. These correlations explain the changes in ranking of the mix energy among the polymer samples as the viscosity ranking changes. The change in viscosity ranking probably represents differences in carbon black reinforcement and in polymer breakdown. Based on these results, simple viscosity measurements were sufficient to discriminate between the SBR samples tested.

In the case of EPDM mixing, however, the polymer Mooney viscosity test results correlated very poorly with the early mixing energies. The lowest-viscosity EPDM polymer was found to require the greatest amount of early mixing energy. The flow index had a high correlation with the early mixing energy. The MPT power coefficient also had a high correlation with the early mixing energy.

The higher shear stresses associated with MPT tests of EPDM polymers involved slippage at the shear rates used in the mixing study. The average power term had the best correlation with EPDM polymer mixing energy because a low value of n, in this case, corresponded with slip.

The EPDM masterbatch Mooney viscosities and MPT shear stresses had a high correlation with the end of cycle mix energy, just as was the case with SBR mixes. The EPDM masterbatches also slipped at the shear rates typical in mixing, however, and the power law constant, K, had a low correlation with the end-of-cycle mix energy. When K is combined with the flow index to form the average power coefficient, however, this coefficient had a high correlation with the end of mix energy. Figure 5.26 shows the correlation of K with end of mix energy, and Figure 5.27 shows how the MPT power coefficient had a closer correlation.

The rheological behavior of the materials contained in an internal mixer is a key to determining power requirements in mixing. Mixing uniformity is often controlled by specifying the mix cycle on the basis of power integrator values at important stages. Often the compounder is faced with the need to change sources of polymer or to specify current polymers by a processability test.

If the polymers are as uniform as the SBR samples in Sezna's experiment, the Mooney Viscometer may be an adequate polymer process control tester. If variations in the flow behavior are significant, as in the case of the EPDM polymer samples tested in his experiment, a more sophisticated processability tester is necessary. The capillary rheometer used in this study is an example of a more complete processability tester. The mixing model using the flow response measured by a capillary rheometer predicted mixing energy variations in these examples. As with any model, experience must be applied to explain all of the variables involved.

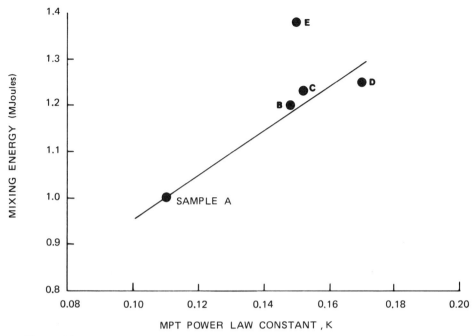

Figure 5.26 End of cycle mixing energy versus capillary power law constant for different EPDM masterbatches. [Data of Sezna (1984).]

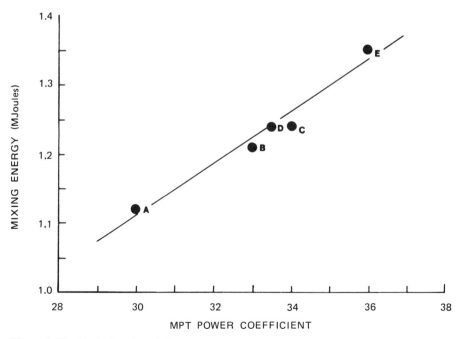

Figure 5.27 End of cycle mixing energy versus processability tester power coefficient EPDM masterbatches. [Data of Sezna (1984).]

Table 5.2 Values of k and n for Gums and Compound Polymers

Mooney viscosity (1' + 4') at 127°C	MWD	$\overline{M}_z/\overline{M}_w$	Gum		Compound	
			k	n	k	n
53	3.1	16	2223	0.26	402.9	0.29
72	3.8	30	2296	0.26	804.8	0.24
73	3.4	12.4	3125.9	0.23	323.9	0.31
60	2.2	3.1	1844	0.23	572.1	0.26
60	2.2	3.1	1844	0.23	148.6	0.16

Cheremisinoff (1987) and Sezna (1984) have shown that k and n constants for compound of EPDMs as measured in the MPT properly describe the rheological characteristics of extrusion. By relating k and n to MWD, some generalizations for selecting polymers for certain extrusion applications can be made. Table 5.2 reports typical k and n values for both gum and compound samples, along with their MWD's. Data are shown for four samples in the same compound formulation, and one sample in two widely different compounds. In scanning the table, it is found that n is nearly constant despite significant variation in MWD and that it appears to be more of a function of the specific compound. The k values, however, vary, implying a k-MWD and k-M_w dependency.

Figure 5.28 shows a plot of compound k values versus M_z/M_w averages for several polymers with approximately the same M_w/M_n. This illustrates the fact that compound viscosity can be altered significantly by varying the moments of the MWD. In this example, the importance of the high end of the distribution is illustrated. The lower k value signifies lower compound viscosity, which should result in a faster extrusion for the compound formulation studied.

As already discussed, the processing chracteristics of polymers depend on contributions from three main areas: compositional and molecular weight properties, and the presence of long-chain branching. In terms of composition, when the ethylene content in an EPDM is generally above 55 wt%, measurable levels of crystallinity can be detected using differential scanning calorimetry (DSC) techniques. Crysallinity can play an important role in extruded articles. Toughness and green strength in compounds are generally derived from crystallinity, which can establish good feeding characteristics in cold-feed extruder operations. In addition, com-

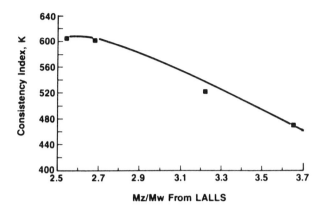

Figure 5.28 Dependency of polymer flow index from MPT on M_z/M_w polydispersity.

pound stocks derived from crystalline-like polymers tend to be more thermoplastic, which enhances mass flow. Finally, extruded profiles will typically have better size control and dimensional stability [see Keller (1985) and Keller and Cheremisinoff (1986)].

Narrow-MWD polymers tend to demonstrate higher throughput rates in extrusions, with extrudates typically having smooth surfaces. Note, however, that the absence of high-molecular-weight ends results in a compound stock having little guts, and hence these materials generally feed poorly. Narrow-MWD polymers are sometimes preferred because they cure faster than broader polymersr, and hence the poor feed qualities of the narrow-MWD polymers are overcome somewhat via introducing semicrystallinity. In contrast, broad-MWD polymers have better extruder feeding characteristics; however, they tend to show higher die swell. Polymers with low levels of long-chain branching tend to be more shear thinning, and hence will exhibit a lower viscosity when subjected to stress. The net result is faster extrusion. Hence, from a processing standpoint, a controlled amount of branching can be favorable.

The presence of branching can be detected by coupling the techniques of gel permeation chromatography (GPC) with laster light scattering in a multi-detector apparatus for molecular weight characterization measurements. This technique enables comparison of linear model molecules with their branched counterparts. At a given elution time,

$$g = \left(\frac{MW_{linear}}{MW_{branched}}\right)^{1+\alpha} \tag{84}$$

where α, the Mark-Houwink exponent for the solvent used, can be related to the degree of branching. The technique is described by Krevelen (1972), Kresge et al. (1984), and Scholte (1983), among others. A crude branching index can therefore be defined from the ratio of weight-average molecular weights as measured by GPC and low-angle laser light scattering (LALLS).

To illustrate the coupling of molecular properties to rheology and then to the ultimate extrusion process, 15 polymers were synthesized in the Exxon laboratories and a capillary rheometer was used to determine the flow and elastic responses of their compounded stocks over a range of shear rates anticipated in the extruder. All stocks were prepared in the same applications formula using IRB No. 5 carbon black at about 30%. Shear rates ranged from 30 to 180 sec^{-1}, and measurements were made at three temperatures (75, 100, and 125°C). These temperatures covered the wall temperature ranges anticipated for the extruder. Table 5.3 provides a summary of the power-law-model rheological coefficients obtained at 100°C, along with the molecular weight and branching characteristics of each polymer. A stepwise regression of k values (using log values) as a

Table 5.3 Summary of Data for Extrusion Study

Polymer	k	n	M_w/M_n	M_z/M_w	$M_w \times 10^{-5}$	Branching index
A	589.5	0.25	2.22	1.8	1.7	1.17
B	605.9	0.29	2.69	2.53	1.94	1.01
C	602.73	0.25	2.78	2.67	1.895	1.04
D	661.53	0.25	2.87	2.69	2.002	1.16
E	983.24	0.25	3.51	2.73	2.525	1.3
F	822.49	0.23	2.87	2.77	2.005	1.07
G	832.6	0.24	4.04	2.92	3.402	0.96
H	957.4	0.22	3.83	3.11	3.1	0.94
I	523.46	0.26	2.7	3.22	1.587	0.83
J	607.88	0.25	2.85	3.53	1.899	0.92
K	842.9	0.21	3.64	3.64	2.774	0.98
L	472.2	0.3	2.67	3.65	1.837	0.78
M	323.9	0.32	3.35	4.17	2.086	0.9
N	415.2	0.27	3.12	4.7	1.661	0.94
O	458.8	0.26	3.24	5.76	2.059	0.84

function of M_w/M_n, M_z/M_w, M_w, and branching index was performed, and the results are tabulated in Table 5.4. As shown, k correlates with the weight-average molecular weight distribution (M_z/M_w). Usually when branching indices $\geqslant 0.9$, it is believed that no branching exists, and hence the lack of k dependency is not surprising. The correlation derived from this empirical fit of the data is

$$k = \exp(6.386)\left(\frac{M_z}{M_w}\right)^{-0.558}(M_w \times 10^{-5})^{0.923} \tag{85}$$

Figure 5.29 shows a parity plot of the calculated k value versus measured, indicating that the correlation is good to within $\pm 20\%$.

The power law exponent, n, in this example appears to be nearly constant (refer to Table 5.3). For EPDMs it appears to be a strong function of branching.

The consistency index varies with temperature according to the Arrhenius-type relationship

$$k = \alpha \exp\left[\frac{E}{R(T - T_0)}\right] \tag{86}$$

where T and T_0 are the measurement and reference absolute temperature, respectively, E the activation energy of flow, and R the universal gas law constant.

To summarize, we can relate the MWD of a polymer through an empirical relation [equation (85)] to its power law coefficients, and the rheology-temperature relation can be determined by equation (86).

It must be noted that the elastic response of the test stocks in a capillary rheometer must also be measured in terms of a Bagley correction experiment, which is calculated by comparing capillary pressures at the same shear rate for two or more dies with different L/D ratios. The entrance correction term, g, can be computed from

$$g = \frac{P_2(L/D)_1 - P_1(L/D)_2}{P_1 - P_2} \tag{87}$$

where P_1 is the pressure at one shear rate using a die with $(L/D)_1$, and P_2 is the pressure at same shear rate using a die with $(L/D)_2$. This correction accounts for the additional pressure developed as the stock is accelerated into a smaller die opening. Using this entrance correction, the true shear stress at a die opening is

Table 5.4 Spreadsheet Regression Results Run Summary

Proc #	Proc Name	Result
1	STEP	ok

Proc 01: Step

Dep. Var: LLN(K)	R-Square:	67.037
File: REGRESS	adj R-Sq:	61.543
Range:	St. Error:	0.199
3.00 Variables	F-Stat:	12.202
15.00 Observations	Dur/Wat:	1.811
	Det.:	1.000

Stepping Report: adj R-Sq.
Step 1: Add LN(Mw)   -38.7333
Step 2: Add LN(Mz/Mw)   61.5430
Step 3: Add LN(INDEX)   61.4801
Step 4: Del LN(INDEX)   61.5430

Var Name	Coeff	t-Stat	St. Err	Beta
Intercept	6.386			
LN(Mz/Mw)	−0.558	−2.951	0.189	−0.489
LN(Mw)	0.923	3.936	0.234	0.652

Actual k-Value	Calcltd. k-Value	Residual	Percent Deviation
589.50	697.70	−108.20	−18.35
605.90	651.78	−45.88	−7.57
602.73	618.93	−16.20	−2.69
661.53	648.42	13.11	1.98
983.24	796.73	186.51	18.97
822.49	638.78	183.71	22.34
832.60	1010.44	−177.84	−21.36
957.40	895.31	62.09	6.49
523.46	473.32	50.14	9.58
607.88	530.67	77.21	12.70
842.90	740.11	102.79	12.19
472.20	505.15	−32.95	−6.98
323.90	527.35	−203.45	−62.81
415.20	399.74	15.46	3.72
458.80	435.10	23.70	5.16

Processability Testing

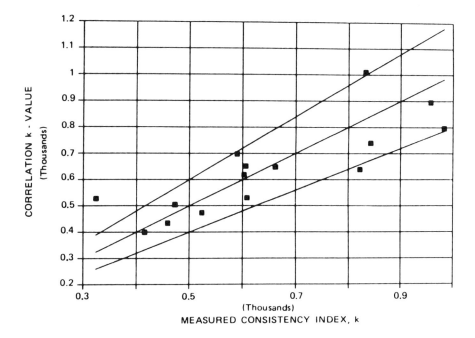

Figure 5.29 Parity plot of K-index correlation.

$$\tau_w = \frac{P}{4(L/D + g)} \tag{88}$$

Figure 5.30 shows a plot of g versus the effective shear rate, $\dot{\gamma}$, generated in the capillary die.

We may now relate the preceding information to extruder operation by developing one-dimensional extruder expressions in terms of the configuration of the apparatus. The extrusion data reported in the discussions that follow were generated in a 2.5-in.-diameter single-screw extruder that was 69 in. in length and equipped with an adjustable slit die. The extruder was equipped with three-zone heating, and although conditions were set to maintain close to isothermal operation, conduction loss calculations indicated that the barrel was typically 10 to 15°C cooler than the die. Head pressures and temperatures were measured with dual thermocouple-transducer sensors. Linear mass rates were measured using the weight-time method at near steady-state conditions, as indicated by machine torque measurements. Relaxed die swell measurements were made on extrudate samples.

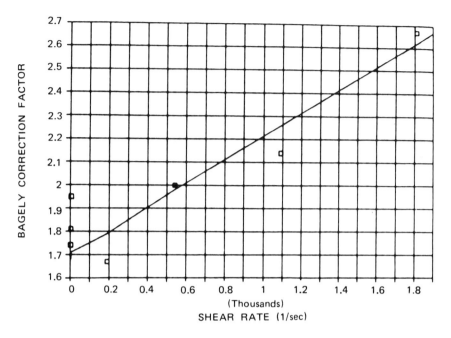

Figure 5.30 Bagely correction term as a function of shear rate.

The shear rate through the slit die can be approximated from the analysis of flow between parallel plates to give

$$\dot{\Upsilon} = \frac{Q}{WH^2} \qquad (89)$$

where Q is the volumetric flow rate, W the die width opening, and H the die height opening. The shear stress at the die is

$$\tau = \frac{P}{4(L/D + g)} \qquad (90)$$

Using the definition that $\tau = k\dot{\Upsilon}^n$ in equation (90) and solving for the extruder head pressure gives

$$P = 4\left(\frac{L}{D} + g\right)k\left(\frac{Q}{WH^2}\right)^n \qquad (91)$$

Processability Testing

To solve equation (91) for P and Q, a second expression is needed. Considering the cold-feed extruder as a melt-conveying screw of simple geometry, the apparent flow through the apparatus then reflects two types of flow: drag flow (i.e., no restriction at the die, Q_d) and pressure flow, Q_p. Q_d represents the maximum theoretical output of the extruder, which is reduced by the head pressure developed at the die (Q_p). Therefore, the actual throughput is

$$Q = Q_d - Q_p \tag{92}$$

The drag flow contribution can be approximated simply by

$$Q_d = AN \tag{93}$$

where N is the screw speed and

$$A = \frac{\pi^2 D^2 H'}{120} \sin\phi \cos\phi \tag{94}$$

where D is the barrel diameter, H' the minimum flight depth, and ϕ the helix angle of screw (17.6°). For the extruder used in the experiments, $A = 2.87 \times 10^{-4}$ m³ · min/sec.

The pressure flow contribution can be estimated from

$$Q_p = B \frac{P}{\eta_s} \tag{95}$$

where

$$B = \frac{\pi D H'^3}{12L} \sin^2\phi \tag{96}$$

and where P is the head pressure and η_s is the viscosity of the stock in the barrel. For the extruder, $B = 2.62 \times 10^{-6}$ m³.

The viscosity of the stock in the barrel is based on the channel apparent shear rate,

$$\dot{\Upsilon}_a = \frac{\pi D N}{60 H'} \tag{97}$$

which at the mean barrel wall temperature is

$$\eta_s = k\dot{\Upsilon}_a^{n-1} \tag{98}$$

where k depends on temperature and the molecular weight and molecular weight distribution of the polymer.

The output of the extruder with a die can now be estimated by

$$Q = AN - B\frac{P}{\eta_s} \tag{99}$$

Equations (99) and (91) can be solved by the method of successive approximations. The iterative calculation procedure is as follows:

1. Input data are the MW (molecular weight) and MWD (molecular weight distribution) and extruder conditions (barrel and screw temperature, screw speed). The rheological coefficient k can be estimated from the empirical equations (85) and (86). The power law exponent, n, in this example is assumed a constant, reflecting the average value for the samples.
2. Compute the channel shear rate from equation (97) using the specified screw speed, N. Using the appropriate k value, obtain an estimate of η_s from equation (98).
3. Make an initial estimate of Q and:
   a. Compute the die shear rate [equation (89)].
   b. Compute the head pressure at the die [equation (91)].
   c. Use equation (99) to compute the head pressure based on N, Q_{est}, and η_s for the barrel section.
4. If the computed head pressures agree to within some specified tolerance (e.g., 2%), the estimate of Q is correct. If not, select a new Q and repeat the calculations until convergence is achieved.

Model predictions of the pressure drop for polymer A (Table 5.4) are compared to measurements in Figure 5.31 in terms of a plot of head pressure versus screw speed. As shown, the model predictions are in good agreement with experimental data for the majority of experiments. The scatter in measured head pressure and the low values at higher screw speeds suggest problems with the transducer accuracy.

Model predictions of the linear extrudate rate for three of the polymers listed in Table 5.4 are shown in the parity plot of Figure 5.32. As shown, the model provides good predictions for polymers widely different in molecular weight distribution. In fact, the more accurately the molecular weight distribution is characterized, the more precise are model pre-

Processability Testing

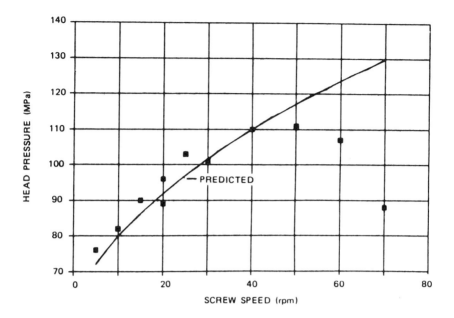

Figure 5.31 Model predictions of head pressure compared to data.

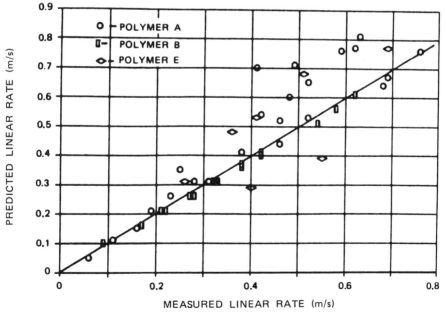

Figure 5.32 Parity plot showing that model properly predicts linear throughput.

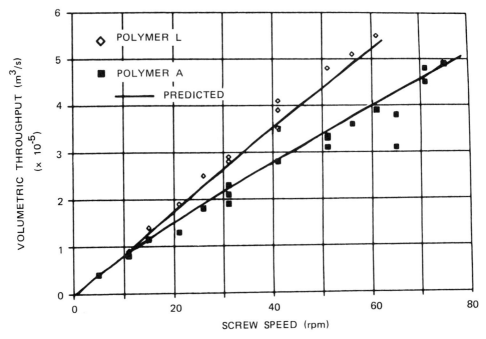

Figure 5.33 Model predictions of volumetric throughput for two polymers.

dictions. Figure 5.33 shows model predictions of the volumetric throughput as a function of screw speed for two polymers. The comparison to measured values in this figure shows that the simple model is capable of quantitative predictions.

The manufacturer of extruded articles not only must define and control his throughput capacity, but must also control the article's dimensional stability or swell. In rubber applications, dies are usually cut to accommodate the article's tolerances based on the characteristics of a single polymer supply or fixed blend production. The reason for this is largely cost; that is, dies, particularly complex geometries such as crosshead dies, are simply too expensive to retool to accommodate different polymer supplies. Consequently, a common problem among elastomer manufacturers in attempting to qualify their products as a second source of supply or even in introducing new products to extruded article manufacturers is to provide a polymer having proper swell characteristics. Again, capillary rheometer data can provide some guidance as a predictive tool in designing polymer structures and/or in selecting polymers for a particular ap-

Processability Testing

plication. A very simple, empirical approach is to correlate capillary rheometer swell measurements with a power law relation:

$$\text{swell} = k'\dot{\Upsilon}^{n'} \tag{100}$$

Table 5.5 provides tabulated data on the swell indices k' and n', along with molecular weight properties of the various polymers tested. An additional parameter, IV (intrinsic viscosity), is also tabulated. The intrinsic viscosity is a solution viscosity normally measured in a solvent such as decolin and can be viewed as a crude measure of molecular weight. A multivariable regression of the data in Table 5.5 reveals that k' is directly proportional to IV:

$$k' \sim \text{IV}^{1.357} \tag{101}$$

The results of the regression are summarized in Table 5.6, and the correlation is shown in Figure 5.34. The correlation offers no surprise; as expected, higher-molecular-weight polymers generally tend to swell more (i.e., higher k' values). Index n' does not show any dependency to any of the

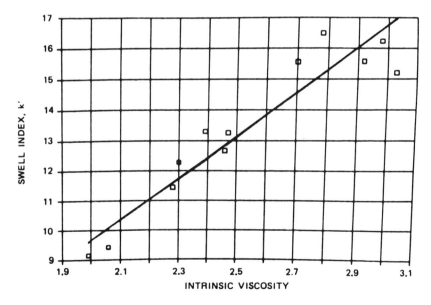

Figure 5.34 Relationship of swell index to intrinsic viscosity for 100°C data.

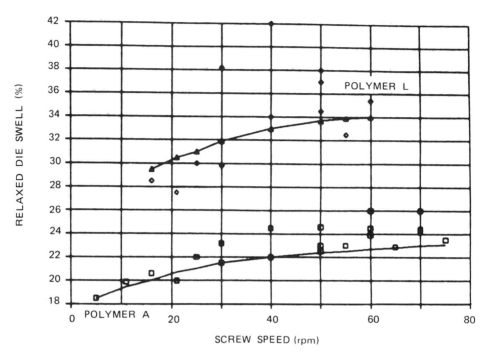

Figure 5.35 Extrudates swell increases with screw speed. Solid lines are predictions; symbols are measured values.

parameters tabulated in Table 5.5, and hence an average value representative of the polymers studied can be used.

Using the effective shear rate for the extruder die, equation (101) was applied directly to estimating the extrudate swell, and comparisons to experiments are shown in Figure 5.35. The simple relationship provides a good qualitative prediction of the extruder operation and for polymer A actually provides quantitative predictions. Figure 5.36 shows a parity plot of predicted and measured extrudate swells for several of the polymers listed in Table 5.5. Although the relationship of equation (100) does not provide highly accurate predictions, it in fact provides a good relative comparison between different polymers. As a first-pass approach to selecting a polymer for an intended extrusion application, equation (101) certainly provides some guidance.

Table 5.5 Summary of Swell Index Data Generated in Capillary Rheometer

Polymer	n'	k'	M_w/M_n	M_z/M_w	M_w	Index	IV
A	0.08	11.44	2.22	1.80	1.70	1.17	2.28
B	0.12	14.13	2.69	2.53	1.94	1.01	
C	0.06	12.67	2.78	2.67	1.90	1.04	2.46
D	0.05	13.26	2.87	2.69	2.00	1.16	2.47
E	0.08	15.20	3.51	2.73	2.53	1.30	3.03
F	0.05	15.56	2.87	2.77	2.01	1.07	2.70
G	0.05	16.23	4.04	2.92	3.40	0.96	2.98
H	0.05	16.49	3.83	3.11	3.10	0.94	2.78
I	0.13	10.74	2.70	3.22	1.59	0.83	
J	0.05	12.28	2.85	3.53	1.90	0.92	2.30
K	0.06	15.57	3.64	3.64	2.77	0.98	2.92
L	0.09	14.58	2.67	3.65	1.84	0.78	
M	0.10	13.30	3.35	4.17	2.09	0.90	2.39
N	0.10	9.15	3.12	4.70	1.66	0.94	1.99
O	0.11	9.43	3.24	5.76	2.06	0.84	2.06

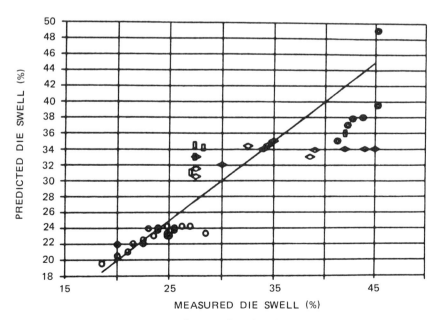

Figure 5.36 Parity plot comparing predicted and measured swells for several polymers.

Table 5.6 Spreadsheet Regression Results

	Proc 01: Step		
Dep. Var: LLN(K')	R-Square:		90.341
File: REGRESS	adj R-Sq:		89.376
Range:	St. Error:		0.065
2.00 Variables	F-Stat:		93.534
12.00 Observations	Dur/Wat:		2.858
	Det.:		1.000

Stepping Report:	adj R-Sq.
Step 1: Add LN(Mw)	52.8098
Step 2: Add LN(Mz/Mw)	71.0405
Step 3: Add LN(IV)	87.0506
Step 4: Del LN(Mw)	88.4285
Step 5: Del LN(Mz/Mw)	89.3755
Step 6: Add LN(INDEX)	89.2526
Step 7: Del LN(INDEX)	89.3755

Var Name	Coeff	t-stat	St. Err	Beta
Intercept	1.329			
LN(IV)	1.357	9.671	0.140	0.950

Actual k'	Calcltd. k'	Residual	Precent Deviation
11.44	11.56	−0.12	−1.03
12.67	12.81	−0.14	−1.13
13.26	12.88	0.38	2.83
15.20	17.00	−1.80	−11.85
15.56	14.54	1.02	6.56
16.23	16.62	−0.39	−2.42
16.49	15.13	1.23	8.27
12.28	11.70	0.58	4.75
15.57	16.17	−0.60	−3.85
13.30	12.32	0.98	7.36
9.15	9.61	−0.46	−5.03
9.43	10.07	−0.64	−6.80

PRINCIPLES OF TORQUE RHEOMETRY

Torque rheometers are multipurpose instruments well suited for formulating multicomponent polymer systems, studying flow behavior, thermal sensitivity, shear sensitivity, batch compounding, and so on. The instrument is applicable to thermoplastics, rubber (compounding, cure, scorch tests), thermoset materials, and liquid materials.

When the rheometer is retrofitted with a single-screw extruder one can measure rheological properties and extrusion processing characteristics to differentiate lot-to-lot variance of polymer stocks. It also enables the process engineer to simulate a production line in the laboratory and to develop processing guidelines.

The torque rheometer with a twin-screw extruder is considered a scaled-down continuous compounder. It allows the compounding engineer to develop polymer compounds and alloys. It also permits the formulation engineer to assure that the formulation is optimum.

The torque rheometer is essentially an instrument that measures viscosity-related torque caused by the resistance of the material to the shearing action of the plasticating process.

Torque can be defined as the effectiveness of a force to produce rotation. It is the product of the force and the perpendicular distance from its line of action to the instantaneous center of rotation (Figure 5.37). By the definition, torque can be expressed as

$$M = FR \qquad (102)$$

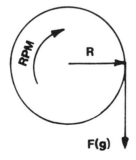

Figure 5.37 Definition of torque.

The units of torque are g·m, dyn·cm, or lb-ft. Power is the time rate at which work is done:

$$P' = FV \tag{103}$$

expressed in units of hp, watts, and ft-lb/sec.

From the two relations above, we can relate torque and power by the following expression:

$$M(g \cdot m) = 725{,}995 \frac{hp}{rpm} \tag{104}$$

The first torque rheometers were "mechanical" instruments, comprised of a dynamometer system with lever arm, weight, and dashpot. The lever arm and weight mechanically operated the torque sensing system of the instrument. As such, instrument calibration was done by manually adjusting the levers and weights, which was a very cumbersome procedure. In addition to the awkwardness of calibrating the rheometer, the instrument consisted of several modular systems for temperature controlling and parameter recording. Although it was an excellent instrument, its two largest disadvantages were the amount of time required for calibration and the large space needed for the unit and auxiliary systems.

The second generation of the torque rheometers were "electronic" instruments. These systems were semiautomatic and compact, with almost every necessary module built into the unit. Figure 5.38 illustrates the torque measuring principle with the electronic torque rheometer. Power was transmitted from the input to the output gears of a gearbox through an idle gear mounted on an output gear shaft, while the other was suspended on a gearbox housing. A transducer was mounted between the swing arm and the gearbox housing. When a load was applied against the gearbox output shaft, it caused a resistance to the power transmitting parts. The resistance on the output gear shaft along with a driving force on the input gear shaft, forced an idler gear downward. This downward pull on the idler produced the strain in the transducer. The strain was converted into an electrical signal that was registered by the torque circuit in the electronic torque rheometer.

The prevalent design is a microprocessor-controlled torque rheometer. The microprocessor-controlled torque rheometer is a unique system and offers more versatility. The system consists of two basic units: an electromechanical drive unit and a microprocessor unit. Advantages over the electronic torque rheometer are:

Processability Testing

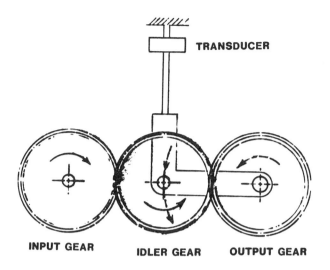

Figure 5.38 Principle of torque rheometer measurement.

It can control two pieces of equipment simultaneously. An example of this would be feeders or postextrusion systems. It allows the user to regulate feed rate or down stream equipment.

The system can drive equipment, control and monitor parameters, and has the additional feature of data acquisition capabilities.

The unit displays all parameters graphically or numerically on-line during testing.

The rheometer also has capabilities to plot any parameter in the off-line mode.

Most units are capable of monitoring and recording three pressures and three melt temperatures, as well as torque, totalized torque, and speed.

The system has automatic calibration capabilities for the torque load cell and pressure transducers in any range.

Every parameter that is being measured is recorded for later reference.

Figure 5.39 illustrates a block diagram of the microprocessor-controlled torque rheometer.

As noted very early in our discussions, most plastics and elastomeric products are not pure materials but rather, mixtures of the basic polymer with a variety of addities, such as pigments, lubricants, stabilizers, antiox-

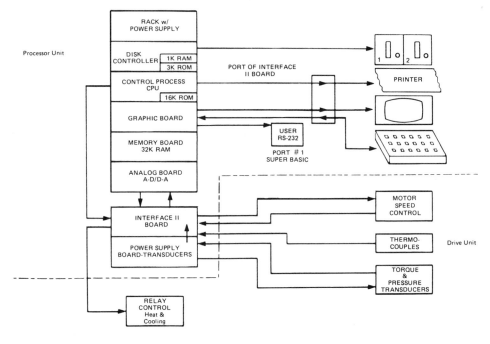

Figure 5.39 Schematic diagram of Haake-Buchler System 40 microprocessor.

idants, flame retardants, antiblock agents, cross-linking agents, fillers, reinforcement agents, plasticizers, UV absorbants, foaming agents, and others. All these additives must be incorporated into the polymer prior to fabrication. Some of the additives take a significant portion of the mixture; others, only minute amounts. Some are compatible; others are not. Depending on the quality of resin and additives and homogenization of the mixtures, the quality of the final product will be varied. Therefore, developing a quality resin and additives that meet with desired physical and mechanical properties of the product and quality control associated with them play an important role in the plastic industry. In the development of formulations and applications for new polymers, the torque rheometer is an invaluable instrument. Common practice is to equip the torque rheometer with a miniaturized internal mixer (MIM) to simulate large-scale production at the bench. The mixer generally consists of a mixing chamber shaped like a figure eight with a spiral-lobed rotor in each chamber. A totally enclosed mixing chamber contains two fluted mixing rotors that revolve in opposite directions and at different speeds to achieve a shear action similar to a two-roll mill.

In the chamber, the rotors rotate in order to effect a shearing action on the material mostly by shearing the material repeatedly against the walls of the mixing chamber. This is illustrated conceptually in Figure 5.40. The rotors have cheverons (helical projections) which perform additional mixing functions by churning the material and moving it back and forth through the mixing chamber. The mixture is fed to the mixing chamber through a vertical chute with a ram. The lower face of the ram is part of the mixing chamber. There is usually a small clearance between the rotors, which rotate at different speeds (e.g., gear ratio 3:2, 2:3, and 7:6 for different mixers) at the chamber wall. In these clearances, dispersive mixing takes place. The shape of the rotors and the motion of the ram during operation ensure that all particles undergo high intensive shearing flow in the clearances.

There are three sets of interchangeable rotors available on the market. They are roller, cam, and sigma rotors designs, although there are many other designs as illustrated in Figure 5.41. Normally, roller rotors are used for thermoplastics and thermosets, cam rotors are for rubber and elastomers, and sigma rotors are for liquid materials. Banbury rotors are used with miniaturized Banbury mixer for rubber compounding formulation.

Plastic materials with two or more components being processed should be well mixed so that the compounded material provides the best physical properties for the final product. There are distinctive types of mixing processes. The first involves the spreading of particles over position in space (called distributive mixing). The second type involves shearing and spreading of the available energy of a system between the particles themselves (dispersive mixing), as illustrated in Figure 5.42. In other words, distributive mixing is used for any operation employed to increase the randomness of the spatial distribution of particles without reducing their sizes. This mixing depends on the flow and the total strain, which is the product of shear rate and residence time or time duration. Therefore, the more random arrangement of the flow pattern, the higher the shear rate; and the longer the residence time, the better the mixing will be.

A dispersive mixing process is similar to that of a simple mixing process, except that the nature and magnitude of forces required to rupture the particles to an ultimate size must be considered. Essential to intensive mixing is the incorporation of pigments, fillers, and other minor components into the matrix polymer. This mixing is a function of shear stress, which is calculated as a product of shear rate and material viscosity. Breaking up of an agglomerate will occur only when the shear stress exceeds the strength of the particle.

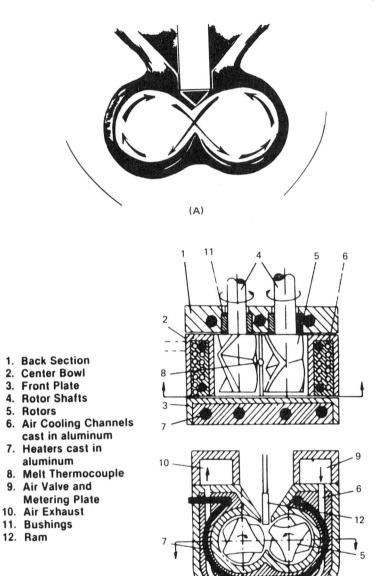

1. Back Section
2. Center Bowl
3. Front Plate
4. Rotor Shafts
5. Rotors
6. Air Cooling Channels cast in aluminum
7. Heaters cast in aluminum
8. Melt Thermocouple
9. Air Valve and Metering Plate
10. Air Exhaust
11. Bushings
12. Ram

Figure 5.40 Mixing action in a MIM.

Processability Testing

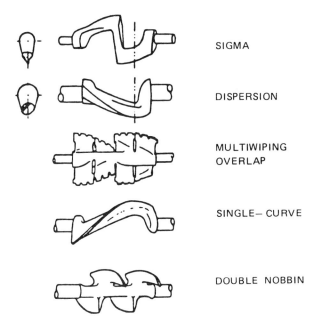

Figure 5.41 Types of mixer rotor configurations.

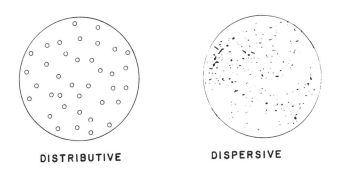

Figure 5.42 Distributive and dispersive mixing.

Commonly used mixing procedures in the industries are *dump mixing,* where all ingredients are added to the mixer at once; *upside-down mixing,* where solid additives are added first followed by the polymer; and *seeding mixing,* where small amounts of a previously well-mixed batch is added to the new batch. Dilutants, on the other hand, are added as late in the cycle as possible.

To illustrate how torque rheometry can be applied to mixing and extrusion processing studies, we must first examine the constitutive equation applicable to this system. Over a limited shear-rate range, the flow behavior can be governed by the power law equation [equation (16)]. The temperature dependence of consistency is given by the Arrhenius expression, restated here as

$$K = K_0 e^{\Delta E/RT} \tag{105}$$

Due to the complex geometry of the mixer rotors and since rotors turn at different speeds, the local wall shear rate varies throughout the chamber. However, we can state that

$$\dot{\gamma} = C_1 N \tag{106}$$
$$\bar{\tau} = C_2 M \tag{107}$$

where M is torque, N is rpm, C_1 and C_2 are constants, and $\dot{\gamma}$ and $\bar{\tau}$ are mean shear rate and mean shear stress averaged over the surface of the chamber. From the above the following relationship between torque and rpm can be obtained:

$$M = K e^{\Delta E/RT} N^n \tag{108}$$

where $K = K_0 C_1^n / C_2$.

Figure 5.43 shows a typical data plot from a torque rheometer. The plot (Figure 5.43A) shows both the instantaneous torque and rotor speed over time for one experiment. The shear sensitivity of the test material can be derived from a cross-plot of torque and rotor speed, shown in Figure 5.43B. Since torque is proportional to shear stress and rpm is proportional to shear rate, as discussed earlier, the following relationship between torque and rpm may be obtained:

$$\log M = \log K + n \log N \tag{109}$$

Therefore, $n = d \log M / d \log N$.

Processability Testing

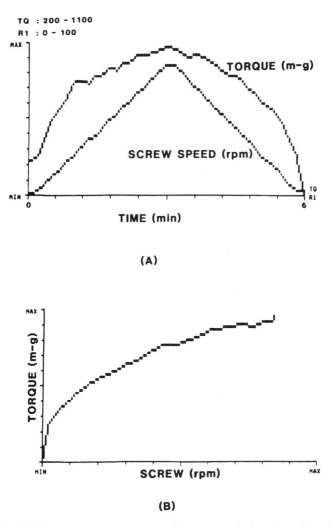

Figure 5.43 (A) Instantaneous torque and rotor speed plotted against time; (B) cross plot of torque and rotor speed showing materials shear sensitivity.

An important aspect of mixing studies on the torque rheometer is the temperature dependency of the mixing process. In general, the viscosity of a polymer decreases as temperature increases, and vice versa. Properties of the material also change depending on temperature. Hence knowledge of the temperature dependency of viscosity is important. Large variations of viscosity for a certain range of temperature means that the material is thermally unstable (i.e., requires large activation energy). This kind of material has to be processed with accurate temperature control. Figure 5.44 shows the torque versus temperature curve obtained from the microprocessor-controlled torque rheometer to see the temperature dependency of viscosity-related torque on an EPDM material. These data can be applied to the Arrhenius equation described earlier.

$$\ln \eta = \ln K' + \frac{\Delta E}{R} \frac{1}{T} \tag{110}$$

Since viscosity is proportional to the ratio of torque and rpm, the following relationship may be obtained:

$$\ln \frac{M}{N} = \ln K' + \frac{\Delta E}{R} \frac{1}{T} \tag{111}$$

Hence Figure 5.44 defines the temperature sensitivity of the system and equation (111) enables quantification. It is often difficult to explain flow

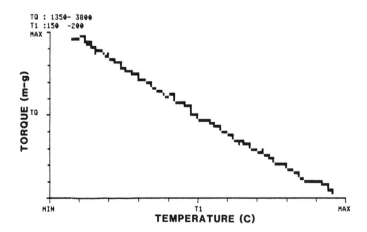

Figure 5.44 Cross plot of torque and temperature from rheocord.

Processability Testing

behaviors of materials obtained from the torque rheometer due to melt temperature changes induced by frictional heat while the material is being mixed. Therefore, it is necessary to compensate the temperature effect of the material. Temperature-compensated torque can be obtained from an integrated form of the Arrhenius equation:

$$\ln \frac{M}{M'} = \frac{\Delta E}{4.57G} \left(\frac{1}{T} - \frac{1}{T'} \right) \tag{112}$$

where M is the measured torque at T (K) and M' is the calculated torque at reference temperature T'.

An important parameter to assess in comparing polymer or mix prototypes is work. The unit work is defined as the work energy required to process unit volume or unit mass of material. It can be calculated from the totalized torque, which is obtained directly from the rheometer. Totalized torque is defined as the energy required to process a certain material for a certain period of time at given conditions. It is therefore simply the area under the torque-time curve. The totalized torque can be converted to work energy by

$$\hat{W} = \frac{\hat{W}_t}{V_b}$$

$$= \frac{2\pi N \int_{t_1}^{t_2} M(t)\, dt}{V_b}$$

$$= 61.59 N \frac{\text{TTQ}}{V_b} \tag{113}$$

where the unit work is expressed in dimensions of J/cm³ and where

$\hat{W}_t$ = total work energy
$M(t)$ = torque
N = rpm
V_b = charged sample volume
TTQ = totalized torque

The MIM along with the torque rheometer can be used to simulate and optimize a variety of processing applications/problems. Some typical but by no means inclusive examples, are in studying fusion characteristics, examining stability and processability, color or thermal stability testing, examining the gelation of plastisols, in developing criteria for the selection of blowing agents for foam products, compound formulation optimization, studying the scorch and cure characteristics of rubber compounds, and in studying the cure characteristics of thermosets. Specific examples of each are discussed below.

Lubricants play in important role in processing and in the properties of the final product. The lubricants also effect the fusion of the polymer materials. That is, internal lubricants reduce melt viscosity, while external lubricants reduce friction between the melt and the hot metal parts of the processing equipment and prevent sticking, controlling the fusion of the resin. Figure 5.45 illustrates the results of an experiment aimed at studying the fusion characteristics of PVC. The level of external lubricant used in the formulation affects the fusion time between points L and F on the curve. The higher the level of external lubricant in the formulation, thel onger the fusion time will be.

If an unnecessarily high level of external lubricant is used in the formulation, it will take a longer period of time to melt the material in processing, which results in reducing production, increasing energy consumption, and poor products. Meanwhile, if too low a level of external lubricant is used, the material will melt too early in the processing equipment, which may result in degradation in the final product. Therefore,

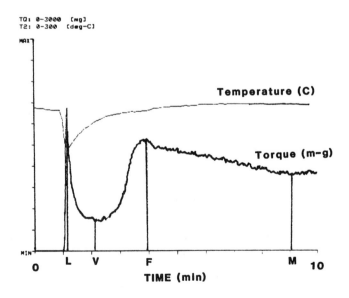

Figure 5.45 Results of PVC fusion study.

Processability Testing

selecting the optimum amount of external lubricant is a must for improvement of processing and for good-quality products.

Stability testing of materials enables formulation engineers to select the best stabilizer and enables process engineers to choose optimum processing parameters, which, in turn, reduces scrap. Figure 5.46 illustrates a case of material degradation indicated by the rise in torque due to a cross-linking reaction such as PVC, rubber, and thermoset curing. It also provides information on the processability of the material under testing. Processability time is determined by the time from loading (point L) to the onset of degradation (point D) of the material. The longer the time, the better the processability. If the material does not cross-link when it degrades, the torque values keep decreasing until structural breakdown occurs.

Another example of the use of a torque rheometer in processing studies is in the area of color or heat stability. Usually, heat-sensitive polymers will degrade if they are exposed to heat for prolonged periods of

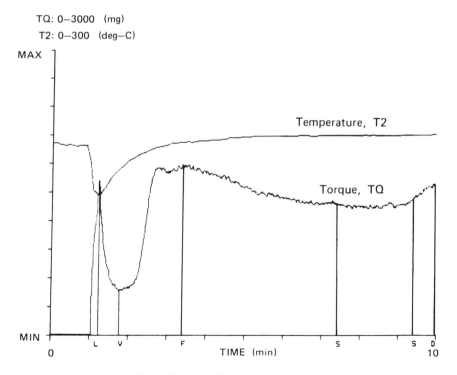

Figure 5.46 Results of material stability study.

time. The sign of degradation can easily be detected by color changes and odors in the material. Determining color stability of materials is helpful to see how stable the material is in processing and how long the processing time (residence time) should be to obtain a high-quality product that has the best mechanical properties. Also, a series of tests can be conducted to assess the best antioxidant packages to be used for a particular polymer. This can be done by simulating shear conditions, causing autooxidation of a polymer. Then by repeating the shearing conditions but with polymer samples doped to either different levels of an antioxidant or with different types of antioxidants, the best formulation can be selected on the basis of color changes. Plastisols are dispersions of homopolymers or copolymers of vinyl chloride in PVC plasticizers; that is, the very small particles are kept in suspension by swelling of the polymer particles from the solving action of the plasticizer. Upon heating, the plastisol particles swell with plasticizer to become a gel. Further heating fuses the ingredients into a homogeneous melt which becomes a continuous solid upon cooling. Figure 5.47 shows the gelation reaction of a plastisol and the amount of gas evolved from a blowing agent used in the plastisol formulation at an elevated temperature. Torque is observed to increase rapidly in the first gelation, due to the high rate of gelation reaction of the plastisol under heat. The reaction rate slows down in the second gelation because most of the reaction has been completed at this stage. After the second gelation, the material turns into a homogeneous melt. The torque value decreases

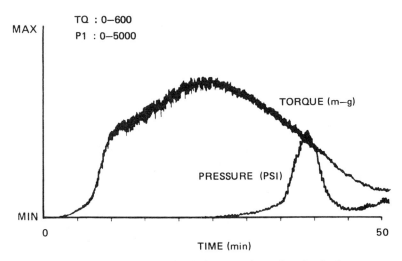

Figure 5.47 Results of gelation reaction of a plastisol.

because of increasing temperatures. Foam products can be produced by adding chemical blowing agents to the formulation. The blowing agents are inorganic or organic materials that decompose under heat to yield gaseous product. Proper selection of a blowing agent is important in foam products. If the proper blowing agent is not chosen, it might react with other ingredients used in the formulation, which results in retarding the blowing agent from activation. The level of blowing agent used in the formulation is also important. If the level of the agent used is too high, too many unnecessary cells form in the foam product and it will collapse easily. If the level is too low, the product does not have enough cells, so it would decrease the cushionability of the product. Figure 5.48 illustrates evolution of gas from the activation of blowing agents used in an EPDM. Figure 5.48A provides information on the temperatures at which the two blowing agents start to activate. Figure 5.48B provides an idea of torque levels when they activate. By studying a series of prototypes, one can assess how good a particular blowing agent is as well as determine whether or not the ingredient reacts with other additives in the formulation during processing.

One of the most frequently applied tests is in the study of additive incorporation and compounding. All of the additives used in a formulation must be incorporated in the major component, and the components should be in a stable molecular arrangement. Figure 5.49 illustrates a test result for incorporation of minor components to the major component as well as homogeneous compound after the additives are incorporated. The test was performed with an EPDM rubber and reblended additives based on ASTM D3185. The EPDM was loaded into the mixer and mixed for 30 sec. Preblended additives (carbon black, zinc oxide, sulfur, and stearic acid) were added at 30 sec. Torque values immediately dropped sharply and increased as the additives incorporated. When the ingredients were fully incorporated, a second torque peak was observed and finally stabilized when the material was homogeneously compounded. The second peak is called the "incorporation peak." If hard fillers are added to the polymer, torque increases sharply and generates the second peak. This can be seen when carbon black is incorporated. The time from the addition of the minor components resulting in the incorporation peak is referred to as the "incorporation time" and is critical to standard batch compounding operations with rubbers.

The final mixing application study we will consider addresses characterization of the scorch and cure characteristics of rubbers. Cross-linkable thermoplstics, rubber, and thermoset materials change their physical properties usually from liquid to solid (cure) by chemical reaction, by heat, or by catalyst during processing. It is important for process engineers to

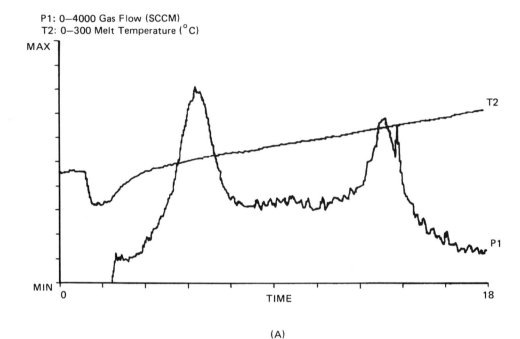

(A)

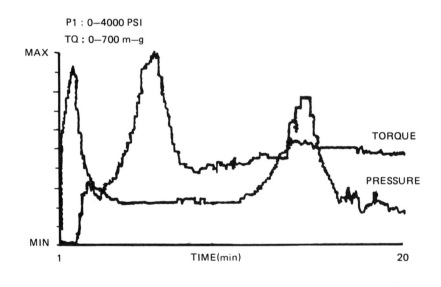

(B)

Figure 5.48 (A) Temperature response and (B) torque response for blowing agent experiment.

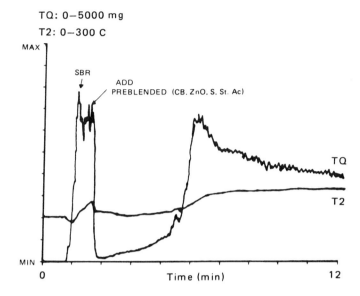

Figure 5.49 Development of incorporation time.

know how long it takes to cure, and how high a torque is generated as well as how long it will be in stable flow. This information would help engineers predict how the material should be processed to provide the best processability and properties. Figure 5.50 illustrates test results of the cure characteristics of a rubber (EPDM). Once the rubber is introduced to the heated mixer, it generates a sharp increase in torque referred to as the "loading peak." After it is loaded, the material starts to soften, which results in decreasing torque. It reaches minimum torque (maximum flow) and starts to scorch prior to the curing reaction taking place. When it cures, the torque value increases and torque decreases after the curing reaction is over. The cycle time is considered the time from loading to cure peak.

To this point, emphasis has been on applications testing where the torque rheometer has been retrofitted with a MIM. Another common practice is to incorporate a screw extruder. Solids conveying, melting, mixing, and pumping are the major functions of polymer processing extruders. The single-screw extruder is the most widely used machine to perform these functions. The plasticating extruder has three distinct regions: solids conveying zone, transition (melting) zone, and pumping zone,

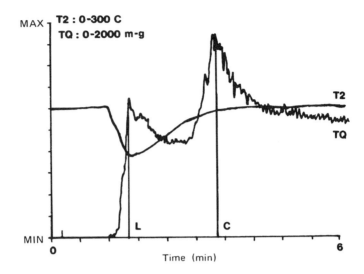

Figure 5.50 Curing of an elastomer.

shown in Figure 5.51. The unit can be fed polymer in the particulate solids form or as strips, as in the case of rubber extrusion. The solids (usually in pellet or powder form) in the hopper flow by gravity into the screw channel, where they are conveyed through the solids conveying section. They are compressed by a drag-induced mechanism, then melted by a drag-induced melt-removal mechanism in the transition section. In other words, melting is accomplished by heat transfer from the heated barrel surface and by mechanical shear heating. Refer to Cheremisinoff (1987) for a detailed discussion of extrusion principles.

Mixing can be carried out either in solid state or in molten state and is achieved through the application of shear to the material. The purpose of the pumping zone is to force the molten polymer through a die to shape the commercial product or for further processing.

Solids conveying is one of the basic functions in the screw extruder. The polymer particles in the solids conveying zone exert an increasing force on each other as the material moves forward, and voids between the pellets are gradually reduced. As the particles move toward the transition section, they are packed closely together to reach a void-free state, and form a solid bed that slides along the helical channels. The solids conveying mechanism is based on the internal resistance of a solid body sliding over another (friction) generated between the plug and barrel surfaces and the screw. This type of flow is known as a drag-induced plug flow. The fric-

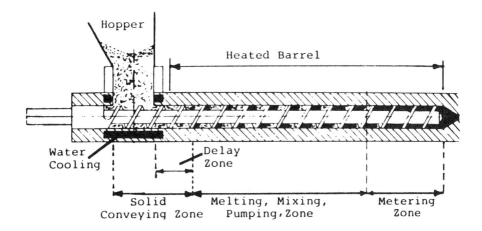

Figure 5.51 Essential elements of a single-screw extruder.

tional force between the barrel surface and the solid plug is the driving force for the movement of the plug; the forces between the screw and the plug retards the motion of the plug in the forward direction. Figure 5.52 illustrates the melting mechanism of material in the extruder. Melting is taking place due to mechanical and thermal energy, which are transformed into heat. The plug that is formed in the solid conveying zone generates friction in contact with the heated barrel surface and in contact with the screw. Both of these frictional processes result in frictional heat generation, which raises the material temperature, which in turn, exceeds the melting temperature or softening point of the polymer. This will convert the frictional drag into a viscous drag mechanism. This creates a melt film between the hot metal and solid bed. As the plug moves forward, the melt portion increases and forms a melt pool, which gets larger and larger. The conveying mechanism in this zone is one of viscous drag at the barrel surface determined by the shear stresses in the melt film and frictional (retarding) drag on the rest of the screw and the flights. Solid and molten material coexist in the melting zone of the extruder. Figure 5.53 conceptually illustrates the melting procedure in single-screw extruders, where the sequence of events is as follows:

1. The sequence commences in the solid conveying section (delay zone).
2. The transition is begun (formation of melt film).
3. The melt pool is formed.

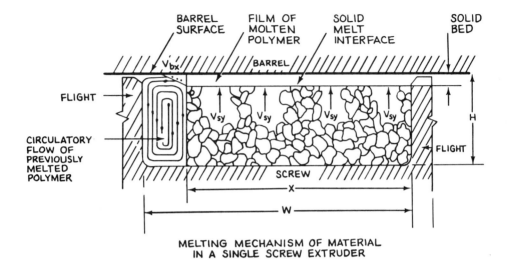

Figure 5.52 Melting zone in an extruder.

4. Melting continues and the width of the solid bed decreases as the channel depth continues to decrease as it progresses down the transition.
5. The plastic continues down the metering section to the discharge.

An ideal screw has a zero solid profile at the end of screw, that is, solid bed profile = $X/W = 0$. Melt conveying occurs in two distinctive regions. One is downstream of the melting zone (after the completion of melting) and the other is in the melt pool, which is an extension of the solid bed profile. In the metering section or at the end of the screw, the polymer transforms totally into a melt. At this point, the solid bed profile, which is the ratio of the length of the solid bed to the screw lead length, has to reach zero, so that only melted polymer comes out of the extruder and die. To create a homogeneous melt, mixing screws, barrier screws, and sometimes two-stage screws are often used.

Figure 5.53 Sequence of events during the melting phase of extrusion.

Processability Testing

A standard single-screw extruder can exhibit good mixing properties within limits. Mixing is achieved through the application of shear to the material and constant orientation of the flow pattern. The screw can be modified to improve mixing by adding a mixing section following the metering section. The mixing section is especially designed to break up flow patterns. Figure 5.54 illustrates some of the mixing devices used frequently in screw design.

Mixing sections have no pumping action and the metering section behind it pumps the melt through this section. Mixing always results in heating the melt by mechanical work, so many polymer melts can degrade. Therefore, the amount of mixing often must be limited.

The solids conveying section establishes the effectiveness of solids feeding or conveying. The depth of the feed section should be determined based on the bulk density of the material and the strength of the remaining screw diameter to withstand the maximum torque of the drive.

It is desirable to utilize the maximum flight depth possible while maintaining the mechanical strength of the screw, especially for low-bulk-density materials. A flight land of one-tenth of the screw diameter is normally used. If the tip of the flight is too narrow, the flight may crack during use due to weakness caused by erosional force.

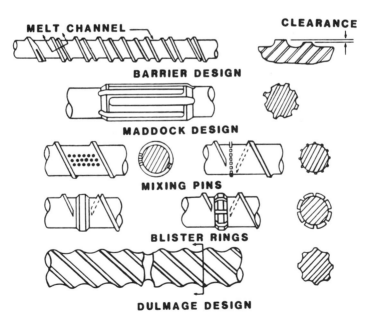

Figure 5.54 Mixing devices used in single-screw extrusion.

The length of any zone (solid conveying, melting, melt conveying) depends on operating conditions, screw geometry, and physical properties of the polymer being processed. Longer feed sections are usually desirable for a material that has a higher melting point. Long feed sections are favorable for crystalline materials because the solid bed is rigid and compressibility is low. Shorter feed sections can be utilized for amorphous polymers because these materials do not have a heat of fusion and have higher compressibility. Additional flights (only in the feed section) or grooving the barrel in the feed section would be helpful for efficient solid conveying.

In the transition section, melting takes place and high pressures can be developed. The compression ratio (ratio of feed depth to metering depth) is commonly used terminology for describing the transition of the screw from the feed section to the metering section. The higher rate of compression in the transition section can increase the rate of melting due to higher pressure. The length of the transition section is affected by the length of the feed section. If the screw has a long feed section, there will be more molten polymer in the feed section; therefore, the solid polymer will be at a higher temperature by the time it reaches the transition section. Consequently, the polymer is more easily deformed and a shorter transition section, or greater compression ratio, can be used.

It is difficult to balance the first stage output with the second stage output. If the first stage delivers more than the second stage pumps through the die, the result is vent flow. If the second stage tends to pump more than the first stage delivers, the result is surging of output and pressure. This can be adjusted by controlling the feed or by valving the output. In the venting section, maintaining zero pressure drop is the most important factor. A flight depth similar to the feed depth or approximately 15% greater than the first metering section is usually adequate. A multiple flighted configuration can be used for the vent section to increase the amount of surface exposure and the efficiency of devolatilization. The vent section should have at least three flights, more if possible.

It is important for the process engineer to choose a proper ratio of length and diameter of screw (L/D ratio of screw) and ratio of flight depth in the feed section and in the metering section of screw (compression ratio) for the polymer to be processed optimally. Normally, short L/D ratio screws may be chosen for heat-sensitive materials to lessen the chance of degradation by shortening the residence time. High-compression-ratio screws can be chosen for a high percentage of regrind, powders, and other low-bulk-density materials. Low-compression-ratio screws can be used for engineering and heat-sensitive materials. The compression ratio of such screws should be reduced by deepening the metering section, not by making the feed section shallower.

Simulation of the extrusion process in the laboratory is one of the most important applications of the torque rheometer in conjunction with single-screw extruders. Figure 5.55 illustrates simulations of widely used extrusion processes in the industries.

It is important for the process engineer to know the rheological properties of a material since the properties dominate the flow of the material in extrusion processes and also dominates the physical and mechanical properties of the extrudates. Therefore, it is also important to measure the properties utilizing a similar miniaturized extruder in the laboratory so that a process engineer knows the flow properties in the system by simulating the production line. Also, it is desirable to know the flow properties of a material to be processed in the range of shear rates of equipments to be used. Figure 5.56 illustrates typical shear-rate ranges encountered in different processing equipment.

The figure also shows that the rheological properties measured at lower shear rate ranges cannot be applicable to high-shear-rate ranges, and vice versa, because the viscosity of material A is lower than that of material B at low-shear-rate ranges (1 to 10 sec^{-1}), but it is the opposite at the high-shear-rate ranges (above 100 sec^{-1}).

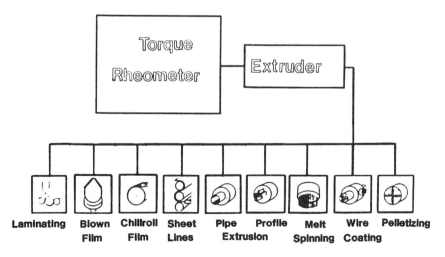

Figure 5.55 Application of torque rheometry to studying extrusion operations.

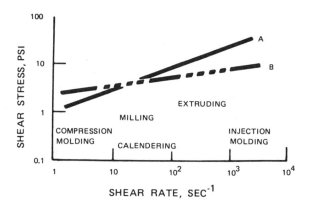

Figure 5.56 Shear ranges encountered in standard rubber processing operations.

In the discussions that follow, application of the torque rheometer to single-screw extrusion studies include the use of two simple geometry dies: a capillary die and a slit configuration. Discussions are limited primarily to steady isothermal flows.

With a capillary die, a pressure drop occurs at the entrance of the capillary because fluid particles entering the tube from the reservoir are accelerated to their final steady flow velocity since the cross section of a capillary is generally much smaller than that of a reservoir section. The energy consumed in this process causes the pressure to drop more rapidly in this region than in the steady flow region, as illustrated in Figure 5.57. The pressure drop is usually considerably greater for viscoelastic polymer melts than for Newtonian fluids.

The apparent shear rate (Γ_w) at the wall and resultant apparent shear stress (τ_w) can be obtained by using the following formulas [see Han (1971)]:

$$\Gamma_w = \frac{4Q}{\pi R^3} \tag{114}$$

$$\tau_w = \frac{R \, \Delta P}{2L} \tag{115}$$

Therefore, apparent viscosity (η_{app}) may be calculated as

$$\eta_{app} = \frac{t_w}{\Gamma_w} \tag{116}$$

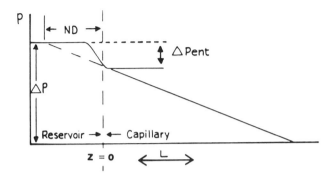

Figure 5.57 Pressure profile in a capillary die.

where

Q = volumetric flow rate, cm³/s
R = radius of capillary, cm
L = length of capillary, cm
ΔP = pressure drop, dyn/cm²

The true shear stress (τ_w^*) and true shear rate ($\dot{\gamma}_w^*$) can be obtained with correction of entrance effects in the capillary.

$$\tau_w^* = \frac{R\,\Delta P}{2(L + ND)} \tag{117}$$

$$\dot{\gamma}_w^* = \frac{3}{4}\Gamma_w + \frac{\tau_w^*}{4}\frac{d\Gamma_w}{d\tau_w^*} \tag{118}$$

$$\eta^* = \frac{\tau_w^*}{\dot{\gamma}_w^*} \tag{119}$$

where ND is the fictitious length obtained from the Bagley correction method described earlier.

To obtain viscosity date from a slit die, three pressure transducers along the die must be used. Since the slope of pressure drop versus length of die (L) curve in the fully developed region is used to calculate shear stress, entrance effect correction is not necessary in slit rheometry. The viscosity of materials can be calculated as follows:

Apparent shear rate (Γ_w):

$$\Gamma = \frac{6Q}{wh^2} \tag{120}$$

True shear rate (γ_w^*) with Rabinowitsch corrections:

$$\gamma_w^* = \frac{\Gamma_w}{3}\left(2 + \frac{d\ln\Gamma_w}{d\ln\tau_w^*}\right) \tag{121}$$

True shear stress (τ_w^*):

$$\tau = \frac{h}{2L}\Delta P \tag{122}$$

where

h = opening of slot, cm
w = width of slot, cm
L = length of slie die, cm
Q = volumetric flow rate, cm^3/s
ΔP = pressure drop, dyn/cm^2

The processing characteristics of polymer prototypes can easily be tested in the laboratory by simulating the production line and postanalyzing the data. A miniaturized single-screw extruder in conjunction with a torque rheometer can characterize the processing by measuring total shear energy (mechanical energy), specific energy, residence time, shear rate, and specific output.

Total shear energy (TE) is the energy introduced to the polymer material by the motor drive during processing. This is calculated from the measured torque multiplied by the screw speed (N):

$$\text{TE} = \text{torque} \times N \times 9.087 \times 10^{-3} \tag{123}$$

These data give information about total mechanical energy required to process the material.

Specific energy (SE) is defined as an energy required to process a unit mass of material. This is calculated from total shear energy divided by the total mass flow rate:

$$\text{SE} = \frac{0.0167\,\text{TE}}{\dot{m}} \tag{124}$$

where $\dot{m}$ is the total mass flow rate. These data give information about viscous dissipation heat built up in the system based on screw speeds.

Residence time is the time required for a material to reside in the extruder before it comes out of the die. It can be obtained from the volume of screw (V) divided by the volumetric flow rate (Q):

$$t = \frac{V}{Q} = \frac{L}{\pi DN} \tag{125}$$

Maintaining uniform residence time is important to obtaining quality product. If it is not maintained, the material will sometimes degrade because of excess exposure to heat. The shear rate in the metering section of the screw can be obtained as follows:

$$\dot{\gamma} = \frac{\pi DN}{h} \tag{126}$$

where D is the diameter of screw, N the rpm, and h the gap between screw and barrel.

Specific output (SO) is defined as the mass flow rate per unit rpm of screw and is calculated from the total mass flow rate ($\dot{m}$) divided by the screw rpm (N) used:

$$SO = \frac{\dot{m}}{N} \tag{127}$$

This information provides information on the uniformity of solids conveying, melting, and pumping mechanisms of the screw being used.

The definitions above provide a basis for scaling up the performance characteristics of polymers in extrusion operations and allow the product development engineer to select materials logically for an intended application. The use of a torque rheometer and single-screw extruder for product development is illustrated by the two examples described below.

The first case we consider is the application of torque rheometry to the assessment of lot-to-lot variations in the production of an extrusion-grade rubber. The torque rheometer simulates a great detail of the extrusion process and as such can often provide information that the polymer fabricator may not observe in many quality control tests. Table 5.7 summarizes the gum and compound properties of samples of the same EPDM retrieved from five different times during one day's production. The column on the rightmost side of the table reports the ranges of each property that the customer will accept for fabricating. In scanning the table, it is clear that every one of the lots meets the customer's criteria.

Compound samples were prepared in masterbatches in a carbon-black formulation typically used for hose extrusion. Each compound was then processed in a single-screw extruder (1.25-in. diameter, $L/D = 20$) at one typical speed and temperature, simulating the customer's extrusion

Table 5.7 Properties of Polymers Tested in Lot-to-Lot Variations Study

Lot in time (hr)	1100	1210	1320	1440	1530	Customer acceptance range
Mooney viscosity (1 + 8) at 127°C	73	69	72	77	73	65–75
Ethylene content (wt %)	61	62	60	60	58	56–64
Diene content (wt %)	5.2	5.0	5.1	5.4	5.1	5–55
Mooney scorch at 132°C (min)	12.0	12.3	11.3	11.5	11.5	>10
Rheometer at 160°C						
MH (in.-lb)	64.9	58.2	65.3	61.3	62.1	>55
Rate (in.-lb/min)	23.9	21.6	26.4	24.6	22.6	>20
Cure data						
Shore A hardness	67	67	66	66	67	60–70
100% modulus (MPa)	5.7	5.1	5.4	5.4	5.3	>5.0
Tensile (MPa)	13.0	11.9	12.1	12.2	11.8	>11.0
Elongation (%)	240	240	220	220	220	>200

process. Figure 5.58 shows time plots of the instantaneous pressure and torque values recorded by a torque rheometer coupled to the extruder. Figure 5.58A shows the head pressure and torque responses, and part B plots the upstream and head pressure data for the same samples. It is clear from these two plots that the lots from the latter portion of the production are distinctly different from the earlier lots. The lower torque and head pressure responses of these three samples suggests that these materials suffer in extrusion rate.

The low-pressure profile further substantiates that these samples are in fact not properly filling the effective barrel volume of the extruder, particularly in the compaction and transition zones. This of course translates into slower extrusion rates, values of which are reported in Figure 5.58B. As shown, the extrusion performance of the suspect polymers is nearly 50% of the earlier lots. To the product development engineers, the obvious question to ask is: Why? The answer in this case lies with the apparent levels of crystallinity for each polymer. Figure 5.58B also reports heat of fusion data as obtained by DSC on gum samples. The poorest extruding samples are those having the lowest heats of fusion, even though the melt section and die of the extruder were maintained at 115°C, well above the melting range for crystalline factions of the polymer. Crystallinity, however, determines (along with molecular weight) cold green strength properties which are important to the feed, compaction, and transition zones of the extruder, where temperatures are generally much lower. In simple terms, a low-green-strength material essentially has no "guts," and we can therefore envision the poor extrusion samples experiencing either slipping and/or the strips of feedstock pulling apart, creating stick-slip flow near the entrance region of the extruder. This is substantial by noting that the fluctuations in the instantaneous torque response is much greater for the latter lot samples (refer to Figure 5.58A).

The investigation must now shift to the synthesis of these polymers to determine why lots varied. Recall from Table 5.7 that the levels of ethylene were quite similar, with small variations simply not enough to explain the significant differences observed in the processing evaluations. Possible speculations as to the causes for different levels of crystallinity are the presence of long-chain branching in the latter portion of the run or perhaps ethylene composition variations over the molecular weight distribution. For our purposes, however, we need go no further into the evaluation.

The final illustration involves an exercise in evaluating/developing a good-processing, fast-extruding rubber, with the additional property of having an ultrasmooth surface appearance. This material is earmarked for automotive parts such as weatherstrip seals and door and trunk seals, where the surface aethetics of the final extruded part are an important consideration to car manufacturers. This product development problem

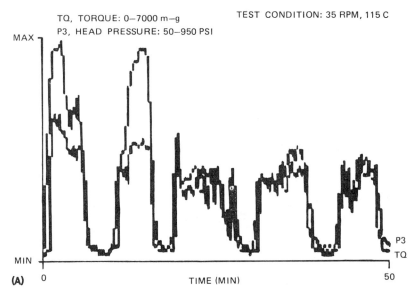

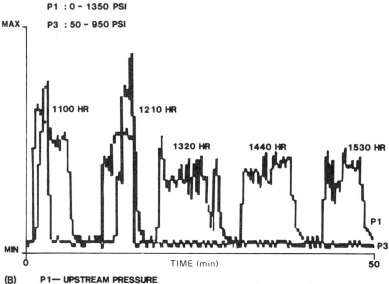

	Heat of Fusion (cal/gm)	Cmpd. Green Strength (MPa)	Extrusion Rate (gm/min)
1100 HR	1.4	3.1	51
1210	1.2	2.9	55
1320	0	0.4	35
1440	0	0.5	20
1530	0.2	0.5	33

Figure 5.58 (A) Torque and head pressure response plots; (B) upstream and head pressure response plots.

Processability Testing

develop a general-purpose grade. That is, we want a rubber material that is versatile enough or has a wide enough processing/property latitude to be used by different customers using different formulations and extruders. In other words, the product must be of a robust design to be soft to a wide variety of customers making weatherstrips who may use significantly different processing conditions.

The present study was performed on several EPDM polymers ranging in molecular weight and composition. The properties of these polymers are reported in Table 5.8. The terpolymers were compounded in a high-quality weatherstrip formulation.

The compounded polymers were strip fed the sam emini-extruder described earlier. Extrusion tests were conducted using a standard tube die with a floating mandrel (15.9-mm-OD tube × 12.7-mm-OD mandrel). The L/D of the die was 65, which was considered adequate for minimizing exit effects in studying swell characteristics of the materials. The discharge was collected on a conveyor belt equipped with variable-speed drive control. The speed of the takeup section was controlled to match the linear speed of the extruded tube.

Running die swell measurements were obtained using a Brabender laser-optical swell tester, positioned 10 mm from the die exit. Extruded tubes were cured at 180°C for 3 min in a high-velocity 6-m Dawson hot air oven. Relaxed wall thickness and tube distortion measurements were then obtained. The skin appearance of the extruded tube was measured by means of a photographic technique. Sections of the extruded tube were razor cut to obtain a flat wall section, and a photograph of the surface at 14 × magnification was made both prior to and after oven curing for each screw speed. A qualitative surface rating scheme based on an elaboration of the standard Garvey die rating system was developed for comparison between samples. Figure 5.59 shows the surface classification system devised. The extruder was operated adiabatically, at a constant wall temperature of 89°C, with three strip heaters located axially. Screw speeds were varied between 20 and 200 rpm, which translated into shear rates ranging from 10 to 400 sec^{-1} in the die.

To interpret the relative surface and swell responses of each sample, we must first understand what information the torque rheometer provides. Recall that torque rheometer extrusion data provide a relative comparison of the flow characteristics and processability of the polymer materials. Screw torque represents the total extrusion process work energy and is therefore a summary of what happens to the material as it is formed downstream by the action of the screw through the steps of feeding, fusion, melting, pumping, and metering, right through forming in the die. Figure 5.60 shows that a unique relationship exists between the specific energy of

Table 5.8 Properties of Polymers Studied

Polymer identification	Symbol on plots	Mooney viscosity (1' + 4') at 125°C	Oil (phr)	C_2 (wt 5)	ENB content relative to polymer A
A	○◐	50	50	50	High
B	□	67	—	57	0.98
C	●	71	—	69.9	1.07
D	△	66	—	66	0.99
E	▽	72	75	70	1.00
F	▲	93	—	62	1.04
G	▽	92	—	62	1.00
J	△	36	30	58	0.97
K	○	88	—	60.5	0.94
L	○	103	—	63	0.79
M	○	98	—	62	0.90
N	■	98	—	62.6	0.89
O	◇	88	—	58.2	0.88
P	◆	78	—	61.1	0.91

Processability Testing

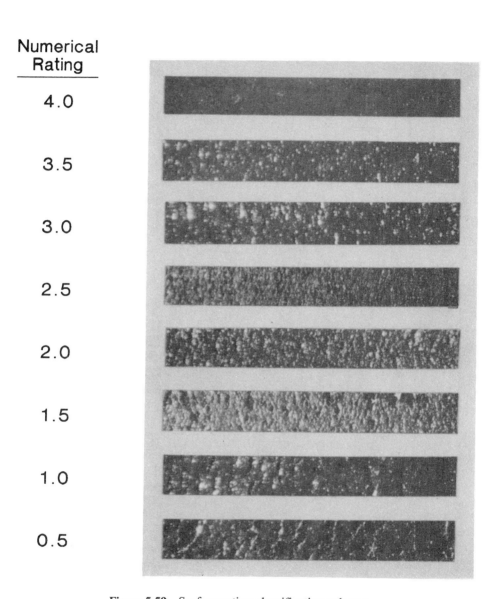

Figure 5.59 Surface rating classification scheme.

requires an understanding of both the parameters of extrusion, and polymer molecular and composition properties important to contributing to a particular rheological response of the compound, resulting in a smooth-skinned part. In addition to the raw material selection (i.e., the rubber), the degree of dispersion in compounding, extrusion, continuous vulcanization media, profile complexity, and postprocessing operations all influence surface and general product property performance. However, discussions here will be limited to applying torque rheometry to studying the extrusion process itself. One of the constraints of this type of problem that should be kept in mind from the point of view of the polymer manufacturer is that in product development the objective is usually to extrusion (i.e., energy per unit mass of flow) and screw speed for a particular screw geometry. The specific energy varies with the level of shearing along the extruder barrel, and all polymers, regardless of their viscoelastic response, will follow the curve shown in Figure 5.60.

From a productivity standpoint, the critical parameter is the machine capacity, or more specifically the line speed at which articles can be

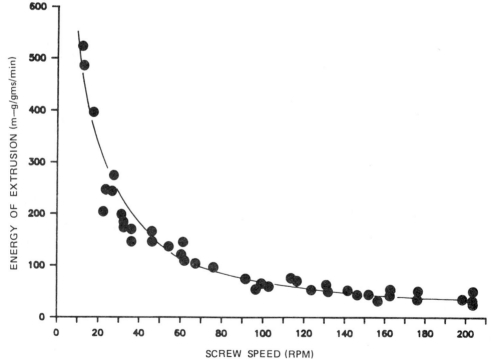

Figure 5.60 Specific energy versus screw speed for metering screw used in study.

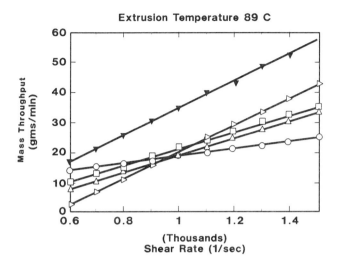

Figure 5.61 Comparison of mass throughput performance for polymers studied.

formed. The weatherstrip extrusion process requires continuous vulcanization in line with extrusion. It is desirable to have the vulcanization time dictate maximum line speed, not the extrudability of the material. The extrusion productivity of EPDM compounds depends on the rheological response of the compound over the various stages of the extrusion operation, as well as the screw and die configurations. For a fixed apparatus geometry, the rubber's extrusion productivity depends on the molecular and compositional properties of polymer: in particular, its molecular weight, the molecular weight distribution, the degree of crystallinity, and the extend of long-chain branching. The polymers studied cover a reasonably wide range of molecular weight distributions and composition for EPDM and consequently were observed to display significant variation in extrusion throughput. Figure 5.61 provides a comparison of the mass throughput curves for polymers studied, indicating an almost twofold difference in throughput performance over the extremes tested.

In this study, we are most interested in the quality of the extrudate in terms of its swell and tolerances, degree of distortion or shape retention, and the relative smoothness of the surface. Interest must therefore be focused on the region of the die in order to describe properly what immediately happens at the discharge, recognizing, however, that upstream stages of the extrusion process establish conditions of head pressure and uniformity of feeding through and forming in the die cavity. This is best illustrated by the plots shown in Figure 5.62. Figure 5.62A and B show linear dependence of machine torque and head pressure to screw speed.

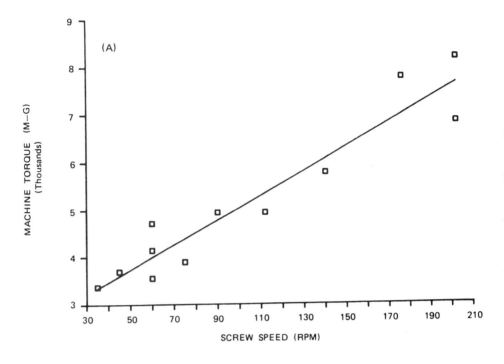

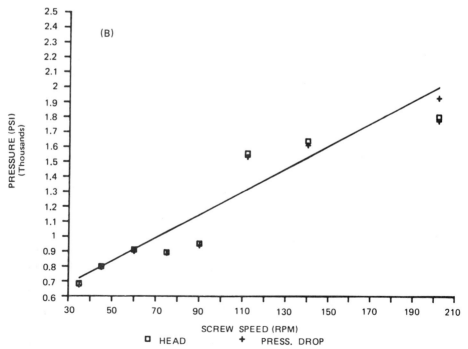

Processability Testing

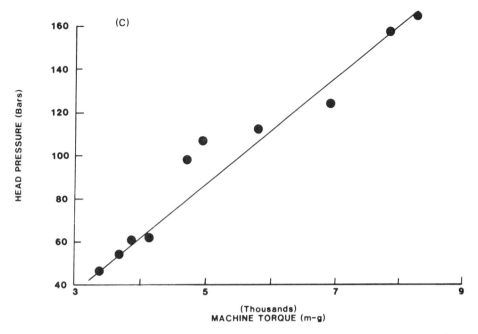

Figure 5.62 (A) Machine torque screw speed dependence; (B) head pressure and pressure drop across barrel dependency on screw speed; (C) head pressure torque relation for one polymer.

Figure 5.62B further illustrates that the upstream pressure has a dramatic effect on die and head conditions in terms of the pressure drop across the metering section of the barrel. As screw speed is increased, the higher pressure drops are indicative of greater compaction and metering capacity. For this particular polymer, the available screw volume is more effectively utilized at higher screw speeds. In other words, a relation between machine torque or specific energy exists, which is unique for the properties of the rubber being processed. This is illustrated in Figure 5.62C.

The rheological properties of the polymer in the die cavity depend on the extent of flow shear and stresses induced in the melt state. Using the standard definitions for flow through an annulus, the following expressions can be applied to estimating the effective shear rate and melt state stress in the die cavity (Han, 1976):

$$\dot{\gamma} = \frac{6Q}{\pi d' h^2} \quad \sec^{-1} \tag{128}$$

$$\tau = \frac{\Delta P R}{2l \times 10^5} \quad \text{Pa} \tag{129}$$

where

Q = volumetric flow rate, cm³/s
d' = medium diameter of annulus, cm
h = gap of annulus, cm
l = die length, cm
ΔP = pressure drop, Pa

Shear stress and shear rate can be related through the power law relation and the apparent melt viscosity in the die is

$$\eta_a = \tau/\dot{\gamma}$$

Using these formulas, the power law coefficients k and n can be evaluated. The power law index, n, reflects the relative shear sensitivity of the melt state, where as $n \rightarrow 1$, the material behaves more Newtonian. This plays an important role in establishing swell and surface properties.

Surface appearance (i.e., relative smoothness, texture, and glossiness) can depend on several factors: the compound ingredients, extent of ingredient dispersion during mixing, degree of compatability between materials, propensity for melt fracture during extrusion, and postextrusion of operations (i.e., vulcanization media and stock nerviness). In this study efforts were given to examining the extrusion process. Figure 5.63A shows a plot of the surface appearance rating (SAR) immediately after extrusion (i.e., before curing) versus mass throughput for several polymers. The plot suggests that definite relationships exist between production capacity and skin appearance, which are polymer-property dependent. In other words, each polymer has a characteristic operating line, whereby surface appearance is a function of production capacity. This is expressed by the curves shown in Figure 5.63B in terms of a plot of SAR versus specific output. Output is defined as the mass flow rate at a given screw speed. Figure 5.63B affords a basis for comparing the extrusion performance of different polymer stocks in terms of throughput efficiency and surface appearance. It does not, however, explain the mechanisms responsible for surface deterioration.

Since Figure 5.63A shows the SAR to be throughput dependent, it is obvious that this is a shear-rate-dependent phenomenon. Figure 5.64 shows that surface deterioration takes place at increasingly higher shear rates. The plot shows SAR values both before and after oven curing. Both sets of data follow identical trends, but the oven cured sets are lower, illustrating that the vulcanization stage can greatly affect the article's quality.

The variation in surface appearance between samples over identical shear rates can be attributed to significantly different temperature respon-

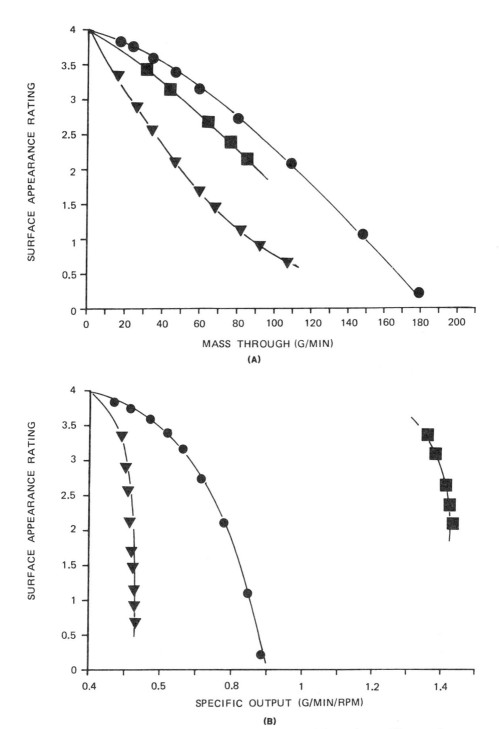

Figure 5.63 (A) Surface appearance as a function of throughput; (B) operating curves for surface appearance.

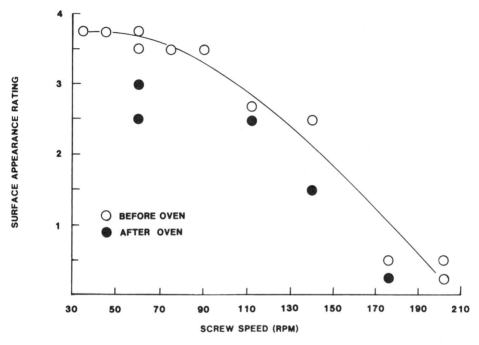

Figure 5.64 Surface appearance as a function of die shear rate.

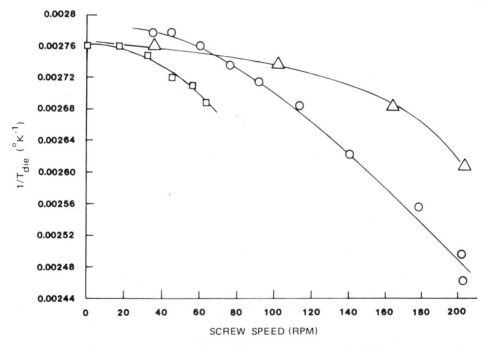

Figure 5.65 Melt temperature response curves for die flow.

ses through the die. Figure 5.65 shows a plot of the inverse of melt temperature in the die versus shear rate for the various polymers. As shown, the temperature responses are considerably different, with several samples showing greater temperature rise than others. This means that polymer samples have significantly different activation energies and hence viscous heat rises, which accounts for the relative differences in surface smoothness. Figure 5.66 illustrates that, in fact, shear rate translates into viscous heat, as recorded by the difference in temperature between melt state the die cavity wall, and hence accurately describes the deterioration of surface properties at increasingly higher shearing.

The actual mechanism responsible for surface smoothness can be attributed to prevulcanization in the die. That is, at successively higher shear rates, a portion of the polymer undergoes some level of curing, which changes its viscoelastic and surface properties upon exiting the die. Figures 5.67 and 5.68 illustrate this for one polymer sample. For each screw speed, a sample of extrudate was run in an oscillating disk rheometer at 3° arc, 6 min, 182°C, to determine what extent, if any, the sample scorched. Figure 5.67 shows a plot of the cure rate (i.e., the slope of the torque/time response curve from the curemeter) versus the viscous heat tempreature rise corresponding each sample's rpm. As shown, cure rate is shown to slow down at higher viscous heats (i.e., shear rates), in-

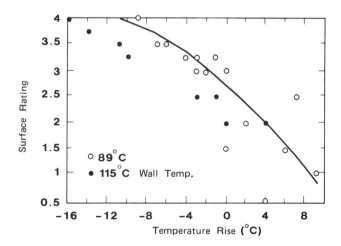

Figure 5.66 Surface appearance shown to worsen at higher levels of viscous heating.

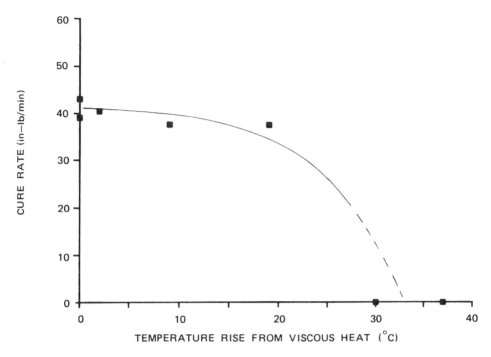

Figure 5.67 Cure rate decreases after polymer undergoes successively higher viscous heating.

dicating that a definite portion of the polymer has vulcanized within the die cavity.

Figure 5.68 further illustrates the same phenomenon but in terms of the average bulk density of the extrudate. Figure 5.68A shows a plot of stock density after extrusion versus the viscous heat temperature rise. The increasing density in a reflection that the polymer is reacting or vulcanizing in the die at successively higher shear rates. Figure 5.68B simply illustrates that this can also be observed by the torque measurements on the rheocord.

As noted earlier, the postvulcanization stage plays an important role in establishing the finished article's properties. Depending on the method used, surface appearance and article tolerances can be controlled, within limits. In the present experiments surface properties were shown to deteriorate after oven curing, simply implying that a severe test was imposed. This does illustrate, however, that surface response is also a func-

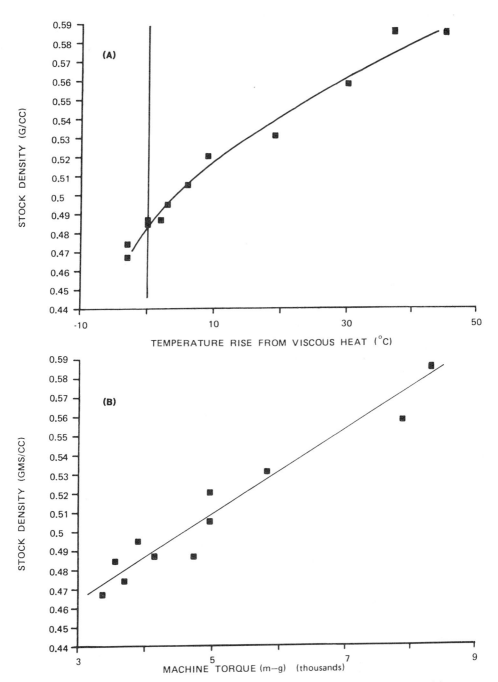

Figure 5.68 (A) Stock density increases due to higher set properties resulting from greater levels of viscous heating in die; (B) torque response from Rheocord accurately reflects more dense stock.

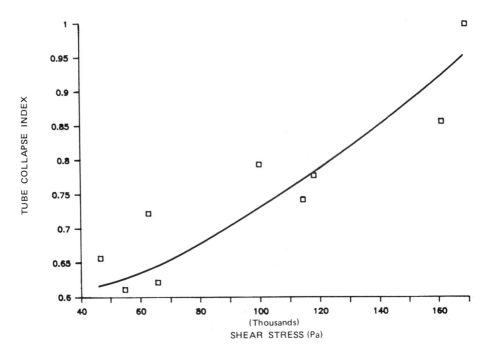

Figure 5.69 Plot of tube collapse versus shear stress.

tion of a dimensionless heat transfer group that accounts for conduction and residence time during the postcuring stage.

It is important to realize that surface response not only depends on the viscoelastic properties of the polymer and vulcanization conditions, but on the shape and tolerances to which the article is controlled. It was shown earlier that the SAR is a function of elastic stresses of the polymer. The same is true for tube collapse but from a different behavioral aspect. SAR tends to decrease as stresses increase. In contrast, tube collapse (defined as the height-to-width ratio of the outer tube dimensions) is shown to increase at higher stresses. Figure 5.69 shows a plot of τ_w versus tube collapse after oven curing for one polymer sample. A cross plot of surface appearance versus tube collapse is shown in Figure 5.70 for the same polymer, demonstrating an inverse relationship. We may conclude that although high elastic response is beneficial to controlling tolerances and article shape, it can result in a worsening of surface properties. A balance between these two response parameters must be devised when attempting to introduce new polymer stocks in an extrusion application.

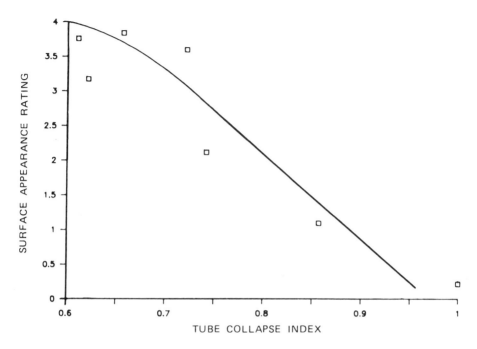

Figure 5.70 Plot of tube collapse versus SAR.

In closing this section, we reflect on the few examples illustrating application of torque rheometry to processability testing. Both this instrument and the capillary rheometer are critical laboratory items that play a major role in new products testing, assessing fabrication quality and in applications development of new materials. Specific procedures have not been published in the ISO and ASTM codes for torque rheometry; however, the application of the apparatus should really be structured toward the specific problem posed to the product development engineer.

There are, however, standardized extrusion tests that provide relative comparisons of samples and are employed on a regular basis with product qualification studies. One of the most widely used is the Garvey die extrusion test. Details of the Garvey die and the rating system used to quantify the extrudate quality are shown in Figure 5.71. The purpose of the test is to provide a quantitative rating of the extrudate dimensions, surface appearance, sharpness of edge appearance, and the porosity and swell properties of the extruded compound.

Without too much thought, analogous rating systems can be devised for assessing the relative processing performance of polymers on other

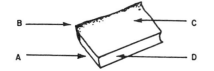

GRADING NUMBER	DEFINITION
A	SWELLING & POROSITY
B	SHARPNESS & CONTINUITY OF 30° EDGE
C	SMOOTHNESS OF THE SURFACES
D	SHARPNESS & CONTINUITY OF THE CORNERS

DIE PROFILE (ORIGINAL PROFILES GENERATED ON 1/2 ROYLE TUBER AT TEMP. OF 180°F)

EXTRUSION RATING BASED ON FOUR SEPARATE GRADINGS, EACH RANGING FROM 1 = VERY POOR TO 4 = EXCELLENT. THE PERFECT RATING IS 16, AND 12 IS SATISFACTORY FOR MOST PRODUCTION EXTRUSIONS PROVIDED THE 30° EDGE RATES AT LEAST 3.

Figure 5.71 Details of Garvey die and of extrudate rating system.

Table 5.9 Qualitative Rating System for Assessing Laboratory Mixing and Milling Performance[a]

Polymer	Mixing (type dump)	Polymer appearance on mill	Rolling bank appearance	Mill adjustment needed	Tacky
A	3	4	3	N	N
B	3	4	4	N	N
C	4	4	3	N	N
D	3	4	3	N	N
E	3	3	3	Y	Y
F	3	4	3	N	N
G	3	4	3	N	N
H	2	3	3	Y	Y

[a]Key to ratings:

Rating	Mixing	Mill appearance	Rolling bank
4	Very good	Smooth	Good
3	Good	Slightly rough	Slight bagging
2	Fair	Very rough	Medium bagging
1	Poor	Crumbly	Heavy bagging

Processability Testing

processing equipment. An example of this is shown in Table 5.9 for the operations of mixing and milling. Such observations clearly cannot be used from the standpoint of scaling-up product performance quantitatively, but they do provide a basis of screening potential processing problems among polymer candidates.

THE OSCILLATING DISK RHEOMETER

A practical processing problem in rubber fabrication is that concerned with minimizing the risk of premature vulcanization. This phenomenon is referred to as "scorching." For an elastomer to be of engineering and/or consumer interest, it must have sufficient scorch resistance; that is, it must be capable of displaying sufficient time delay in vulcanization to enable processing (mixing, milling, forming, molding, etc.) of the final article. In other words, the process of cross-linking must be controllable within the limitations of the processability requirements. One requirement, then, in selecting a polymer-compound formulation system for a particular end use is that the processing time and scorch resistance be compatible.

Scorch resistance can readily be studied in the laboratory and ideally should be measured at the intended processing temperature. By using a Mooney viscometer, one can measure this time from a sharp increase in

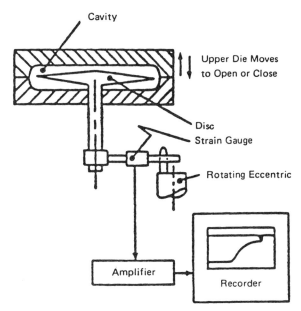

Figure 5.72 Oscillating disk rheometer.

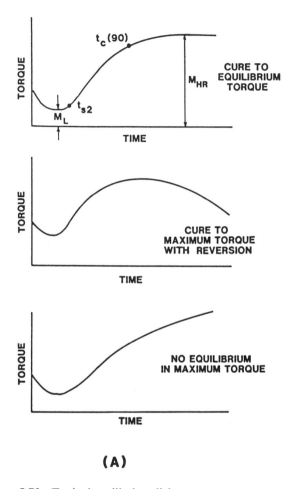

Figure 5.73 Typical oscillating disk curemeter response curves.

the compound stock viscosity. A rheometer or curemeter, as shown in Figure 5.72 can supplement this evaluation by providing a measure of the extent of cure. The curemeter illustrated is based on the principle of oscillating disks: a sample of material is cured in a cavity, while a disk, embedded in the sample is in oscillatory rotation covering an archlike motion. The resistance to the disk's oscillation is measured as a function of time, producing the curemeter response curves shown in Figure 5.73. This type of rheometer chart characterizes both the induction time and the rate

Processability Testing

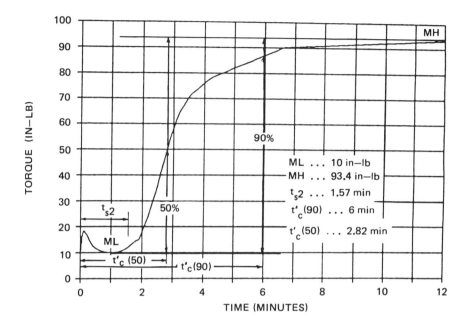

(B)

of cure (i.e., the vulcanization rate). Mooney viscosity measurements over time provide the processing safety guideline in terms of scorch resistance. The latter measurement is usually reported as the time in minutes for a prespecified change in Mooney units. In the rubber industry, scorch times are typically reported as minutes for a five-point rise in Mooney viscosity, or over a three-point rise.

Other key features referred to in curemeter plots are the ts_2, $t(90)$, M_L, M_H, and the cure rate. t_{s2} is known as the 2-min scorch time and t_c (90) is the time to 90% of cure completion. Both are often taken as an indication of scorch. M_L and M_H are the minimum and maximum torque readings, and $M_H - M_L$ is often interpreted as the cure state (i.e., the effectiveness of cure; the higher the value, the tighter the cure). Finally, the slope of the rising portion of the curve (or tangent slope) is the cure rate (i.e., how fast the material vulcanizes) in units of torque per time (in.-lb/min).

NOTATION

A	area
A, B	rheological constants in Ellis model

A	width
b	dimension
C	applied stress
C_A	concentration of component A
C_P	specific heat
C_1, C_2	constants
C_r	Couette viscosity correction term
D	diameter
D_{AB}	binary molecular diffusivity for two-component system
D_b	dimensionless Debora number
d'	medium diameter of annulus
E	flow activation energy
F	force
F'	normal thrust
f'	shear stress function
G	shear modulus
g	branching index
H	die height opening; height or distance
H'	minimum screw flight depth
h	height
IV	inherent viscosity
K	power law viscosity coefficient
k_{PV_o}	constant characteristic of material
k	thermal conductivity
l	effective bob length
l_{eff}	length
M	torque
MW	molecular weight
$\dot{m}$	mass throughput
N	Bagley friction correction term
N_f	rotational velocity
n	power law exponent
n''	slope of torque-rpm curve
P	pressure
P'	power
Q	volumetric flow rate
q	heat flow
R	universal gas law constant
R_i, R_o	inner and outer radii
r	radial distance
SE	specific energy
SO	specific output

T	absolute temperature
TE	total shear energy
t	time
V	velocity
V_b	volume
W	work
W, w	width
x, y, z	coordinates

Greek Symbols

α	Mark-Houwink exponent; thermal diffusivity
β	effective slip coefficient; viscosity pressure coefficient
Γ	effective shear rate
$\dot{\gamma}$	shear rate or ratio of applied stress to shear modulus
δ_c	channel thickness
η	apparent viscosity
$\eta(\dot{\gamma})$	viscosity function
θ	temperature
θ	angle
λ	material relaxation time
μ	Newtonian viscosity
μ_a	apparent viscosity
μ_0	viscosity at zero shear rate
μ_∞	viscosity at infinite shear rate
ρ	density
Ψ	torque
τ	shear stress
τ	stress tensor
τ_y	yield stress
Υ	torque
ϕ	helix angle of screw; angle
$\Psi_1(\dot{\gamma})$	first normal stress difference coefficient function
$\Psi_2(\dot{\gamma})$	second normal stress differences coefficient function
ω	vorticity tensor

REFERENCES

Azbel, D. S., an N. P. Cheremisinoff, *Fluid Mechanics and Unit Operations,* Ann Arbor Science, Ann Arbor, Mich., 1983.

Bagley, E. B., *J. Appl. Phys., 28,* 62A (1954).

Bird, R. B., W. E. Stewart, and E. N. Lightfoot, *Transport Phenomena,* Wiley, New York, 1960.

Calderbank, P. H., and M. B. Moo-Young, *Trans. Inst. Chem. Eng. (London)*, *37*, 26–33 (1959).
Chapman, F. M., and T. S. Lee, *SPE J.*, *36*, 37 (1970).
Cheremisinoff, N. P., *Fluid Flow: Pumps, Pipes and Channels*, Ann Arbor Science, Ann Arbor, Mich., 1982.
Cheremisinoff, N. P., *Polymer Mixing and Extrusion Technology*, Marcel Dekker, New York, 1987.
Criminale, Erickson, and Filby, *Arch. Rat. Mech. Anal.*, *1*, 410–417 (1956).
Furuta, I., V. M. Lobe, and J. L. White, *J. Non-Newt. Fluid Mech.*, *1*, 207 (1976).
Goldstein, C., *Trans. Soc. Rheol.*, *18*, 357 (1974).
Han, C. D., *J. Appl. Polym. Sci.*, *15*, 2576, 2579, 2591 (1971).
Han, C. D., *Trans. Soc. Rheol.*, *18*, 103 (1974).
Han, C. D., *Rheology in Polymer Processing*, Academic Press, New York, 1976.
Keller, R. C., paper presented at the meeting of the Rubber Division, American Chemical Society, Los Angeles, Apr. 24, 1985.
Keller, R. C. and N. P. Cheremisinoff, *Automot. Polym. Des.*, *6*(3), 3–20 (1987).
King, R. G., *Rheol. Acta*, *5*, 35 (1966).
Kreiger, I. M., and S. H. Maron, *J. Appl. Phys.*, *25*, 72 (1954).
Kresge, E. N., C. Cozewith, and G. VerStrate, paper presented at meeting of the Rubber Division, American Chemical Society, Indianapolis, Ind., May 8, 1984.
Krevelen, D. van, *Properties of Polymers*, Elsevier, New York, 1972.
Lee, B. L., and J. L. White, *Trans. Soc. Rheol.*, *18*, 467 (1974).
Lobe, V. M., and J. L. White, *Polym. Eng. Sci.*, *19*, 617 (1979).
Meissner, J., *Rheol. Acta*, *8*, 78 (1969).
Metzner, A. B., and J. C. Reed, *Am. Inst. Chem. Eng. J.*, *1*, 434 (1955).
Middleman, S., *Trans. Soc. Rheol.*, *13*, 123 (1969).
Minagawa, N., and J. L. White, *J. Appl. Polym. Sci.*, *20*, 501 (1976).
Mooney, M. J., *Rheology*, *2*, 231 (1931).
Nakajima, N., and E. R. Harrel, *Rubber Chem. Technol.*, *52*, 9 (1979).
Scholte, Th., in *Developments in Polymer Characterization*, Vol. 4, J. Dawkins, ed., Applied Science, New York, 1983.
Sezna, J. A., *Elastomerics*, *16*(8) (1984).
Skelland, A. H. P., *Non-Newtonian Flow and Heat Transfer*, Wiley, New York, 1967.
Thomas, D. G., *Ind. Eng. Chem.*, *11*(18), 55 (1963).
Wales, J. L. S., *Rheol. Acta*, *8*, 38 (1969).
Wales, J. L. S., and H. Janeschitz-Kriegel, *J. Polym. Sci.*, *5*, 781 (1967).
Weissenberg, K., *Proc. First Int. Rheol. Cong.* (1948).
Zakharenko, N. V., F. S. Tolstukhina, and G. M. Martenev, *Rubber Chem. Technol.*, *35*, 236 (1962).

6
Quality in Product Development

INTRODUCTION

W. Edward Deming was one of the first to recognize the importance of statistical methods to manufacturing, not just from the production phase, but upstream during the initial stages of conceptual through R&D, then through plant operations. The Japanese are generally accepted as the leaders in application of scientific methodology to quality control. This is exemplified by a 50% loss in U.S. market shares to Japanese products over the past decade in product categories such as automobiles, cameras, color television sets, computer chips, and machine tools (see Wheelwright, 1984). There are however, socioeconomic differences between the two countries which in modern times have given the Japanese an advantage over American manufacturing, often not cited by proponents of the Deming philosophy [Deming's concepts and principles to quality control are detailed in a major text (Deming, 1986).] In addition, ineffectual political protection and regulation of foreign trade has contributed significantly to the deterioration of U.S. market shares. Scientific methodology and statistical techniques are by no means new concepts in this country, and in fact, were introduced to Japan as late as post-World War II by people like Deming. Whether it is a major contributor or not to the widening U.S.-foreign market shares, a perceived difference between past practices in the United States and Japan lies in the fact that quality control techniques in this country are more often centered around inspecting to get rid of bad quality. In other words, the philosophy traditionally employed has been to apply the techniques of process control charts and inspection at the end of the production line. Today, the emphasis is moving upstream, with the intent of building good quality into the design of products and processes.

In this chapter, the principles of quality control as applied to product development and product testing are discussed. The chapter is presented as an overview, highlighting concepts important to establishing and implementing effective laboratory support programs aimed not only at developing new polymers, but in servicing customers with their technical needs.

PHILOSOPHY OF QUALITY AND ITS RELATION TO THE MARKETPLACE

Over two decades ago the Japanese first began selling higher-quality products in several U.S. markets. Japan's success in many industries confirms what quality philosophers have said for years: Customers will buy quality. Thompson et al. (1985) have reported that market share for excellent-quality companies is growing at 6% per year, while market share for low-quality companies is shrinking at 2% per year. In direct research on whether customers will buy quality, the 1985 ASQC/Gallup survey concluded that American consumers "would be willing to pay a substantial premium to get the quality they desire in consumer products." Although the ASQC/Gallup survey covered consumer products in general, the role of quality in buying decisions is also being confirmed in a variety of specific industries.

Of course, the fact that customers buy on the basis of quality is not strictly limited to the consumer marketplace. Suppliers to most industries confirm that quality is also a key variable in the buying decisions of corporate purchasing agents.

These facts present a compelling challenge to the developers of new products. The company that can differentiate its new products and services from others on the basis of quality will have a major advantage in competing for customers. But how does a company differentiate its new products and services on the basis of quality? Quality is not a feature that can simply be added to a new product or service at some point in the process.

Quality must be planned in, and designed in, from the earliest phases of the product/service realization process. We can start by concentrating on the very beginning of the design phase of a product: the marketing/development interface and the creation of new product quality objectives that drive the design effort.

To achieve competitive advantage, an organization must determine the quality characteristics of interest to customers in the targeted market. This focusing effort begins at the marketing/development interface. The goals are to define what quality means in the targeted market and to capture that definition in a set of specific objectives. Each function in the product production process can then measure its activities according to those objectives.

Quality is a frustrating concept. It seems intuitively obvious and easy to visualize, but it is very hard to define. This uncertainty feeds the myth that quality cannot be defined, and therefore cannot be planned in the development of a new polymer. Under this myth, quality is not even an issue at the marketing/development interface; the strategy is: We do our best, see what the customer thinks, and react quickly to any deficiencies in quality. This approach to quality is costly and it does not yield competitive advantage. Instead, it is a "catch-up" philosophy.

Crosby defines quality as "conformance to requirements," while Juran (1986) defines it as "fitness for use." In practice, quality must encompass both of these concepts. New product quality objectives must be developed that capture the customers' expectations (fitness for use) and translate these into requirements for the company's operations (conformance to requirements).

Let's consider a specific industry, automobiles. It is difficult to define quality. Both a Mercedes and a Ford can be considered high-quality cars, even though they are not the same. One must realize that quality is not a singular characteristic but rather many characteristics that contribute to product or service quality in the eyes of the customer.

Table 6.1 lists the various characteristics, or dimensions, of quality. Examples of the various dimensions are listed in Table 6.2. These dimensions of quality support a broad understanding of what quality might mean and provide a framework for planning for quality at the marketing/development interface for a new product or service. This framework for defining quality is comprehensive. All the words or phrases that are used to define the concept of quality can fit under one of the nine dimensions. Note that the various dimensions of quality are somewhat independent. A product or service can be good in one dimension but average or poor in another. This comprehensive framework, with its independent dimensions of quality, illustrates how different products can be considered high quality. Each of these products or services has achieved a level of excellence in one or more of the specific dimensions that make up the quality composite.

While all the dimensions contribute to the customer's impression of overall quality, being the quality leader in a particular market does not

Table 6.1 Dimensions of Quality

Dimension	Definition
Performance	Primary product or service characteristics
Features	Added touches, "bells and whistles," secondary characteristics
Conformance	Match with specifications, documentation, or industry standards
Reliability	Consistency of performance over time
Durability	Useful life
Serviceability	Resolution of problems and complaints
Response	Characteristics of the human-to-human interface like timeliness, courtesy, professionalism, etc.
Aesthetics	Sensory characteristics liek sound, feel, look, etc.
Reputation	Past performance and other intangibles

Table 6.2 Examples of Dimensions of Quality

Dimension	Product Example: Stereo Amplifier	Service Example: Checking Account at Bank
Performance	Signal-to-noise ratio, power	Time to process customer requests
Features	Remote control	Automatic bill paying
Conformance	Workmanship	Accuracy
Reliability	Mean time-to-failure	Variability of time to process requests
Durability	Useful life (with repair)	Keeping pace with industry trends
Serviceability	Ease of repair	Resolution of errors
Response	Courtesy of dealer	Courtesy of teller
Aesthetics	Oak-finished cabinet	Appearance of bank lobby
Reputation	*Consumer Reports* ranking	Advice of friends, years in business

necessarily mean being the best in every aspect. Consider U.S. and Japanese automobiles in the 1970s in terms of various dimensions of quality. Japanese cars were not the best in every dimension. For example, in the 1970s Japanese cars tended to score low in acceleration and safety. Furthermore, the unavailability of replacement parts made them hard to service, and they had low scores on corrosion resistance tests. *Fortune* magazine (June 2, 1980) summarized the quality strengths of the Japanese: "The Japanese... excel in the quality of fits and finishes... materials that look good... flawless paint jobs. Japanese cars have earned a reputation for reliability.... Technically, most Japaneses cars are fairly ordinary...."

Japanese cars of the 1980s were not the best in every dimension, yet they were cited for high quality. Apparently, those dimensions of quality in which they did excel—reliability, conformance, and aesthetics—were the most important dimensions to customers. Thus the Japanese achieved a huge competitive advantage through quality. In this example, reliability, conformance, and aesthetics should not be considered as a generic set of dimensions of quality. Volvo buyers are probably looking for durability, while Corvette buyers might stress performance over reliability. This leads to an important conclusion: Products or services can be differentiated on the basis of quality by focusing attention on a small subset of the dimension of quality. The key lies in picking those dimensions that are most important to the target market.

MARKET RESEARCH ON QUALITY

Most firms invest a lot of time and effort at the marketing/development interface in defining the performance and feature objectives. Marketing planners use various market research tools, such as focus groups, questionnaires, customer surveys, competitive analysis, product tear-downs, and advertising claims analysis to define those objectives. These techniques can also be used at the marketing/development interface to set objectives for the other dimensions of quality.

Using the tools of market research, which are normally used only to define performance and feature characteristics, a company can systematically determine how a new product can become the quality leader in the marketplace. For example, in planning the development of a new line of antiskid plastic decking materials, a study might be made to answer several querries.

How important are product features in the target market?
What product features are most important to the target customers?

How do the target customers define what constitutes a feature decking or industrial flooring?

Among competing products, how do the target customers determine which decking material has the best set of features? What "cues" do they use to make judgments on product features during the buying process?

Who are the competitors, what product features do they offer, and what perceptions do the target customers have of those product features?

What product features would make the target customers notice us among a sea of competitors?

Questions like these can be answered using the tools of market research. The answers could then be used to develop a list of features of the new plastic material to increase its chances of succeeding in the marketplace. Such a list is a valuable input to the design team. It helps them focus their efforts on applying the design tools and technologies that would bring about the desired set of features. The results of this market research might also be used by other functions in the product/service production processes, such as purchasing, manufacturing, or field support.

While market research studies are routinely used to define performance and feature objectives, the questions above could just as easily be asked of any of the other dimensions of quality listed in Table 6.1. When we substitute the dimensions "reliability," "conformance," "durability," and so on, for the word "features" in the questions above, we have outlined a market research study on quality. Such a study provides a comprehensive definition of what quality means in the target market. More important, it is the basis for creating quality objectives that:

Identify and define the most important dimensions of quality in the target market.

Focus the quality efforts of the design team and other downstream functions.

Differentiate the new product or service in the marketplace on the basis of quality.

Give the new product a competitive advantage.

Figure 6.1 shows the process for conducting a market research study on quality at the marketing/development interface. The process begins with internal planning, progresses through customer- and competitor-oriented data collection, and ends with the development of new product quality objectives.

The objectives of the first phase are to appoint and train a cross-functional team, develop background materials and a plan for data

Quality in Product Development

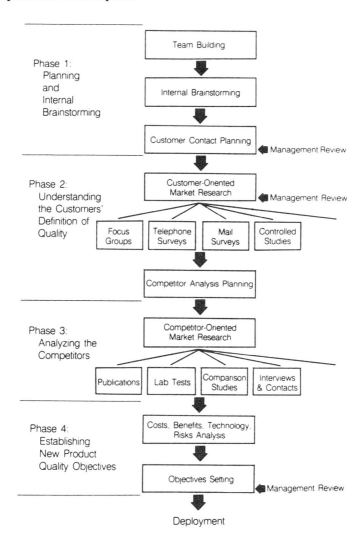

Figure 6.1 Market research study approach to quality.

gathering, and secure management's commitment to proceed. To begin the effort, a group of managers from various functions in the product/service production process should be assembled. This group should include the marketing manager, the development manager, the quality manager, someone who is familiar with market research techniques, and someone who has routine contact with customers (e.g., the sales or service manager).

This group should brainstorm to list all the possible meanings of the various dimensions. The various meanings of reliability could include time to first failure, time between failures, number of failures on a single unit, and so on. These various meanings could lead to different strategies in product design, the purchasing of component parts, or the activities of the field support network. Therefore, it is important to identify the most appropriate definition for reliability—the one that matches the customers' needs and expectations—and set objectives that drive the decisions and operations of the other functions in the organization. The same is true of each of the other dimensions. They all have many possible meanings that require different design, purchasing operations, delivery, or support strategies. Emphasis should be given to identifying the meaning that matches the target customers' expectations and to set appropriate objectives to drive the other functions. This is the basic goal of the market research study.

The brainstorming session also helps the market research study designer understand the choices and distinctions as the customers make them. The session should identify differences in interpretations between the various functions in the company—which can cause departments to work at cross purposes.

The results of the brainstorming session should be reviewed with people in the company who have routine contact with potential customers. These informal reviews help refine the lists and provide initial perceptions of how customers might rank and define the dimensions. The group then identifies the issues that must be resolved through customer contact and develops a detailed plan (complete with cost estimates) for a market research study on quality. Finally, this plan should be reviewed with top management to obtain their commitment.

The objectives of the second phase of the effort are to determine the key dimensions of quality from the customers' point of view, find any differences between that view and the internal view, find any differences between various groups of customers, and develop a plan for competitive analysis. Many techniques can be used in this phase to gather information from target customers and resolve the issues identified in the planning phase.

Focus groups are 1- to 2-hour discussion sessions with 8 to 10 carefully chosen participants and a trained leader. The open format allows the leader to probe the issues, but has the potential pitfall that the results might not be truly representative of the target population as a whole. For this reasons, market research studies almost always include a series of focus groups, and often a larger follow-up survey to confirm results. Telephone questionnaires are another market research tool. Although the ability to probe and explore issues is somewhat limited compared to a focus gorup, phone interviews can be given to a larger sample of the target population.

Mailed questionnaires are less costly than phone interviews, but they must be severely limited in length and flexibility, and require a lot of time. Furthermore, a large number of unreturned questionnaires can jeopardize the credibility of the results.

Controlled studies can provide insight into more subjective dimensions such as "aesthetics" or "response." The cost of these studies will vary widely depending on the sophistication of the testing. Controlled studies are also useful for research into the "cues" that customers use during the buying process to evaluate technical dimensions such as "performance," "durability," or "conformance."

In addition to these techniques, sales call, installation visits, trade shows, industry publications, and industry experts are other sources of information about customers' expectations. The key lies in knowing exactly what you are looking for and setting up some systematic means of getting it. The choice of techniques used in this phase should be guided by the experience of the team member familiar with market research techniques, subject to budget, time, and security constraints. Regardless of the techniques used, the study report should identify the customers' views of:

The relative importance of the dimensions
The exact meanings of the most important dimensions
The points of differentiation (i.e., what it will take to get us noticed)

The conclusions should highlight and explain any differences between customer groups and differences between what the customers said and the opinions of those inside the company. Finally, the group should develop and present to management a plan for determining where the competition stands on the key dimensions of quality.

In phsae 3, the goal is to determine where the major competitors stand on the key dimensions of quality. Besides the customer-oriented market research that has traditionally been limited to performance and feature attributes, many organizations perform ongoing competitor analysis in the other dimensions of quality: conformance, reliability, durability, response, and so on. Competitor analysis is often done in support of the concept of world-class quality—the strategic notion that a company should position itself among the top three competitors in the world, in all aspects of its products and services. In practice, companies who embrace this noticon are almost always forced to make subjective decisions and trade-offs, even though they possess reams of quantitative data on competitors. The problem reduces to establishing, and evaluating the potential payoffs, for the various investments that must be made to position a product or service among the top three in the world. Knowing where competitors stand does not necessarily tell you what is important in the marketplace. Competitor analysis should build on customer-oriented research and focus mainly on the key dimensions of quality, as defined by the target customers. While the broad-based competitor analysis used by many companies helps establish minimum quality objectives for a variety of dimensions, the focused competitor analysis will yield a higher return on investment and be more strategically important.

There are a variety of techniques that can be used to benchmark competitors on the key dimensions on quality. Trade publications and special industry reports can contain useful, low-cost information. Product teardowns, laboratory tests, and controlled-comparison studies are more costly, but more detailed, techniques. Finally, valuable information can be obtained simply by asking specific questions. Customers and third-party dealers are often surprisingly willing to talk about their experiences with competing products. Furthermore, the competitors themselves may be more than happy to describe their techniques and results in some area in return for similar information. In other caess, consulting firms and third parties must be used to gather detailed data on competitors while protecting the identities of each.

The goal of the final phase is to establish the new product quality objectives that will focus and drive the design effort and other functions. The market research done in the preceding phase has identified, defined, and quantified the key dimensions of quality; the relative positions of competitors; and the points of differentiation in the customers' eyes. Setting the quality objectives is a strategy decision involving cost and technology constraints. It would be a gross oversimplification to suggest taht all objectives be set to exceed the best competitor and the customers' points of differentiation. There is obviously a cost associated with achieving a given level on any dimension. Furthermore, in some cases there may be basrriers of technology with associated cost and time penalties. Thus the process for establishing the objectives must be iterative and experience based. In practice, the cross-functional team that has been guiding the process might pursue the following course:

Quality in Product Development

1. For each key dimension, establish the cost baseline by evaluating the level of performance that the company could achieve in the new product or service using the technologies and techniques that it already has. This might be done through comparisons to similar products or services that the company already offers.
2. Estimate the costs and benefits of moving from this baseline level to new product quality objectives with the level of the average competitor, the level of the best competitor, and the point of differentiation identified by the customes.
3. Estimate the lead times and risks associated with any new technologies or process techniques that would be needed to reach each of these three objective levels.
4. Finally, considering all the key dimensions simultaneously, estimate the benefits, costs, lead times, and risk associated with various combinations of new product quality objectives for the overall product. Select the combination that acheives the company's objectives, subject to cost and lead-time constraints, with an acceptable level of risk from new technologies and process techniques.

W. Edwards Deming writes that "the consumer is the most important part of the production line." Kaoru Ishikawa states that "the first step in quality control is to know the requirements of the customer." The key question is: How can a company be close to the customer? How can a company know the requirements of the customer? How can customer input become part of the production line?

Several other questions surround this issue. What exactly are customers? Are they the outside users of a company's product or service, or are there internal customers as well? Should anyone listen to "internal" customers?

Should a company's quality improvement efforts involve the customer? Who defines quality? What input should customers have in the development of a product or service? Who should talk to customers?

To address these questions, Arbor, Inc. (1987) organized a special team consisting of experts from the fields of quality control, market research, and organizational development and developed the so-called "customer window" concept, which is a method of identifying customers, gathering customer data, and using these data to deliver a quality product. The customer window concept is based on three premises:

1. Everyone in an organization has customers. The customer may be the ultimate user (external customer) or someone within the organization (internal customer). A customer is anyone to whom someone provides service, information, or a product.

2. Everyone—not just the marketing department—can benefit by becoming more customer oriented.
3. Quality is defined by the customer. To improve quality, find out what the customer wants but isn't getting—and then, whenever possible, provide it.

Figure 6.2 illustrates the customer window model. The customer window grid is based on quadrant analysis, which is a market research tool. The grid divides product features into four groups or quandrants:

1. The customer wants it—and gets it.
2. The customer wants it—and does not get it.
3. The customer does not want it—and gets it.
4. The customer does want it—and does not get it.

For example, if a new product is being designed, potential attributes or features of this product can be placed in the appropriate quadrant of the customer window grid. Which quadrant? This can be answered by conducting customer research. Knowing where a feature or attribute falls can guide the decision-making process. Effort should be devoted to those features that a customer wants/needs but is not getting. If the customer wants it and is getting it, there is no problem; the feature is meeting the customer's needs. If the customer is getting a feature and does not want or need it, it may be possible to eliminate or alter that feature, or to educate the customer on how that feature can be helpful. If it is not delivered and is not important, there is little reason to be concerned about it.

Cary et al. (1987) recommend viewing the planning, development, and delivery of a product or service as a cycle of continuous improvement. Shewhart's (1939) plan-do-check-act cycle is an important element of the customer window. The concept of studying customers complements the ongoing cycle of planning, implementing, measuring, and improving. Plans are more realistic and accurate if customers have input. During implementation, customer reaction is important and can indicate potential problems with a new product. Customer feedback can be measured on an ongoing basis to identify areas for improvement. Figure 6.2B illustrates the customer study cycle. When addressing quality, it is important to focus on the quality of design as well as quality of conformance.

Managers who obtain information from both internal and external customers tend to make better decisions and can offer quality based on customer perceptions. If quality is defined by the customer, it makes sense to find out what the customer thinks. To communicate with customers, principles of market research must be used. In such analyses one should clearly identify and define customers (segmentations and profiles). A

Quality in Product Development

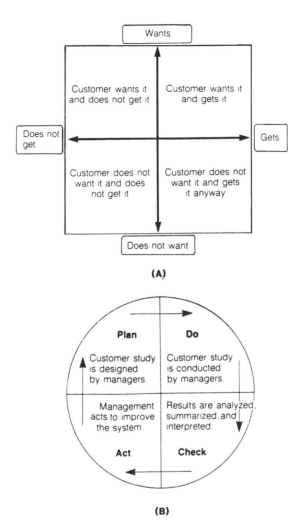

Figure 6.2 (A) Customer window model; (B) customer study cycle.

major part of this study involves collecting customer data (interviews, questionnaires, focus groups, etc.). Customer data should be analyzed and summarized in a clear format (statistical analyses, quadrant analysis, etc.).

Traditionally, these techniques have been applied mainly to "customers"—the outside user of a company's product or service. The application of certain market research tools can help managers and departments better serve their "internal" customers as well and can improve on day-to-day decision making.

Some customer research requires professional assistance. Advertising tests, new product introductions, and other major undertakings may require professional market research. However, there are numerous situations in any organization when information from customers would improve quality decision making and productivity and eliminate wasted effort. Unfortunately, decisions are frequently made with little or no direct customer study.

Even armed with a definition of customer-driven quality, many organizations cannot take action to improve their systems. Organizational change may be viewed as threatening to individuals for a myriad of reasons. Over the past several decades, a growing body of applied research has evolved to address the process of organizational change. For the beneficial applications of customer studies, a number of organizational development approaches were integrated into the customer window approach.

Organizational change can be difficult; the change process requires careful planning and attention. The change process can be organized into three phases: unfreezing (communicating the need for change, creating awareness, and motivating); changing (providing new patterns of behavior, which might include a new procedure, new equipment, or new methods); and refreezing (reinforcing the desired change). This analogy is effective in pointing out how people become frozen in or bound by old behavior patterns.

Customer-driven quality improvement may represent a drastic change for some organizations. It may require educating management on the importance and impact of using customer information in quality improvement projects. Some of the customer window tools and techniques may be new to managers who are being asked to use them. And to reinforce ("refreeze") this new approach, organizations should measure, document, and publicize the impact of improvement projects and changes in customer attitude/satisfaction.

Another helpful concept involves the "depth of intervention." How far are the individuals and groups willing to go? Is the change something an external consultant wants but something that is not generally desired by the organization? How is the change being positioned? It is often helpful to communicate with employees about the personal benefits of a customer orientation. The benefit of customer-oriented research is that it can bring all employees closer to their internal and/or extern customers. Information produced by customer research can be shared with every level of the organization.

Professional market and customer research has become a complex and highly technical field. In product development a method called conjoint analysis is often used to predict which combinations of product features will be most attractive to customers when developing a new product.

Although complex questions require complex techniques, there are simpler and more direct methods that are often appropriate for day-to-day decisions. Line managers, quality engineers, and management teams can conduct simple customer studies. Following are the steps involved in constructing a customer window analysis:

1. Identify, segment, and profile the customer base.
2. Develop research questions, such as those based on gaps in the profile.
3. Define the sample and collect data using an appropriate collection format.
4. Analyze and summarize data.
5. Take action.

To start a customer study, first identify and segment the customer base. Who are the internal customers? Who are the external customers?

Determining the segments comes through discussion, experimentation, or even initial customer research. If the segments are determined through research and statistical analysis, they are empirically determined segments. If customers are segmented based on assumptions or unmeasured perceptions, the resulting segments are termed "a priori" segments.

Usually, different customers want different things. Each segment can be profiled by answering such questions as:

What are the demographics of this customer? (Characteristics such as company size, location, product; organization level, department; etc.)
What is important to this customer?
What do they like about our product?
What do they dislike?

Who is competing for this customer's business?
How could this customer be better satisfied?
How does this customer define quality?
What is the size of this segment?
What percentage of the total business does this segment represent (dollars and/or time/effort)?

The result will be a matrix containing a series of data about the customer. If there are a priori data, some of the assumptions will be right on target. Some will be off by a little, and some will be dead wrong. There may be other pieces of information that will be unidentifiable without collecting customer data.

Segmenting and profiling internal customers can be beneficial as well. Just as companies position different products to different market segments, managers can profit from recognizing and meeting the needs of different internal customers.

After customers are identified, segmented, and profiled, certain pieces of information may be missing. It may also be important to verify some of your perceptions about the customers, or to get customer input on planned changes. The next step is to develop research questions to verify or obtain information. When developing the research questions:

1. Clarify goals and objectives. What information is needed, and why? Each question should contribute to the end purpose.
2. Avoid loaded, biased questions. Ensure that the data gathered are accurate and unbiased.
3. Don't ask "double-barreled" questions (two questions phrased as a single question). To ensure accurate responses and to make things clear for the person being questioned, make sure that there is no confusion about what is being asked.
4. Remember the customer window grid: ask "how well" and "how important." Check how important something is to the customer and how well the product or service delivers (or will deliver) this attribute. The combination of these responses will help to develop a plot of the data on the appropriate quadrant of the customer window grid.

Market researchers employ several analysis methods. If the customer research has produced an undecipherable mass of data, a marketing professional can provide assistance. If goals were carefully established and a manageable amount of data was collected, proceed to the next step—condensing the data into a usable form. Any data collected about customers' perceptions about the importance and current delivery for certain attributes can be plotted on the customer window grid. Other data analysis tools, such as Pareto charts, can help summarize customer information and indicate key issues to be addressed.

Customer research should be a driving force for new product development quality improvement teams, quality circles, productivity improvement efforts, and managerial decision making. Customer data can aim these efforts in the right direction, and can prevent overengineering and wasted effort.

For companies that have problem-solving or quality improvement teams in place, customer research can be a source of guidance and motivation. Analysis of customer research can guide a manager or organiation in repositioning products or services to meet internal and external customer needs.

It is important to emphasize that product design embraces all those activities undertaken to implement the functionality specified for a product. It is assumed that the product's specification was developed with the participation of marketing personnel, for only by using feedback from customers and marketing personnel can a company be reasonably confident that a newly designed product will find enough buyers to be profitable. Its worthwhile highlighting cost-effective augmentations to the processes typically used. First, the inclusion of one or mor emanufacturing engineers to a design team from the start of the project is important. Also, providing a means for design and manufacturing engineers to evaluate and improve the manufacturability of a design is essential. These two issues are often viewed from the point of being prevention costs.

Involving manufacturing engineers from the beginning of a product development project has benefits for both the product itself and for the manufacturing system that will produce the product. Manufacturing engineers help improve the product's inherent manufacturability. The product design engineers, in expressing to the manufacturing engineers what will be needed in the manufacturing processes to achieve the desired functionality of the product, will be providing lead time for the development and debugging of any new synthesis processes or capacity enhancement needed.

From the quality assurance viewpoint, using reviews and assigning a quality/reliability engineer early in a design project should be included in a good design process. In addition, the use of an "experimental design" approach such as the Taguchi method (discussed later) would undoubtedly be of great benefit in the development of more robust (less marginal) designs for a given cost or less costly designs for a given robustness.

An important question to ask is: What limits 100% quality of conformance; that is, what keeps some units of a product from conforming to the design specification? In many cases the problem is poor design. To plastics fabricators this can refer to both the parts they make and how the parts fit together. Some parts may be designed with features that are difficult to fabricate repeatedly or with tolerances that are unnecessarily and unbuidably tight. Some parts may lack details for self-alignment or features that prevent insertion in the wrong orientation. In other cases, the design parts may be so fragile or so susceptible to contamination that a fraction of the parts will inevitably be damaged in shipping or internal handling. Sometimes a design, due to lack of refinement, simply has more parts than are really needed to perform the desired function(s), so there is a greater chance of assembly error. Thus problems of poor design may show up as errors, poor yield, damage, or functional failure in fabrication, assembly, test, transport, and end use. There are basically three areas in which a product's design affects quality and cost: at the supplier's plant, in the manufacturer's own plant, and at customer locations.

A common cause of quality problems at a supplier is incomplete or inaccurate specification of the item to be provided. The supplier provides items that conform to the specification it received, but unfortunately, the products or parts do not satisfy the design intent because of some specification error of commission, omission, or transmission. This problem often occurs with custom parts (as opposed to stock parts) due either to weakness in the design process, engineers who do not follow set procedures, or sloppiness in the procurement and purchasing process. The greater the number of different parts in a design and the more suppliers involved, the more likely it is that a supplier will receive an incomplete or inaccurate part specification.

Another common design flaw that leads to supplier quality problems is a specification that is too tight relative to the inherent variability in the supplier's manufacturing process, causing some items to fall out of the specification tolerance band. This can often occur with raw material suppliers to polymer fabricators. Portions of a product to separate conforming from nonconforming units, the purchaser has two problems: It must pay, directly or indirectly, for the supplier's yield losses, and it must defend against errors the supplier may make in screening out the nonconforming portions of the production that the process will continually generate.

To the end user, if part fabrication is done in the plant, many of the same problems as described above for a supplier can occur. In addition, there will be problems in the area of assembly and test. For example:

Designs with numerous parts may cause part mix-ups, missing parts (because among so many parts, a missing part is not as noticeable), and more test failures.

Quality in Product Development

If some parts are very similar but not identical, the chances of an assembler using the wrong part in a given location are increased.

Parts without details to prevent insertion in the wrong orientation may be assembled improperly.

Complicated assembly steps and/or tricky joining processes may lead to incorrect, incomplete unreliable, or otherwise faulty assemblies.

Designs that require adjustments during or after assembly increase the chance of errors.

The designer's failure to consider the conditions to which parts will be exposed in the assembly process (e.g., temperature, humidity, vibration, static electricity, finger oils, and dust) may lead to subtle weaknesses in some units or unit failures during testing.

Hence the product design can directly affect assembly yields, the amount of scrap and rework, and manufacturing overhead expense.

The survivability of units during shipment to customers depends somewhat on the design of the product, and to a great degree on its packaging. Two additional design considerations that affect customer satisfaction—the perceived quality of the product—are the product's ease of use and the number of options of the product (i.e., the number of varieties in the model family). From the quality standpoint, a design should be so simple that correct assembly and use of the product are foolproof and should have as few options as possible.

There is an obvious correlation between product quality and product cost. Product cost includes the manufacturing cost, the expenses associated with product warranties, and engineering redesign costs. Furthermore, short-run loss of orders due to specific product problems and long-run loss of orders due to a general lowering of reputation, although these costs are hard to estimate, represent product costs due to poor quality.

Many of the characteristics of a product's design that affect conformance quality have an associated effect on manufacturing cost. Table 6.3 lists the principal components of manufacturing cost. The items are grouped by three main categories:

1. Labor directly and indirectly applied to actual production of products
2. Materials for products and manufacturing processes
3. Overhead chargeable to manufacturing operations

In many companies, all but direct manufacturing labor and product material cost are counted as overhead.

Table 6.3 Components of Manufacturing Cost

Labor
- Direct Manufacturing
- Indirect Manufacturing (includes supervisory and manufacturing-related engineering support labor)

Materials
- Product
- Process
  —Consumables
  —Equipment (Expensed, Capitalized)

Overhead
- Labor (includes purchasing, accounting, facilities, etc.)
- Nonproduction materials
- Equipment
  —Engineering (Expensed, Capitalized)
  —Office (Expensed, Capitalized)
- Utilities
- Facilities (includes buildings, land, and taxes thereon)
- Cost of capital
- Corporate charges (e.g., allocations of some corporate headquarters expenses to the manufacturing operation)

GENERAL PRINCIPLES OF QUALITY CONTROL

Producing good quality and increasing productivity as low cost is achieved by learning about processes. Stewhart (1939) has stated that "The three steps [specification, production, and judgment of quality] constitute a dynamic scientific process of acquiring knowledge. . . . Mass production viewed in this way constitutes a continuing and self-corrective method for making the most efficient use of raw and fabricated materials." He never intended quality control to be just passive inspection. The essential elements in discovering something are first, a critical event, one that contains significant information, and the presence of a perceptive observer present. These two things, a critical event and a perceptive observer, are essential.

The use of quality control charts enables informed observations to be noted and acted on. Knowledgeable individuals will examine such charts and ask "What happened at point x?" That is, they look for an assignable cause. With a quality control chart, data from the process are not buried in a notebook, a file drawer, or a computer file. In practice the charts are put

Quality in Product Development

up on the wall where the people who know about the process can observe, ask what happened, and take action. Thus people working with a system can slowly eliminate the bugs and create a process of continuous, never-ending improvement.

Ishikawa (1976) created a set of tools that helps people get at the information being generated in processes, referring to them as "The Seven Tools": check sheets, the Pareto chart, the cause-and-effect diagram, histograms, stratification, scatter plots, and graphs (in which he includes control charts).

Consider some of these tools and how they might be used to examine a problem. Suppose that in one week's production of gold balls, 75 are defective. That is informative, but not of much help for improving the process. Suppose that all the defective balls are sorted into those having defects because of cracks, because of scratches, because the dimensions were not right, and so on, and listed on a check sheet. The results can then be displayed using the Pareto chart, as shown in Figure 6.3A. In this instance, 42 of the 75 balls were discarded because of cracks. Clearly, cracks are an important problem. Pareto charts focus attention on the most important things to work on. They separate the vital few from the trivial many.

The next step is to figure out what could be causing the cracks. The people who make the balls should convene around a blackboard and make a cause-and-effect diagram listing possible causes, and causes of causes. Somebody may believe that the inspection process itself produces cracks. Another questions whether the gauges for measuring surface cracks are set right. Someone doubts whether inspectors agree on what a crack is. It is suggested that the source of cracks could be in the process of assembly. It is pointed out that two types of balls, A and B, are made; do cracks occur equally in both? The foreman believes that the reason for cracks may be that the hardening temperature is sometimes not set correctly.

The final cause-and-effect diagram resembles the skeleton of a fish and is often called a "fishbone" chart. Its main function is to facilitate focused brainstorming, argument, and discussion among people about what might cause the cracks. Often, some of the possibilities can be eliminated right away simply by talking them over. Usually, there will be a number of things left and it will be unknown which items cause which problems. Using data coming from the process can often help sort out the different possibilities with the seven tools.

For example, suppose that the cracks have been sorted by size. The histogram, shown in Figure 6.3B, can then be used to show how many cracks of different sizes occurred. This is of great value for conveniently

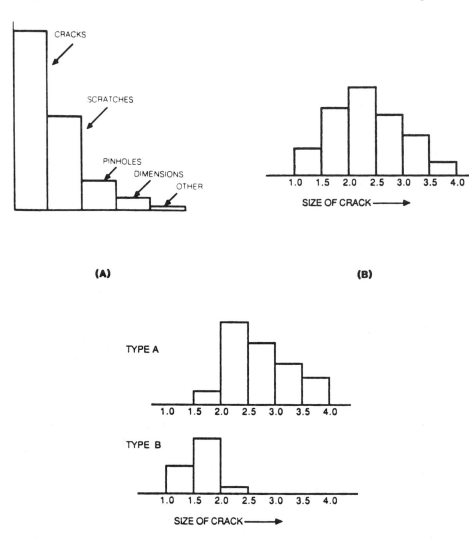

Figure 6.3 (A) Pareto diagram; (B) histogram of defects; (C) stratification of defect data.

summarizing the crack-size problem. But much more can be learned by splitting, or stratifying, the histogram in various ways. For example, the fact that there are two types of balls provides a chance for stratification. In Figure 6.3C the original histogram has been split into two, one for balls of type A and one for those in type B. It is clear not only that most of the cracks are in balls of type A, but also that the size of the type A cracks is on the average larger and that the spread is also larger. Now the question is: What's special about type A balls? If we know who inspected what, the data could also be stratified by inspector.

Scatter diagrams can also be helpful. If hardening temperature varies significantly, the crack size might be plotted against the hardening temperature. A control chart recording the oven temperature can tell how stable the temperature is over time and whether a higher frequency of cracks is associated with any unusual patterns in the temperature recording.

Clearly, these seven tools are based on common sense. By helping people to see how often things happen, when they happen, where they happen, and when they are different, the work force can be trained to be quality detectives producing never-ending improvements.

Informed observation is not enough in itself, because no amount of fine-tuning can overcome fundamental flaws in products and processes due to poor design. Directed experimentation is needed to ensure that we have excellently designed products and processes to begin with.

In the past, quality control was thought to concern only the downstream side of the process, with an emphasis on control charts and inspection schemes. Today the emphasis is moving upstream, focusing attention on getting a product and a process that are well designed so that seldom is anything of unsatisfactory quality produced. This reduces the need for inspection and provides economic gains. This push for moving upstream, strongly recommended by pearson (1935), is embodied in the Japanese approach. Direct experimentation is needed to reach this goal.

Statistical experimental design was invented in teh early 1920s by R. A. Fisher in England. In her biography, Joan Fisher Box (1978) describes how in 1919 R. A. Risher went to work at a small and at that time, not very well known agricultural research station near London called Rothamsted. The workers at Rothamsted were interested in finding out the best way to grow wheat, potatoes, barley, and other crops. They could have conducted their experiments in a greenhouse, carefully controlling the temperature and humidity, making artificial soil that was uniform, and keeping out the birds and insects. By doing so they would have produced results that applied to plants grown under very artificial conditions. However, the results would probably have been entirely useless in deciding what would happen on a farmer's field.

The question, then, was how to run experiments under the imperfectly controlled conditions of the real world. Fisher showed how to do it, and his ideas were quickly adopted worldwide. He showed how experimentation could be moved out of the laboratory. His methods produced relevant, practical findings in many important fields, including medicine, education, and biology. In particular, his ideas were suitably developed for the industrial setting, and many statistically designed experiments have been run in industry based on his original concepts. Statistical design is a potent tool; it is almost invariably successful.

Consider a specific example of designed experiments. Suppose that we want to redesign the gold balls mentioned earlier to eliminate the problem with cracks. The temperature of the rubber before the quenching stage are important factors. Thus the effect on the number of cracks of varying the three factors will be studied: temperature of the rubber before quenching, content of fillers, and the temperature of quenching medium.

Traditional approaches recommend that the scientifically correct way to conduct an experiment is to vary just one factor at a time, holding everything else fixed. Suppose we do that in this situation, starting with the rubber temperature. Since we want reasonably reliable estimates of the effects of the factors, we run four experiments at T_1' rubber and four experiments at T_2' rubber, fixing the filler content to a and using T_1F quenching oil. Suppose that using an accelerated life test the percentage of balls without cracks is as follows:

T_1'	T_2'	Percent difference
72	78	6
70	77	7
75	78	3
77	81	4
Average difference		5

The average difference in the number of cracks between $T_1'°F$ and $T_2'°F$ is 5%. From this it might be concluded that T_2' °F is the better temperature. But is it? Remember, we fixed the carbon content at a% and used T_t °F quenching oil. After these eight trials all that can really be said is that it appears to be better to use the high rubber temperature if the filler con-

Quality in Product Development

tent is $a\%$ and the oil is $T_1°F$. If somebody asks whether changing the rubber temperature would produce the same reduction in cracks with $b\%$ filler or with $T_2°F$ oil, the honest answer is that we don't know. To study the effect of changing the carbon content in the same way will require an additional eight runs. After that, all we could say is that for the particular fixed choice of steel temperature and oil temperature, there is a certain change in response when the filler content is charged. The same difficulty would apply to a third set of eight runs meant to determine the effect of changing the quenching temperature. Thus, after 24 experiments, all we would know would be the effect of each variable at one particular combination of settings of the other two. Additional experiments would be needed to find out more.

Fisher proposed the concept that it was much better to vary all the factors simultaneously in what is called a factorial design. Using such a design, we would run just one set of eight experimental trials to test all three variables, as shown in Table 6.4. One way to interpret the results of such an experiment is to display the data at the corners of a cube, as illustrated in Figure 6.4.

The trials with low rubber temperature are shown on the left side of the cube, and the trials with the high rubber temperature on the right. Low filter trials are on the bottom, and high filler trials on the top. Trials with $T_1°F$ oil are on the front and those run with $T_2°F$ oil on the back. For example, for the third trial in Table 6.4, the rubber tempreature was low (T'_1), filler was high ($b\%$), and the quench temperature was low (T''_1). That takes us to the left top front corner of the cube, where the proportion of balls without cracks is 61%.

Suppose that we want to find the effect of rubber temperature. We assess the effect of changing rubber tempreature by examining the pairs of readings in Figure 6.4 (67,79), (59,90), (61,75), and (52,87). In each pair, one of the readings reflects an experiment run at high rubber temperature and the other experiment at low rubber temperature—but within each pair the other factors remain constant. Therefore, as with the one-at-a-time experiment, four comparisons allow us to calculate the effect of changing only rubber temperature:

$$
\begin{aligned}
79 - 67 &= 12 \\
75 - 61 &= 14 \\
90 - 59 &= 31 \\
87 - 52 &= \underline{35} \\
\text{Average} \quad &+23
\end{aligned}
$$

Table 6.4 Example of Experimental Design

Run	T Rubber Temp.	F Filler Content	Q Quench Temp.	Balls without cracks	Day Run
1	T_1 F	a%	T_1 F	67%	1
2	T_2 F	a%	T_1 F	79%	2
3	T_1 F	b%	T_1 F	61%	2
4	T_2 F	b%	T_1 F	75%	1
5	T_1 F	a%	T_2 F	59%	2
6	T_2 F	a%	T_2 F	90%	1
7	T_1 F	b%	T_2 F	52%	1
8	T_2 F	b%	T_2 F	87%	2

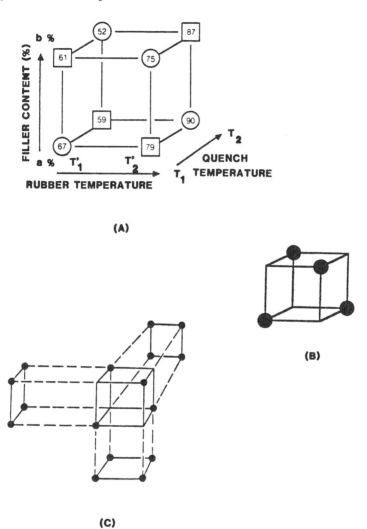

Figure 6.4 (A) Two-level, three-factor designed experiment; (B) half-fraction of an eight-run factorial design; (C) projections of a fractional factorial in two dimensions.

Comparing the four corresponding pairs of numbers along the top-bottom axis and four pairs on the back-front axis permits the same calculations for filler content and quench temperature. This gives an average carbon effect of -5.0 and an average quench temperature effect of $+1.5$. Another way to calculate the same quantities would be to take the average of the four numbers on the left face and subtract that from the average on the right face. Similarly, to calculate the effect of filler, subtract the average of the four numbers on teh bottom face from the average of the four numbers on the top. The difference between them is the effect of filler. Thus, using a factorial design in only eight runs allows us to test each of the three factors with the same precision as in a one-factor-at-a-time arrangement that has three times the number of runs!

To this point we have discussed quantities that represent the averages or main effects of the single factors. If they are denoted by the letters T, F, and Q, we find from the data:

> T (rubber temperature) $+23.0$
> F (filler) -5.0
> Q (quench temperature) $+1.5$

Even more information can be extracted. Examining the individual rubber temperature effect differences again—12%, 14%, 31%, and 35%—the cube in Figure 6.4 shows that the first two numbers represent contrasts on the front of the cube where $T_1°F$ oil was used, while the last two came from the pairs on the back of the cube where $T_2°F$ quenching was used. This indicates that the effect of rubber temperature is different depending on whether $T_1°F$ or $T_2°F$ quenching oil temperature is used. Therefore, rubber temperature (T) and quenching temperature (Q) interact. This $T \times Q$ interaction is calculated as follows:

> Rubber temp. effect with $T_2°F$ oil $= (31 + 35)/2 = 33$
> Rubber temp. effect with $T_1°F$ oil $= (12 + 14)/2 = 13$
> —————————————————
> Difference   20

$T \times Q$ interaction effect $= 20/2 = 10$

Similar calculations can be done to produce the $T \times F$ and $Q \times F$ interactions: $T = 23.0, F = -5.0, Q = 1.5, TF = 1.5, TQ = 1.0$, and $QF = 10.0$. This single eight-run design has not only helped estimate all the main effects with maximum precision but has also helped determine the three interactions between pairs of factors. This would have been virtually impossible with the one-factor-at-a-time design.

The practical engineer might, with reason, have reservations about process experimentation. Processes do not always remain stable or in a

Quality in Product Development

state of statistical control. For example, the quenching media used for quenching deteriorates with use. Rubber may have other properties that vary from batch to batch. The engineer might therefore conclude that the process is too complex for experimentation to be useful. On the other hand, it is important to remember that Fischer conducted his agricultural experiments in an extremely complex, "noisy" environment. The soil was different from one corner of the fluid to the other, some parts of the field were in shadow while others were better exposed to the sun, and some areas were wetter than others. Sometimes birds would eat the seeds in one part of the field. All such influences could bias the experiment results. Fisher used *blocking* to eliminate the effect of inhomogeneity in the experimental material and randomization to avoid confounding with unknown factors.

To understand blocking, suppose in the experiment above that only four experimental runs could be made in one day. The experiments would then have to be run on two different days; any day-to-day differences could bias the results. But suppose that we ran the eight runs in the manner shown in Figure 6.4, where the data shown in circles were obtained on day 1 and those in squares obtained on day 2. There are two runs on the left side and two runs on the right side of the cube that were made in the first day. Similarly, there are two runs on the left side of the cube and two runs on the right side of the cube that were made on the second day. With this balanced arrangement, any systematic difference between days will cancel out when the temperature effect is computed. A similar balance occurs for the other two main effects and even for the interactions.

The beauty of Fisher's method of blocking is that inhomogeneities such as day-to-day, machine-to-machine, batch-to-batch, and shift-to-shift differences can be balanced out and eliminated. Effects are determined with much greater precision. Without blocking, important effects could be missed, or it would take many more experiments to find them.

Blocking almost seems to permit experiments to be run in a nonstationary environment at no cost. However, this is not quite true. From an eight-run experiment one can calculate at most eight quantities: the mean of all the runs, three main effects, three two-factor interactions, and a three-factor interaction. Using this arrangement, one can associate the block difference with the contrast corresponding to the three-factor interaction. Thus these two effects are deliberately mixed up; in other words, the block effect and the three-factor interaction are confounded. In many examples, however, the three-factor interaction will be unimportant.

What about unsuspected trends and patterns that happen within a block of experimental runs? Fisher's concept was to assign the individual treatment combinations in random order within the blocks. In the example above, the four runs made on day 1 would be run in random order, and the four runs made on day 2 would also be run in a random order. In this way the experimenter can avoid the possibility of unsuspected patterns biasing the results.

If several parameters (factors) are to be examined, design might require a prohibitively large number of runs. As an example, to test eight factors each at two levels, a full factorial design would require 2^8 or 256 experiments. It is important to note, however, that significant conclusions can be drawn from experimental designs that employ only a carefully chosen piece (or fraction) of the full factorial design. This is known as the process of fractionation.

Fractionation can be illustrated by starting with Figure 6.4A. Instead of the full eight-run design, suppose only that the four runs marked with circles are run. This arrangement is shown in Figure 6.4B. Imagine a bright light shone along the ball temperature axis. A complete square of four points would be projected for the factors filler and quench temperature. Figure 6.4C shows that a similar phenomenon occurs if the light is shone down the "filler" axis or the "quench temperature" axis. Suppose now that we wanted to test three factors, a Pareto effect was expected; it was fairly certain that, at most, two factors would be important enough to be concerned about. If just the four runs shown in Figure 6.4B are run, we would have a complete 2^2 design in whichever two factors turned out to be important.

The idea can be extended to a much larger number of factors. Table 6.5 is a design that can be used to test seven factors, A, B, C, ..., G, in

Table 6.5 Eight-Run Two-Level Fractional Factorial Experiment for Varying Seven Factors

	Factors						
Run	A	B	C	D	E	F	G
1	−	−	+	−	+	+	−
2	−	−	+	+	−	−	+
3	−	+	−	−	+	−	+
4	−	+	−	+	−	+	−
5	+	−	−	−	−	+	+
6	+	−	−	+	+	−	−
7	+	+	+	−	−	−	−
8	+	+	+	+	+	+	+

only eight runs. The high level of a factor is indicated by (+) and the low level by (−). If any two columns are chosen, there will be a complete set of factor combinations (− −), (+ −), (− +), (+ +) repeated twice in those factors. Thus if the engineer believes that there will be a strong Pareto effect, so that no more than two factors of the seven tested will be important, he can use the design to assure that whichever two turn out to be influential, he will have them covered with a completely duplicated factorial design.

As an even more complex example, Table 6.6 shows an experimental arrangement involving 16 runs and 15 factors. This is a fully saturated fractional factorial. If the experimenter thinks that not more than two factors will be important, these 16 runs can be used to screen as many as 15 factors each at two levels, denoted by + and −, using all 15 columns to accommodate factors. For illustration, consider an experiment by Quinlan (1985) involving quality improvement of speedometer cables. In this analysis, the speedometer cables can sometimes be noisy. One cause is shrinkage in the plastic casing material. Hence an experiment was done to find out what caused shrinkage. The engineers started with 15 different factors—linear material, liner line speed, wire braid type, braid tension, wire diameter, and so on—labeled A through O. They suspected that only a few of the factors listed would be important, so they decided to test 15 factors in 16 runs. For each run, 3000 ft of plastic casing was produced and four sample specimens were cut out and the percentage shrinkage measured. The average shrinkage is tabulated on the right of Table 6.6.

Fifteen effects can be computed from these 16 numbers. A simple way to find out which of the 15 effects really are active is to plot the effects in order of magnitude along the horizontal scale as an ordinary dot diagram. The points are then referred to their respective order of magnitude along with vertical scale of a normal probability plot. To interpret this plot, it is important to realize that effects due to random noise plot on this paper as a straight line, so that points falling off the line are likely to be due to real effects. In this case only factors E and G, the type of wire braid and the wire diameter, turn out to be important. This is a typical example of Juran's vital few (E and G) and trivial many.

If main effects and higher-order effects exist, it will be impossible to tell from a highly fractionated experiment whether an observed effect is due to a main effect or an interaction. The effects are confounded with each other. Nevertheless, fractional factorials often work very well. There are three different rationales for this fact. One is that the experimenter can guess in advance which interaction effects are likely to be active and which are likely to be inert. The third rationale is based on th Pareto ideas and the projective properties of fractionals. The most appropriate use of fractional designs is for purposes of screening when it is expected that

Table 6.6 Example of Sppedometer Cable Noise

Run	H	D	-L	B	-J	-F	N	A	-I	-E	M	-C	K	G	-O	y
1	−	−	+	−	+	+	−	−	+	+	−	+	−	−	+	0.4850
2	+	−	−	−	−	+	+	−	−	+	+	+	+	−	−	0.5750
3	−	+	−	−	+	−	+	−	+	−	+	+	−	+	−	0.0875
4	+	+	+	+	−	−	+	−	−	+	−	+	+	+	+	0.1750
5	−	−	+	+	−	−	−	−	+	+	+	−	+	+	−	0.1950
6	+	−	−	+	+	−	−	−	+	−	−	−	−	+	+	0.1450
7	−	+	−	+	−	+	+	−	+	−	+	−	+	−	+	0.2250
8	+	+	+	−	+	+	−	−	−	−	+	−	−	−	−	0.1750
9	−	−	+	+	+	+	+	+	+	−	+	−	+	+	−	0.1250
10	+	+	−	−	−	−	−	+	−	+	−	−	+	+	+	0.1200
11	−	−	−	−	−	−	−	+	+	+	−	−	+	−	+	0.4550
12	+	+	+	+	−	−	−	+	+	−	+	+	−	−	−	0.5350
13	−	−	+	+	+	−	−	+	−	−	−	+	−	−	+	0.1700
14	+	−	−	+	+	+	−	+	+	+	−	+	+	−	−	0.2750
15	−	+	−	+	−	−	+	+	−	−	−	+	−	+	−	0.3425
16	+	+	+	+	+	+	+	+	+	+	+	+	+	+	+	0.5835

Source: Quinland (1985).

only a few factors will be active. It may be necessary to combine them with follow-up experiments to resolve any ambiguities. Highly fractionated designs should not be used at later stages of experimentation, when the important factors are known.

The objective of experimentation is often the attainment of optimum conditions—for example, achieving a high mean, low variance, or both, of some quality characteristic or property. One method for solving optimum condition problems involves two phases, as described by Box and Wilson (1951). In the first phase the experiment progresses from relatively poor conditions to better ones, guided by two-level fractional factorial designs. Phase 2 begins when nearly optimal conditions have been found. It involves a more detailed study of the optimum and its immediate surroundings. The study in phse 2 tends to disclose ridge systems, much like the ridge of a mountain that can be exploited, for example, to achieve high quality at a lower cost. In phsae 2, curvature and interactions become important. To estimate these quantities, response surface designs were developed that are often assembled sequentially using two-level factorial or fractional factorial designs plus *star points*. These composite designs are not of the standard factorial type, and for quantiative variables, have advantages over three-level factorials and fractions.

When the initial conditions for the experiments are poor, progress can be made in phase 1 with simple two-level fractional factorial designs. The idea of sequential assembly of designs—developing the design as the need is shown—can also be used with fractions. Thus, if after running a fractional design the results are ambiguous, a second fraction can be run selected to resolve those ambiguities.

The Japanese have been successful in building quality into their products and processes upstream using fractional factorial designs and other orthogonal arrays. In particular, Genichi Taguchi emphasizes the importance of using such experimental design in:

Minimizing variation with mean or target
Making products robust to environmental conditions
Making products insensitive to component variation
Life testing

The first three categories are examples of what Taguchi calls parameter design. He also promotes novel statistical methods for analyzing the data from such experiments using signal-to-noise ratios, accumulation analysis, minute analysis, and unusual application of the analysis of variance (ANOVA) discussed in Chapter 8.

Genichi Taguchi is a Japanese engineer who has been active in the improvement of Japan's industrial products and processes since the late

1940s. He has developed a philosophy and a methodology for the process of quality improvement that depend on statistical concepts and tools, especially statistically designed experiments. He wrote a very popular Japanese experimental design text. Many Japanese firms have achived great success by applying his methods. Taguchi has received some of Japan's most prestigious awards for quality achievement, including the Deming Prize.

Although he has given seminars in the United States since at least the early 1970s, it was not until about 1980 that American companies (including AT&T, Ford, and Xerox) began applying his ideas actively. In cooperation with some of Ford's quality organizations, he set up the American Supplier Institute to provide training in statistical methods mainly to Ford and its suppliers.

The heart of Taguchi's approach has to do with a conceptual framework for the process of quality improvement. Taguchi's ideas can be distilled into two fundamental concepts:

1. Quality losses must be defined as deviation from target, not conformance to arbitrary specifications; quality losses must be measured by systems-wide costs—"loss to society"—not local costs at points of defect detection.
2. Achieving high system quality levels economically requires that quality be designed in. Quality cannot be achieved economically through inspection and product screening; the three stages of quality by design are: (a) systems (functional) design; (b) parameter (targeting) design; and (c) allowance (tolerance) design. Systems and parameter design provide the greatest opportunity for quality cost reduction; allowance design oftne entails increased costs to achieve higher quality.

Taguchi builds both a conceptual framework and specific methodology for implementation from the precepts. Quality deteriorates, of course, when a product deviates from its design target and that costs of rework or scrap increase as a product moves through the manufacturing system. This is a simple interpretation of item 1 above. Similarly, few would doubt that designing in quality is cheaper than trying to inspect and reengineer it in after a poor design arrives at the manufacturing floor—or worse yet, after it gets to the customer. This is part of what is meant by item 2 above. Unfortunately, these simple interpretations do not fully capture what Taguchi means in his statements; indeed, some of what Taguchi says contradicts conventional quality control in U.S. industry.

At the heart of Taguchi's philosophy is the concept of the quality loss function [see Sullivan (1984)]. This concept is fundamental to an un-

Quality in Product Development

derstanding of what the methodology tries to do and why it is so successful at doing it. The idea can best be explained by comparison to conventional quality thinking. Figure 6.5 is a graph of the loss function implicitly associated with the idea of being in or out of specifications. The central target, T, represents the ideal level of the design parameter in question as determined by the designer. The specification limits, denoted by $T + S$ and $T - S$, are standard symmetric specification limits; the ideas work equally well for nonsymmetric limits.

On the ordinate is a measure of "value lost" due to deviation of the characteristic from desired levels. The exact way in which this lost value is measured is not important. A percentage scale varying from 0 (no value lost) to 100% (all value lost) is shown. The definition chosen to define this value is important. On this point Taguchi (Taguchi and Wu, 1980) is quite clear: Value should be defined in terms of "the losses a product imparts to the society from the time a product is shipped." This definition clearly emphasizes the customer-oriented emphasis that Taguchi places on quality. This should be contrasted with the producer-oriented definition of quality that is inherent in most cost-of-quality systems in which costs of scrap, rework, and warranty repair are the chief measures of quality performance. As such, quality is defined in terms of the entire system in which the product is manufactured and used, not just in terms of the accounting practices at particular points in the system at which the quality characteristic is being measured. This means that not only must the costs of rework or scrap at various stages of the manufacturing process be considered, but so should the costs of lost productivity due to inefficiencies caused by variation. These include costs of increased maintenance, downtime due to equipped breakdown, excess inventory, excess personnel and paperwork, and time wasted in meetings and unproductive dis-

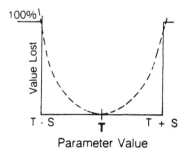

Figure 6.5 Loss function.

cussions with suppliers, dealers, and distributors. Worst of all are the costs to customers incurred because of degraded product performance, reliability, and durability.

The solid line graph in Figure 6.5 indicates what the in/out-of-specs duality defines as the loss. As long as quality is within specifications, everything is fine and no value is lost. As soon as the value ranges out of specifications, the product is not salable, all value is lost, and th eproduct must be either scrapped or reworked. This, of course, is a very gray area often steeped in controversy.

The trouble with this view of quality cost accounting is that it is unrealistic. Specification limits can certainly not be dispensed with. Accountants need a simple system to keep track of production costs, manufacturing efficiencies, material usages, and so on. They therefore must be able to quantify quality measurements by assigning results to one side or the other of their ledger sheets. On the other hand, nature does not work within such rigidness. There is no abrupt change from perfect to useless as an arbitrary boundary is crossed. What really happens is that performance gradually deteriorates as the quality measure deviates farther and farther from the intended target. Indeed, the deterioration continues as one moves beyond the specification limits: Greater than "100%" losses can occur when the degraded product can cause losses beyond the totality of its own value (e.g., when poor performance of one component of an assembly results in the destruction of the whole).

The appropriate loss function paradigm is therefore not the "is it defective or not?" step function, but the smoothly varying dashed curve shwon in Figure 6.5. The exact shape of such a function is usually difficult or impossible to determine. However, in practice, good approximations often exist.

One consequence of these concepts is that achieving zero defects may not be good enough. Figure 6.6A and B give histograms of the distributions of product received from two suppliers. The numbers in each bar of the histogram represent the percentage of product falling within that bar. Supplier A's product all falls within the specifications, while supplier B produces about 4% that is just outside. On the other hand, the distribution of supplier A's product is not centered on the design target, T, and is considerably more variable, on the whole, than B's. It is, in fact, what one typically receives from a 100% inspection screen.

Now consider what the losses are in each case. Figure 6.6C represents a typical loss function that often is a reasonable practical approximation—quadratic loss. This means that the loss incurred when a product parameter deviates from target is proportional to the square of the dis-

Quality in Product Development

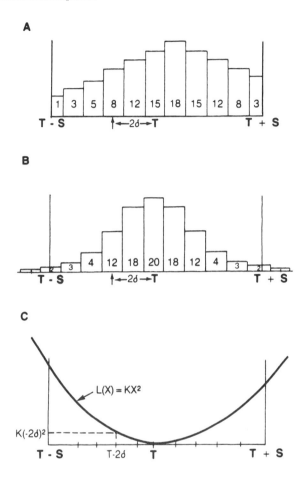

Figure 6.6 Why zero defects is not a practical goal.

tance that the parameter is from target. Algebraically, this can be written as

loss when parameter is distance, D, from target = KD^2

For example, for the cell indicated in the figure, which is centered at a distance 2δ from T, the approximate loss from the 8% (= 0.08) of supplier A's product that it contains is

$$0.08K(2\delta)^2 = 0.32K^2$$

For supplier B, on the other hand, the loss would be

$$0.12K(2\delta)^2 = 0.48K^2$$

If this computation is carried out for all the cells in the figure, it is easy to see that the total loss due to supplier A's product is approximately

$$K\delta^2(0.25 + 0.48 + 0.45 + 0.32 + 0.12 + 0.18 + 0.60 + 0.72 + 1.28 + 0.75) = 5.15K\delta^2$$

The loss for B is approximately

$$K\delta^2(0.36 + 0.50 + 0.48 + 0.36 + 0.48 + 0.18) \times 2 = 4.72K\delta^2$$

The loss due to supplier A is about 9% greater than the loss from B, even though supplier A supplies zero defects while supplier B supplies about 4% defective! Moreover, supplier A may have to bear the additional costs of 100% inspection, while supplier B does not.

When the importance of the loss function concept is fully grasped, it is clear why go/no-go definitions of quality are inadequate. Key quality characteristics must be quantified and measured; quality improvement means that the process/product must be kept on target while making ongoing efforts to reduce the measured variability.

Gauss used the concept of the loss function to develop the theory of least squares in the early nineteenth century, and it has been central to that branch of statistics known as decision theory since the 1940s. Nevertheless, its crucial importance to the management of quality and productivity in industry had not previously been emphasized. The Taguchi method point of view is that it provides both cost justification and a scientific basis for ongoing quality improvement. In addition, it lays the basis for the second of his two fundamentals.

Taguchi's second fundamental concept rests on the importance of achieving high quality economically (i.e., consistency of functional performance). The challenge for the design engineer is to develop design approaches that can meet both cost and quality criteria. Taguchi calls this "systems design"; it is fundamental to achieving quality within the cost constraints of a competitive marketplace. By referring to the concept of the loss function, insight can be gained into how this can be achieved.

Quality in Product Development

The design's consistent ability to achieve its quality, reliability, and durability objectives under these variations is related directly to the concept of the loss function. Variability is ever present in all stages of manufacturing and use. Variability tends to degrade performance, which results in a measurable loss (to society). In certain phases of manufacturing and use, measures can be taken to combat variability. Statistical process control (SPC) is one powerful approach to reducing variability within the procurement/production phase. This is almost always a cost-saving and quality-enhancing approach and is practiced in companies producing world-class quality products. SPC reduces the quality and productivity costs due to assignable causes. It is important to note, however, that SPC may not be enough. The process capability may need to be improved, or more costly process control schemes may be required. These costly avenues can be avoided by making the design insensitive to the variability that would otherwise require these actions to reduce. This is the idea of *robust design,* where Taguchi concentrates his efforts.

The effectiveness of statistical approaches in achieving robust designs, which is usually what is meant by Taguchi methods, follows from the following engineering insight. Taguchi divides the design cycle into three phases (summarized in Figure 6.7):

1. *Systems design.* The fundamental design and engineering concept is established.
2. *Parameter design.* The design nominals—target dimensions, material properties, voltages, and so on—are established.
3. *Allowance (or tolerance) design.* The design tolerances are established.

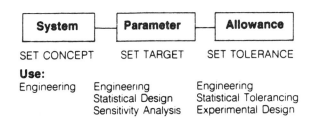

Figure 6.7 Three phases of a design cycle.

Systems design requires engineering expertise; becasue it is a concept-oriented activity, there is little or not role for statistical methods. But in both parameter design and allowance design, the engineer can make decisive use of statisitcal methods to improve design robustness.

With allowance design the principles of classical statistical methods are employed. In embarking on an experimental program the only effective way to conduct the experiments and analyze the results is through the use of multivariable statistical design. Changing only one variable at a time will, as already discussed, almost always fail to give a true and precise picture of what occurs. The reason for this is simple: We are studying a complex system in which the interactions may be important. We cannot learn about interactions unless more than one variable is changed simultaneously.

If interactions were important, then when experimentation is done by changing only one variable at a time, cost will probably be higher and quality lower than if properly conducted statistically designed experimentation has been used. Statistical techniques such as the use of two-level fractional factorial designs, nested designs, variance components analysis, and statistical tolerancing help obtain the needed information. Moreover, these techniques have been applied in complex multivariable studies under research lab, pilot plant, and manufacturing conditions. They have even proven to be of importance in computer simulations.

Allowance design is important, but it may indicate that very tight tolerances must be maintained. Tightening tolerances usually means increasing costs. To get both good performance and good economics, Taguchi emphasizes the importance of the robust design.

Figure 6.8 illustrates the idea of the robust design. The horizontal axis illustrates the setting of some crucial design parameter or combinations of parameters. The vertical axis indicates a performance characteristic. To reduce the effect of any small error in the design parameter on the x-axis or the performance criterion on the y-axis, it would clearly be better to set the nominal at B rather than A. Much more variation in the design parameter could be tolerated without the danger of excessive change in the product's performance. This reflects the nonlinearity of the performance in response to the parameter or parameter combination on the x-axis. Unfortunately, the design will not perform as intended at B, so it appears that this cannot be done. Taguchi then advances a critical hypothesis. Although this is not universally true, there is often a duality among the variables affecting design performance as illustrated in Figure 6.9. Signal variables are design variables that affect performance level linearly over the reasonable range at which they could be set. Control variables, on the other hand, tend to have a nonlinear effect and therefore can be used to

Quality in Product Development

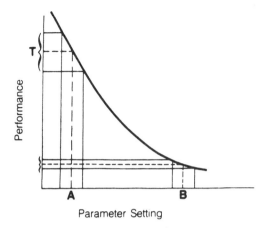

Figure 6.8 Effect of nonlinearity on design robustness.

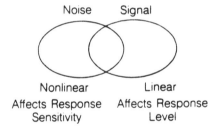

Figure 6.9 Signal-to-noise factor duality.

control the design's sensitivity to various types of noise. When this duality exists, it is usually possible to set the control variables to achieve design robustness—insensitivity to variation in manufacturing or use—and then move back on target by manipulating the signal variables. This two-stage optimization heuristic is at the heart of Taguchi's statistical approaches. This of course can be done in the design of the product; however, it i snot the only approach that Taguchi discusses to improve design robustness. Variations in external (environmental) variables may also influence product performance. Variations in temperature and humidity may affect the yield of chemical processes or the degradation of a polymer. Unlike the variations in the design parameters described above, these sources of variation are not always controllable. However, Taguchi again advocates the use of multivariable experimentation to simulate this variability and, by using appropriate statistical optimization methods, to find settings of controllable design parameters that minimize product sensitivity to this variation.

The Taguchi method is essentially a sensitiivty analysis. Although this is an idea that has been around for a long time, Taguchi has proposed some specific, novel technical approaches. Unfortunately, one of those that is widely advertised—the use of "inner and outer arrays"—generally appears to require so much experimentation that it is impracticable.

That approach seems impracticable because the method advocates the use of large "one-shot" experiments to acquire all needed information. This inevitably results in wasted experimental effort, because the "trivial many" unimportant variables must be included in th experiment, along with the "vital few" important ones (if we knew which was which a priori, we would not have to experiment). Fortunately, this problem has long been recognized by practicing statisticians. The key to dealing with it is to use sequential experimental strategies. As a result:

Minimal effort is spent studying variables that are ultimately found to be unimportant.

Key variables and promising experimental regions can be studied in detail without unreasonable costs.

Box et al. (1978) provides an overview of how such strategies can work. They make the recommendation. "As a rough general rule, not more than one quarter of the experimental effort (budget) should be invested in a first design."

Another aspect of such sensitivity analyses long emphasized by practicing statisticians but largely ignored in Taguchi's methods is the importance of interactions. Figure 6.10 is a plot showing the effect of a two-factor interaction of variables A and B. At low levels of B, small variations in A will have a large effect on functional performance; at high levels of B, A does not have much effect. Clearly, then, controllable parameters such as B, which combine with many other design or external variables to produce important interaction effects, are strong candidates for manipulation to reduce design sensitivity.

Unfortunately, many of the designs that Taguchi proposes—the three-plust level orthogonal arrays, in particular—make it hard to identify such interactions. This has been kn'own for 40 years, and it was precisely for this reason that George Box (1986) developed the approach known as response surface methodology (RSM) in the late 1950s and early 1960s.

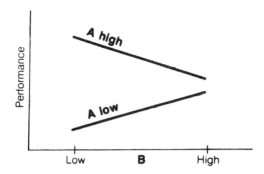

Figure 6.10 Application of interaction to reduce sensitivity.

Implementation of Taguchi methods requires the appropriate use of statistical design and analysis methods. There is a fair amount of controversy over the meaning of "appropriate." However, in most respects the method is neither new nor radical; there is more similarity than disagreement between the Taguchi method and standard statistical approaches. For example, Taguchi advocates the use of multivariable statistical design methods as the only sensible way to study the simultaneous effects of many variables. He rejects as ineffective the experimental practice of varying first one variable while holding others constant to find an "optimum," then varying a second while others are controlled to find its "optimum," and so on. Taguchi also advocates the application of structured statistical analysis procedures rather than unstructured perusal of the data for analysis.

Unfortunately, the TM methods for analysis are nonstandard, and no scientific (i.e., mathematical) justification has yet been provided for their use. This has led, inevitably, to controversy, including those areas mentioned above. Others include the use of signal-to-noise ratios, accumulation analysis, and ANOVAs.

CLOSURE

We have examined the principal ideas important to establishing quality as a key element in the design of a product and in producing and delivering the product to customers. The suggested references given at the end of the chapter will enable the reader to obtain more in-depth discussions. From the standpoint of product testing, we end this chapter by asking four basic questions:

1. What are we testing?
2. Which properties should be measured?
3. How should test(s) be made?
4. What are the problems in testing, and how can they be overcome?

One group of applicable answers to question 1 are quality control, design and scale-up data/information, prediction of service performance, investigating faults and part failures, and assessing the composition of new polymers. These reasons can apply to any material or product; however, it is important to stress that the requirements for each of these reasons is different. It is not logical to address questions 2 through 4 until the reason for testing has been clearly understood. Although this seems to be an obvious point, quite often there is a loss in communication between technology and operating departments. For example, the researcher and production quality control engineer are, more often than not, limited by a lack of appreciation of why the other person is testing and what information is needed from testing.

The first question leads one to two additional queries: "Why are we testing?" and "What must we obtain from the results?" This implies that whatever the reason for testing, the answers to which properties and techniques will depend to some extent on the type and level of information required. One can readily envisage that the same reason for testing may exist but there may be two different levels of need (e.g., in quantity for observing trends only, or in precision of data).

As an illustration let's consider the problem discussed earlier, pertaining to the manufacture of automotive weatherstrip seals from an elastomer such as EPDM. This is an extrusion application of an article that can have a solid and a sponge profile fused together. Figure 6.11 illustrates a simplified profile for this application. What properties of the extrusion operation and postcuring in, say, a microwave train oven do people consider? Certainly, the appearance, possibly the glossiness and staining, and the resilience and weatherability of the article are critical. These product performance parameters can depend on both the operating ranges of the equipment and the polymer feedstock properties. For example, appearance as well as resilience and heat-aging properties are a function of the relative scorchiness of the polymer (in the case of EPDM, its diene level, diene compositional distribution, and diene type), the die design and extruder operating conditions (i.e., wall and screw temperature profiles) and the temperature and residence time in the continuous vulcanization media. Despite not consciously thinking of such performance parameters, it would be reasonable to assume that over 6 months of use, the system will have been adequately assessed for service performance.

Quality in Product Development

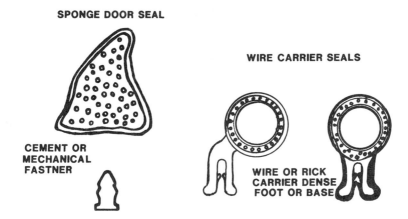

Figure 6.11   Typical sponge weatherstrip profile.

An issue that most often arises is whether to test the final product or to test the feedstock material itself. If we consider the extrusion line from the viewpoint of testing for fitness for purpose, it is clearly reasonable to do testing on the final product. If, on the other hand, the manufacturing train has been adequately proven, and principles of statistical process control properly applied, testing may be shifted to the properties of the neat polymer.

In this example we do not intend to consider the properties or methods in detail, but merely to raise questions of principle, which should be addressed prior to the implementation of any test procedure or quality control program. On the one hand, we have a product that can be tested in finished form and one can readily list the properties of importance for the intended application. One the other, it may be more cost-effective simply to test the feedstock, provided that we can convince ourselves that data generated from the neat polymer adequately reflect the productivity of the process and the finished article's performance. Both of these programs or either one enables us to assess further the entire process for fitness for purpose. The question then becomes: What are the problems? As is often the case, there is not always universal agreement as to the specific test methods to be used, despite some detailed guidelines by ASTM or ISO. In the world of manufacturing and technology support groups for product lines there is a more serious problem related to the human factor—the far-too-lengthy delay in getting together and defining what tests should be used—than to the technical hurdles to designing actual tests or even products.

There is, of course, another problem in the establishment of new manufacturing lines—that the process setup is expected to last for many years and the process qualification tests are short term. Can we apply the rules of extrapolation with confidence? The answer to this constantly occurring question is often so, although many processes can be defined from well-designed performance tests to provide confidence over a reasonable service life.

Consider a second illustration where a polymer is used for jacketing insulation in high-voltage cable applications. The integrity of the cable must be capable of handling many years of service—else a rather costly maintenance situation arises as well as a serious safety issue. In this case the properties of importance are well defined: insulation level, no flaws, adequate tear resistance or strength, adequate deformation resistance, and low emissivity. The test methods and the quality assurance schedules are well established in standards, with the most significant being 100% testing for electrical breakdown strength and current leakage. Initially, we might say that there are no problems in such a test application since the properties are known and the test methods are well defined. However, the tests are realtively slow and expensive. Ideally, quality-control tests should be rapid as well as adequately screening defective production. This in itself implies that there is room for the development of instrumentation and test methods to provide more cost-effective product quality control testing. For a less-safety-critical product one approach would be to accelerate the method under more severe test conditions; however, there would have to be irrefutable evidence that any faster but less directly relevant service test was adequate. At any rate, it would be beneficial to have a test that could be used in the field immediately before use.

In the two examples, we have outlined some of the reasons why tests are conducted as well as some of the problems encountered. These are, however, relatively simple cases in which we are considering only one aspect of, or reason for, testing in each case. In real-world applications, there are inevitably more problems and issues that increase the challenge of implementing quality control practices. Despite this, we can organize the reasons for testing in all types of product performance problems if we view them in relation to the role of standards; that is:

Quality control: a quality control specification and a quality control schedule

Design data: for example, codes of practice for developing and for applying and using data

Prediction of service performance: a performance specification

Identifying/investigating failures: application of quality assurance and performance specifications

Standards can play a role in one form or another in all the reasons for testing and are not necessarily entirely concerned with quality assurance issues. There are different types of standards, each having a particular purpose. This can lead us into a dilema by questioning whether current standards are adequate. It is more than likely that a great many are not. This is particularly true of performance specifications, which are in fact quality control methods.

As noted in the practice, the intent of this book is to review the principles and mechanics of statistical analysis for polymer property testing, with the overall thrust aimed at applying both a methodology and techniques to quality assurance. To accomplish this one must first have an understanding of the conventional test methods and instrumentation, and an appreciation of the philosophy behind testing programs. The first half of this book concentrated the requirements for testing and principal test methods themselves.

As noted above, we can classify the reasons for testing into four main areas: reiterating, these are for quality control, for developing design data, for predicting service performance, and for investigating failures. In quality control, the ideal test should be simple, rapid, and cost-effective. Nondestructive techniques and automation are often attractive. Properly designed tests will accurately relate to the product's performance.

When developing design data, the need is for tests that provide material properties in a form that they can be applied with confidence to a variety of configurations and scale of processes. This requires an understanding of how material properties vary with apparatus geometry, time scale, processing conditions, and so on. In this class of tests speed and optimum economics are usually secondary concerns.

For predicting service performance, the more relevant the test is to service conditions, the better. Again, speed and inexpensiveness are less important issues; however, there is a need for well-established test routines that are not excessively complex. Nondestructive methods are desirable but not a prerequisite.

For investigating failures, a key question to address is what to look for. A properly designed test for this purpose must meet well the requirement of discriminating. Very often there is no need for absolute accuracy.

Test methods can be classified in terms of the various parameters of interest: for example, mechanical, thermal, rheological, dimensional, electrical, chemical, surface, and environmental. These classifications, however, do not necessarily reveal the reasons for testing or what one needs to get from a test. Let's consider, for example, the properties of interst in mechanical tests: strength, stiffness, creep, and so on. Although this subdivision reveals which specific parameters require measurement, it does not define why we are testing, and furthermore, does not establish which method to adopt. If, however, we examine such issues in terms of the fundamental properties or tests, the apparent properties, and functional properties or tests, then we begin to attain an understanding of what is needed from the result and which test method should be adopted. For example, if we consider the property of strength, the fundamental strength of a polymer (or any material) is that measured in such a way that the result can be reduced to a form that is independent of test conditions. In contrast, the apparent strength is that property obtained by a standard method that is based on completely arbitrary conditions. Functional strength is that measured under the mechanical conditions of service. Functional strength might, for example, be measured on the finished product.

We can relate this more generalized classification to the reasons for testing, as well as what we need to derive from a test. In the case of quality control, knowledge of fundamental properties is not necessary; instead, apparent properties are sufficient, and functional properties are most desirable. For design data, clearly fundamental properties are required; however, functional properties can provide a significantly body of information. For predicting service performance, functional properties are certainly most appropriate. Finally, for the study of failures, it is unlikely that fundamental methods would be needed.

It is important to note that many existing measures of such properties as impact strength, tear resistance, or melt flow provide apparent properties. In these cases there is a need for fundamental methods. Tests aimed at dimensional, thermal, and chemical parameters provide fundamental properties. Other types, such as those used for studying the effects of the environment (e.g., weathering, ozone resistance, hydrocarbon resistance, etc.) and those used as design data, do not necessarily have to be based on methods that provide absolute results to relatively minor changes with time. In other words, apparent methods may be sufficient. In the following chapters, the statistical tools applied to data analysis are discussed.

REFERENCES AND SUGGESTED READINGS

Andreasen, M. M., S. Kahler, and T. Lund, *Design for Assembly,* Springer-Verlag, Berlin, 1983.

Anon., Calling Long Distance: User Vote Shows Strong Support for AT&T, *The Wall Street Journal,* Section 2 (Aug. 22, 1986).

Baker, E., and Artinian, H., *Qual. Prog.* (June 1985).

Bolz, R. W., *Production Processes: The Productivity Handbook,* 5th ed., Industrial Press, New York, 1981.
Boothroyd, G., and P. Dewhurst, *Design for Assembly: A Designer's Handbook,* University Massachusetts, Amherst, Mass., 1983.
Box, J. F., *R. A. Fisher: The Life of a Scientist,* Wiley, New York, 1978.
Box, G. E. P., Studies in Quality Improvement: Signal to Noise Ratio, Performance Criteria, and Statistical Analysis, Parts I & II, *Technical Report, 11,* Center for Quality and Productivity Improvement, University of Wisconsin-Madison, Madison, Wis., 1986.
Box, G. E. P., and S. Bisgaard, *Qual. Prog.* (June 1987).
Box, G. E. P., and N. R. Draper, *Evolutionary Operation: A Statistical Method for Process Improvement,* Wiley, New York, 1969.
Box, G. E. P., and N. Draper, *Empirical Model Building and Response Surfaces,* Wiley, New York, 1986.
Box, G. E. P., and C. A. Fung, Some Considerations in Estimating Data Transformations, *Technical Summary Report 2609,* Mathematics Research Center, University of Wisconsin-Madison, Madison, Wis., 1983.
Box, G. E. P., and C. A. Fung, Studies in Quality Improvement: Minimizing Transmitted Variation by Parameter Design, *Report 8,* Center for Quality and Productivity Improvement, University of Wisconsin-Madison, Madison, Wis., 1986.
Box, G. E. P., and S. Jones, *Technometrics, 28*(4) (1986).
Box, G. E. P., and R. D. Meyer, *Technometrics, 28*(1) (1986).
Box, G. E. P., and J. G. Ramirez, "Studies in Quality Improvement: Signal to Noise Ratios, Performance Criteria, and Statistical Analysis, Part II, *Report 12,* Center for Quality and Productivity Improvement, University of Wisconsin-Madison, Madison, Wis., 19 .
Box, G. E. P., and K. B. Wilson, *J. R. Stat. Soc. Suppl., 23* (1951).
Box, G. E. P., L. R. Connor, W. R. Cousins, O. L. Davies, F. R. Himsworth, and G. P. Sillito, in *Design and Analysis of Industrial Experiments,* D. L. Davies, ed., Oliver & Boyd, Edinburgh, 1954.
Box, G. E. P., W. G. Hunter, and J. S. Hunter, *Statistics for Experimenters,* Wiley, New York, 1978.
Burgam, P., *Manuf. Eng.* (May 1985).
Burman, J. P., and R. L. Plackett, *Biometrika, 33* (1946).
Buzzell, R. D., and F. D. Wiersema, *Harv. Bus. Rev.* (Jan.-Feb. 1981).
Cary, M., B. Kay, P. Orleman, W. Robertshaw, G. Ross, D. Saunders, W. Wallace, and J. Wittenbraker, *Qual. Prog.* (June 1987).
Cochran, W. G., *Biometrika, 37* (1950).
Cuthbert, D., *Applications of Statistics to Industrial Experimentation,* Wiley, New York, 1976.
Daetz, D., *Qual. Prog.* (June 1987).
Danforth, D. D., *Qual. Prog.* (Apr. 1986), pp. 15-17.
Daniel, C., *J. Am. Stat. Assoc., 57* (1962).
Daniel, C. *Technometrics, 1*(4) (1959).

Davies, O. L., *Statistical Methods in Research and Introduction,* Oliver & Boyd, Edinburgh, 1947.
Deming, W. E., *Quality, Productivity, and Competitive Position,* MIT Center for Advanced Engineering Study, Cambridge, Mass., 1982.
Deming, W. E., *Out of the Crisis,* MIT Press, Cambridge, 1986.
Department of Defense, Transition from Development to Production, *Directive 4245.7,* U.S. Government Printing Office, Washington, D.C., Jan. 1984.
Department of Defense, Transition from Development to Production Manual, *Directive 4245.7M,* U.S. Government Printing Office, Washington, D.C., Sept. 1985.
Department of the Navy, Best Practices for Transitioning from Development to Production, Publication NAVSO P-6071, U.S. Government Printing Office, Washington, D.C., Mar. 1986.
DeToro, I. J., *Qual. Prog.* (Apr. 1987), pp. 16–20.
Donovan, L., and J. Nichols, *Qual. Prog.* (June 1987).
Flud, L. M., *Competitor Intelligence: Hot to Get It—How to Use It,* Wiley, New York, 1985.
Gallup Organization, Customer Perceptions Concerning the Quality of American Products and Service, Gallup, Princeton, N. J., 1985.
Garvin, D. A., *Sloan Manage. Rev.* (Fall 1984).
Garvin, D. A., Product Quality: Profitable at Any Cost, *The New York Times* (Mar. 3, 1985).
Golomski, W. A., *Qual. Prog.* (June 1987).
Gunter, B., *Am. Stat., 38,* 230 (1984).
Gunter, B., *Qual. Prog.* (June 1987).
Hamada, M., and C. J. Wu, A Critical Look at Accumulation Analysis and Related Methods, *Technical Report,* Department of Statistics, University of Wisconsin-Madison, Madison, Wis., 1986.
Hamada, M., and C. F. J. Wu, *Technometrics, 28*(4) (1986).
Hermann, J. A., and Baker, E. M., *Qual. Prog.* (July 1985).
Hunter, W., *Am. Stat., 35,* 72–76 (1981).
Hunter, J. S., *J. Qual. Technol.* (Oct. 1985), pp. 210–221.
Hunter, W. G., J. O'Neill, and C. Wallen, Doing More with Less in the Public Sector: A Progress Report from Madison, Wis., *Report 13,* Center for Quality and Productivity Improvement, University of Wisconsin-Madison, Wis. 1986.
Ishikawa, K., *Guide to Quality Control,* Asian Productiivty Organization, Tokyo, 1976.
Joiner, B. L., and P. R. Sholtes, *Qual. Prog.* (Oct. 1986).
Juran, J. M., *Managerial Breakthrough,* McGraw-Hill, New York, 1964.
Juran, J. M., *Qual. Prog.* (Aug. 1986), p. 19.
Juran, J. M., *Planning for Quality,* Juran Institute, Wilton, Conn., 1986.
Kackar, R., *J. Qual. Technol.* (Oct. 1985), pp. 176–188.
Kackar, R., *Qual. Prog.* (Dec. 1986), pp. 21–29.
Kane, E. J., *Qual. Prog.* (Apr. 1986).
Liepins, K., *Qual. Prog.* (Nov. 1985), pp. 53–56.

Mehrotra, S., and J. Palmer, in *Perceived Quality: How Consumers View Stores and Merchandise,* Olson and Jacobi, eds., Lexington Books, Lexington, Mass., 1984.
Melan, E. H., *Qual. Prog.* (June 1985).
Melan, E. H., *qual. Prog.* (June 1987).
Minton, P., *Am. Stat., 37,* 284–289 (1983).
Nair, V. N., *Technometrics, 28*(4) (1986).
Parker, K. T., in *Proc. 1984 IAQC Annual Conf. and Exhibition,* Cincinnati, Ohio, Apr. 16–19, 1984.
Pearson, E. S., *J. R. Stat. Soc. Suppl., 2* (1935).
Pignatiello, J., and J. Ramberg, *J. Qual. Technol.* (Oct. 1985), pp. 198–206.
Plsek, P. E., *Qual. Prog.* (June 1987).
Propst, A. L., *Qual. Prog.* (June 1987).
Putnam, A. O., *Harv. Bus. Rev.* (May–June 1985).
Quinlan, J., Paper presented at the 3rd Supplier Symposium on Taguchi Methods, American Supplier Institute, Dearborn, Mich., 1985.
Rao, C. R., *J. R. Stat. Soc. Suppl., 9* (1947).
Schreiber, R. R., *Robotics Today, 7*(3), 45–46 (1985).
Shewhart, W. A., in *Statistical Methods from the Viewpoint of Quality Control,* W. Edwards Deming, ed., Graduate School, Department of Agriculture, Washington, D.C., 1939.
Snedecor, G. W., and W. G. Cochran, *Statistical Methods,* 7th ed. Iowa State University Press, Ames, Iowa, 1980.
Snee, R. D., L. B. Hare, and J. R. Trout, *Experiments in Industry: Design, Analysis, and Interpretation of Results,* 1985 (available through ASQC's Quality Press, Milwaukee, Wis.).
Sullivan, L. P., *Qual. Prog.* (July 1984), pp. 15–21.
Sullivan, L. P., *Qual. Prog.* (June 1987).
Taguchi, G., and Y. Wu, *Introduction to Off-Line Quality Control Systems* Central Japan Quality Control Association, Nagoya, Japan, 1980.
Thompson, P., G. DeSouza, and B. T. Gale, *Qual. Prog.* (June 1985).
Tippett, L. H. C., *J. Manchester Stat. Soc.* (1934).
Wheeler, R. E., *Quality* (May 1986).
Wheelwright, S. C., in *Advances in Applied Business Strategy,* JAI Press, Greenwich, Conn., 1984.
Wu, Y., *Off-Line Quality Control: Japanese Quality Engineering,* American Supplier Institute, Dearborn, Mich., 1982.
Xerox Corporation, *Competitive Benchmarking: What It Is and What It Can Do for You,* Xerox, Stamford, Conn., 1984.
Youden, W. J., Experimental Design and ASTM Committees, *Mat. Res. Stand.* (1961); reprinted in *Precision Measurements and Calibrations,* No. 1, National Bureau of Standards, Washington, D.C., 1969.
Zelen, M., *Technometrics, 1*(3) (1959).

7
Basic Definitions in Probability and Statistics

THE IMPORTANCE OF FREQUENCY DISTRIBUTIONS

Frequency distributions can readily be used in any process operation for two purposes. First, they can be used to determine average conditions or product properties and their variations expected within the current process. Second, they can be used to control current production and/or provide direction in changing product composition to improve performance.

Every process operation has a degree of variability. The key question is whether the degree of variability is the most economical or cost-effective for a particular operation. In this volume we adopt the principle that statistical quality control is a never-ending goal—a continual practice that must be integrated in normal production. This does not mean that one can never bring a process under a state of statistical process control or that product consistency and quality are never the "best." Although one can always argue that "there's room for improvement," this must be tempered by the query "at what cost?" Certainly, processes can be overcontrolled, overchecked, and overrun in analytical cost such that the ends do not finally justify the means. In this regard, statistical quality control methodology must be applied in a logical fashion, whereby the cost of the level of efforts required to implement a program are clearly shown to be outweighed by the return on investment due to reduced operator time in process control, improved sales volumes and margins due to better consistency, and higher quality products compared to the competition.

The use of frequency distributions and measure of variability are among the simplest aspects of statistics. They are tools that provide a first step in assessing process and product control. Uncovering the existence of variability and its underlying causes in any operation is essentially application of the science of statistical quality control. The principles are best illustrated by way of an example. In the following discussions we examine the degree of variability in an elastomer extrusion operation, where we are manufacturing radiator hoses for automobiles. Although there are several product performamance parameters to consider in such an operation, we direct attention only to dimensional aspects of the product for this illustration.

A conventional single-screw extruder often used in rubber applications is a relatively straightforward piece of equipment. Despite this, the number of potential sources for product variability can be quite high. For example, although automatic temperature controls are activated only when there is a difference between the actual and desired temperature, temperature fluctuations are inevitable. Upstream premixing and preheating steps are subject to variations that can effect extruder operations. Despite extraordinary precision in die tooling, two or more presumably identical dies will never be exactly alike. Measuring out charges from a die, whether by weight, volume, linear, and so on, introduce further sources of variability. Lot-to-lot or even bale-to-bale variations in the physical and chemicophysical properties of the feedstock polymer are also inevitable.

In our example we illustrate variability by taking 100 extruded hoses, measuring their wall thickness and recording the readings. If measured to the nearest fraction of a millimeter or less, the variation becomes apparent quite soon. This variation can also be followed by weighing pieces and preparing a plot of the variations by weight instead of dimensions. This is just as easy and reliable as measuring dimensions.

The total sources of product variability, when reduced to their practical minimum, can be grouped together and are referred to as the system's "inherent variability." In simple terms, this value is as good as one can hope for with the current process. The obvious goal to achieve is that the observed variability be as close to the practical minimum. Statistical techniques essentially provide a quantitative target for what this minimum variability should be.

A distribution is normally constructed in the form of a histogram as illustrated in Figure 7.1. The ideal distribution may be wider or narrower than illustrated in Figure 7.1A, but its general characteristics should be unimodal and symmetrical. This is what is known as a *normal frequency distribution.* If the frequency distribution is like the alternate shapes illus-

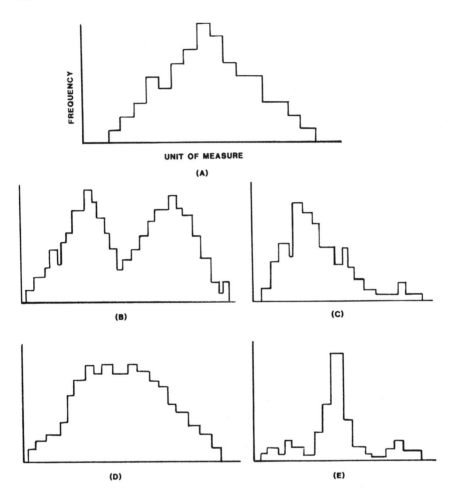

Figure 7.1 Possible shapes of frequency distribution histograms.

trated in Figure 7.1B to E, it is indicative that the process is not performing to its best ability, and that some corrective steps are required to bring the variability back into line.

In many operations, data automation is still not a reality, in which case in-plant histograms can be manually generated by a simple tally. The following table illustrates readings obtained from two extruder lines for our example.

Definitions in Probability and Statistics

Measured tolerance	Frequency	
	Line A	Line B
+0.019		1
+0.018		1
+0.017		
+0.016		
+0.015		1
+0.014		111
+0.013		111
+0.012		++++ 111
+0.011		++++ ++++ ++++ 11
+0.010		++++ ++++ 111
+0.009		1111 1
+0.008		111
+0.007		1
+0.006	1	1
+0.005		
+0.004	1	
+0.003	111	
+0.002	1111	
+0.001	++++ 111	
0.000	++++ ++++ 1	
−0.001	++++ ++++ ++++ ++++ 1	
−0.002	111	
−0.003	1	
−0.004		

In this group of data, the production is out of control since the combined product displays a bimodal frequency distribution (i.e., two peaks, as illustrated in Figure 7.1B). In the case of the data above, the cause for the out-of-control process could be traced to a worn die in extruder line A.

The average of the frequency distribution is computed by multiplying the frequency of occurrence by the recorded dimension, and dividing the sum of these products by the number of units measured. As shown later, the information from a frequency distribution can be used to establish process control limits from which a system of statistical quality control can be developed around such control charts.

THE NORMAL FREQUENCY DISTRIBUTION

The normal frequency distribution (or Gaussian distribution) is illustrated in relation to the histogram of Figure 7.2. To define this curve properly, we need information on:

1. *Its level.* Where is it located on the scale of values?
2. *Its dispersion.* How far along this scale does it spread?

The level of a distribution can be expressed by the *average*, or *arithmetic mean*. This value is derived by summing up all the values of the individual measurements and dividing the total by the number of measurements.

BASIC DEFINITIONS AND THEOREMS OF PROBABILITY

A fundamental concept of probability and statistics is the *set*. A set is a collection of objects which are referred to as *members* or *elements* of the set. Conventional notation denotes a set by a capital letter (e.g., A, B, C) and an element is denoted by a lowercase letter (e.g., a, b, c). Synonomous terms for set are *class, aggregate,* and *collection.*

If an element a is part of the set B, it is denoted as $a \in B$. If a does not belong to B, we write $a \notin B$. If both a and b belong to B, we write $a,b \in B$. For a set to be well defined, one must also establish whether a particular element does or does not belong to the set.

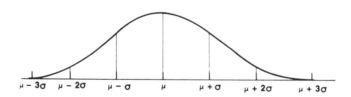

Interval	Area (%)	Interval	Area (%)
$\mu \pm \sigma$	68.27	$\mu \pm 0.674\sigma$	50
$\mu \pm 2\sigma$	95.45	$\mu \pm 1.645\sigma$	90
$\mu \pm 3\sigma$	99.73	$\mu \pm 1.960\sigma$	95
$\mu \pm 4\sigma$	99.994	$\mu \pm 2.576\sigma$	99

Figure 7.2 Normal or Gaussian distribution.

Definitions in Probability and Statistics

Sets can be defined by actually listing their elements or, if this is not possible, by describing some property held by all members and by no nonmembers. The first is referred to as the *roster method* and the latter is the *property method*. For example, the set of parameters characterizing the extrusion quality of a plastic tube can be defined by the roster method as wall thickness, percent swell, or eccentricity, or by the property method as $\{x | x \text{ is a physical dimension}\}$, read "the set of all elements x such that x is a physical dimension," where the vertical "|" denotes "such that" or "given that." As a more general example, the set $\{x | x \text{ is a rectangle in a plane}\}$ is the set of all rectangles in a plane. Note that the roster method cannot be used here.

We now consider *subsets*. If each element of a set A also belongs to a set B, we call A a subset of B, written as $A \subset B$ or $B \supset A$. This is read "A is contained in B" or "B contains A," respectively. Also, for all sets A we have $A \subset A$. Note that if $A \subset B$ and $B \subset A$, then A and B are *equal* (i.e., $A = B$), in which case A and B have exactly the same elements. If A is not equal to B (i.e., if A and B do not have exactly the same elements), we write $A \neq B$. If $A \subset B$ but $A \neq B$, we refer to A as a *proper subset* of B. for example, $\{\alpha, \beta, \gamma\}$ is a proper subset of $\{\alpha, \beta, \omega, \delta, \gamma, \phi\}$. $\{\phi, \gamma, \omega, \delta, \alpha, \omega\}$ is a subset, but not a proper subset of $\{\alpha, \beta, \omega, \delta, \gamma, \phi\}$, because the two sets are equal. A rearrangement of elements does not alter the set.

We summarize the discussions above by stating the following theorem for any sets A, B, and C.

Theorem a. If $A \subset B$ and $B \subset C$, then $A \subset C$.

There are numerous situations where it is useful to restrict attention to subsets of a particular set referred to as the *universal set*. The universal set is denoted by $\mathcal{U}$ and the elements of a space are called *points* of the space. It is also useful to consider a set having no element at all. This is referred to as the *empty set* or *null set*, denoted by $\emptyset$. The null set is a subset of any set.

A universe $\mathcal{U}$ can be represented by a set of points inside a retangular plane. Subsets $\mathcal{U}$, A, and B are illustrated by the shaded regions in Figure 7.3 and are represented by sets of points inside the circles. These diagrams are referred to as *Venn diagrams*. They can be used to provide geometric intuition regarding possible relationships between sets.

The following operations can be performed among sets:

Union. The set of all elements that belong to either A or B or both A and B are referred to as a *union* of A and B. This is denoted by $A \cup B$ and is illustrated by the shaded region in Figure 7.4A.

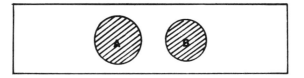

Figure 7.3 Venn diagram.

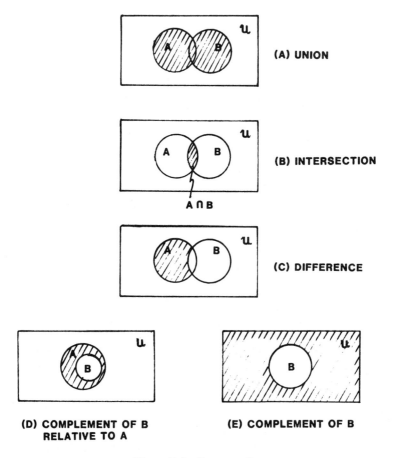

Figure 7.4 Set operations.

Definitions in Probability and Statistics

Intersection. This is the set of all elements that belong to both A and B. The intersection of A and B is denoted by $A \cap B$ and is illustrated in Figure 7.4B.

Difference. This set consists of all elements of A that do not belong to B. The difference of A and B is denoted by $A - B$ and is illustrated in Figure 7.4C.

Complement. If $B \subset A$, then $A - B$ is referred to as the complement of B relative to A. It is denoted by B_A' and is shown by the shaded region in Figure 7.4D. If $A = \mathcal{U}$ (i.e., the universal set), $\mathcal{U} - B$ is simply the complement of B (denoted by B'). This is shown by the dhaded region in Figure 7.4E. The complement of $A \cup B$ is denoted by $(A \cup B)'$.

The following is a summary of important theorems involving sets:

Theorem (b): commutative law for unions. $A \cup B = B \cup A$.

Theorem (c): associative law for unions. $A \cup (B \cup C) = (A \cup B) \cup C = A \cup B \cup C$.

Theorem (d): commutative law for intersections. $A \cap B = B \cap A$.

Theorem (e): associative law for intersections. $A \cap (B \cap C) = (A \cap B) \cap C = A \cap B \cap C$.

Theorem (f): first distributive law. $A \cap (B \cup C) = (A \cap B) \cup (A \cap C)$.

Theorem (g): second distributive law. $A \cup (B \cap C) = (A \cup B) \cap (A \cup C)$.

Theorem (h): difference. $A - B = A \cap B'$.

PRINCIPLE OF DUALITY, RANDOM EXPERIMENTS, AND PROBABILITY

Any true result involving sets is also true if unions are replaced by intersections, intersections by unions, sets by their complements, and if the inclusion symbols, $\subset$ and $\supset$, are reveresed. This is known as the *principle of duality*.

As emphasized throughout this volume, experimentation is the key element to establishing a quality control program. A fundamental principle involved in experimental work is that if we perform experiments repeatedly under very nearly identical conditions, we should in practice obtain results which are essentially the same. There are, however, experiments in which results will not be the same even though conditions are kept nearly identical. These types of experiments are called *random experiments*. For example, if we use an extruder to produce hoses, the result of the experiment is that some of the hoses may be defective. Thus when a hose is extruded it will be a member of the set {defective, nondefective}. Another example is: If we toss a coin, the result of the experiment is that it will come up either "tails" (0) or "heads" (1) (i.e., one of the elements of the set {0, 1}).

A term useful to our discussions is *sample space*. A sample space refers to a set $\mathscr{S}$ that consists of all possible outcomes of a random experiment. Each outcome is referred to as a *sample point*. As if often the case, there can be more than one sample space that can describe outcomes of an experiment; however, there is usually only one that provides the most information. The symbol $\mathscr{S}$ is used to refer to the universal set. A sample space can be portrayed graphically, and hence it is desirable to employ numbers rather than letters to denote outcomes. As an example, if we toss a die, one sample space or set of all possible outcomes is {1, 2, 3, 4, 5, 6}, while another is odd, even {O, E}. The latter convention is inadequate, for example, to assess whether the outcome is divisible by 2.

If a sample space has a finite number of points, it is called a *finite sample space*. If it has as many points as there are natural numbers (e.g., 1, 2, 3, ...) it is called a *countably infinite sample space*. If it has a many points as there are in some interval on the *x*-axis (i.e., $0 \leqslant x \leqslant 1$), it is called a *noncountably infinite sample space*. A sample space that is finite or countably infinite is called a *discrete sample space*. In contrast, one that is noncountably is called a *nondiscrete* or *continuous sample space*.

The term *event* refers to a subset A of the sample space $\mathscr{S}$; that is, it is a set of possible outcomes. An event comprised of a single point of $\mathscr{S}$ is reffered to as a *simple* or *elementary event*. Since events are sets, events, in the mathematical sense, can be treated by conventional set theory. Specifically, the technique known as the *algebra of events* can be applied to the manipulation of sets. By using set operations on events in $\mathscr{S}$, other events in $\mathscr{S}$ can be obtained. If A and B are events, the following set operations can be applied:

$A \cup B$ is the event "either A or B or both." (i)
$A \cap B$ is the event "both A and B." (ii)
A' is the event "not A." (iii)
$A - B$ is the event "A but not B." (iv)

Note that $A \cap B = \emptyset$, the events are mutually exclusive (i.e., they cannot both occur).

With the definitions and theorems above, we may now discuss the concept of probability and important theorems. Any random experiment has associated with it some uncertainty as to whether a particular event will or will not take place. It is convenient to assign a numerical rating between 0 and 1 as a measure of the probability or chance with which we can expect the event to occur. So, for example, if we are sure that the event will not occur, its probability is zero. If the probability is ⅓, we would say there is a 33% chance it will occur and a 67% chance it will not occur.

Estimates for the probability of an event can be obtained by one of two methods. The first method is referred to as the classical or a priori approach. In this method, if an event can occur in h different ways out of a total of n possible ways (all of which are equally likely), the probability of the event is h/n. The second method is referred to as the frequency approach. Here, if after n repetitions of an experiment (where n is very large), an event is observed to occur in h of these, the probability of the event is h/n. An example of the first method is the probability that tails turns up in a single toss of a coin. Since there are two equally likely outcomes (heads or tails), and of these two ways tails can arise in only one way, we deduce that the required probability is ½ or 50%. To illustrate the second method, if the coin is tossed 100 times and we find that it comes up tails 63 times, we estimate the probability of a tail to be $63/100 = 0.63$.

Both methods are in fact very imprecise and can lead to ambiguous predictions. For this reason, mathematicians have devised an axiomatic approach to probability utilizing sets. Consider a sample space $\mathscr{S}$, where if $\mathscr{S}$ is discrete, all subsets correspond to events and conversely. If, however, $\mathscr{S}$ is continuous, only special subsets correspond to events. A real number $P(A)$ can be associated to each event A in the class C of events. In other words, P is considered a real-valued function defined on C. P is referred to as the *probability function,* and $P(A)$ is the probability of the event A. The following axioms must be satisfied for this to be true:

Axiom (a): For every event A in the class C,

$$P(A) \geqslant 0 \tag{1}$$

Axiom (b): For the certain event $\mathscr{S}$ in the class C

$$P(\mathscr{S}) = 1 \tag{2}$$

Axiom (c): For any number of mutually exclusive events $A_1, A_2, \ldots$ in the class C,

$$P(A_1 \cup A_2 \cup \cdots) = P(A_1) + P(A_2) + \cdots \tag{3}$$

Note that for only two mutually exclusive events A_1 and A_2,

$$P(A_1 \cup A_2) = P(A_1) + P(A_2) \tag{4}$$

The following summarizes the principal theorems on probability.

Theorem (i). For $A_1 \subset A_2$, $P(A_1) \leqslant P(A_2)$ and $P(A_2 - A_1) = P(A_2) - P(A_1)$

Theorem (j). For every event A,

$$0 \leqslant P(A) \leqslant 1$$

That is, the probability lies between 0 and 1.

Theorem (k). The impossible event has zero probability:

$$P(\emptyset) = 0$$

Theorem (l). If A' is the complement of A,

$$P(A') = 1 - P(A)$$

Theorem (m). If $A = A_1 \cup A_2 \cup \cdots A_n$, where $A_1, A_2, \ldots, A_n$ are mutually exclusive events,

$$P(A) = P(A_2) + \cdots P(A_n)$$

Theorem (m_1). If $A = \mathscr{S}$ (the sample space), then

$$P(A_1) + P(A_2) + \cdots + P(A_n) = 1$$

Theorem (n). For any two events A and B,

$$P(A \cup B) = P(A) + P(B) - P(A \cap B)$$

Definitions in Probability and Statistics

Theorem (n). If A_1, A_2, and A_3 are any three events,

$$P(A_1 \cup A_2 \cup A_3) = P(A_1) + P(A_2) + P(A_3)$$
$$- P(A_1 \cap A_2) - P(A_2 \cap A_3) - P(A_3 \cap A_1) + P(A_1 \cap A_2 \cap A_3)$$

Theorem (o). For any events A and B,

$$P(A) = P(A \cap B) + P(A \cap B')$$

Theorem (p). In the case that event A must result in one of the mutually exclusive envents $A_1, A_2, \ldots, A_n$, then

$$P(A)\ P(A \cap A_1) + P(A \cap A_2) + \cdots + P(A \cap A_n)$$

Suppose that a sample space $\mathscr{S}$ is comprised only of the elementary events $A_1, A_2, \ldots, A_n$; then from Theorem (m) we note that

$$P(A_1) + P(A_2) + \cdots + P(A_n) = 1 \tag{5}$$

Provided that equation (5) is satisfied, we can arbitrarily assign nonnegative numbers for the probabilities of these simple events. Assuming equal probabilities for all simple events,

$$P(A_k) = \frac{1}{n} \tag{6}$$

where $k = 1, 2, \ldots, n$.

Furthermore, if A is any event comprised of h such simple events, then

$$P(A) = \frac{h}{n} \tag{7}$$

By assigning probabilities, we are essentially posing a hypothesis or mathematical model. The validity or accuracy of this model must be verified to properly designed experiments.

We now turn our attention to the concept of *conditional probability*. Consider two events, A and B, such that $P(A) > 0$. The probability of B, given that A has occurred, is denoted by convention as $P(B|A)$. Since A has taken place, it becomes the new sample space (i.e., it replaces the original $\mathscr{S}$). This becomes the basis for the following statement:

$$P(B|A) \equiv \frac{P(A \cap B)}{P(A)} \tag{8}$$

Statement (8) can also be expressed as

$$P(A \cap B) \equiv P(A)P(B|A) \tag{9}$$

Statement (9) says that the probability that both A and B occur is equal to the product of the probability that A occurs and the probability B occurs (provided that A has occurred). $P(B|A)$ is referred to as the *conditional probability* of B given A (in other words, it is the probability that B will occur given that A has occurred).

The following are important theorems on conditional probability:

Theorem (q). The probability that any three events $A_1, A_2, \ldots A_n$ are mutually exclusive events whose union is the simple space $\mathscr{S}$ (one of these events must occur). Then if A is any event,

$$P(A_k|A) = \frac{P(A_k)P(A|A_k)}{\sum_{k=1}^{n} P(A_k)P(A|A_k)} \tag{10}$$

Bayes' theorem enables the probabilities of the various events $A_1, A_2, \ldots, A_n$ that can cause A to occur, to be determined.

COMBINATORIAL ANALYSIS, COUNTING, AND TREE DIAGRAMS

If the number of sample points in a sample space is not very large, the counting of sample points needed to compute probabilities is not difficult. There are, however, situations where direct counting is impractical. For such cases the method of *combinatorial analysis* can be applied. Let's consider the following example. Suppose that a rubber compounder has two types of carbon blacks and four polymer grades from which to prepare his mix. The compounder has $2 \times 4 = 8$ ways of selecting a carbon black and then a polymer. We can restate this example in a very general manner. If a task can be accomplished in n_1 different ways, and after this a second task can be accomplished in n_2 ways, ..., and finally a kth task can be accomplished in n_k different ways, then all k tasks can be performed in the specified order in $n_1 n_2 \cdots n_k$ different ways. This principle is often applied with a tree diagram, illustrated in Figure 7.5. The diagram in Figure 7.5 graphically shows the various ways in which the rubber compounder selects a carbon black (CB_1, CB_2) and then a polymer (P_1, P_2, P_3, P_4).

Definitions in Probability and Statistics

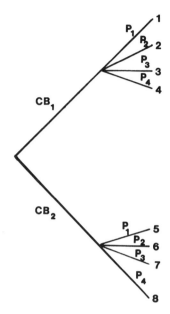

Figure 7.5 Tree diagram.

As a final example, suppose that we wish to prepare an optimized masterbatch mix for a super-soft sponge formulation that is to be used for the manufacture of automotive weatherstrip seals. The compounder has already selected the polymer and will custom design the formulation to achieve desired compound properties such as tensile, resilience, surface texture, and scorch safety. There are three major ingredients in the mix that can be varied to alter these properties significantly. These ingredients are the carbon black, the blowing agent, and the co-catalyst for the vulcanization stage. If there are three possible carbon blacks suitable for weatherstrip formulations, two types of blowing agents, and four types of co-catalysts, let's determine how many different mixes must be evaluated using (1) the fundamental principle of counting and (2) a tree diagram.

1. We can select a carbon black in three different ways and a blowing agent in two different ways. Then there are $3 - 2 = 6$ different ways of choosing a carbon black and blowing agent candidate. With each of these ways we can choose a cocatlyst in four different ways. Thus the number of different mixex that can be prepared for study is $3 \times 2 \times 4 = 24$.

2. Denote the three carbon black candidates by B1, B2, B3; the blowing agent candidates by A1, A2; and the cocatalyst candidates by C1, C2, C3, C4. Then the tree diagram of Figure 7.6 shows that there are 24 different mixes in all. From this tree diagram we can list all these different mixes (e.g., B1A1C1, B1A1C2, ...).

PERMUTATIONS, BINOMIAL COEFFICIENTS, AND STIRLING'S APPROXIMATION

Let's consider that we are assigned n distinct objects and we wish to arrange r of these items in a line. There are n ways to choose the first object, and after this is done, $n - 1$ ways of choosing the second object, ..., and finally, $n - r + 1$ ways of choosing the rth object. It follows by the fundamental principle of counting that the number of different arrangements, or *permutations*, is given by

$$_nP_r = n(n - 1)(n - 2) \cdots (n - r + 1) \tag{11}$$

where it should be noted that the product has r factors. The notation $_nP_r$ refers to the number of permutations of n objects taken r at a time.

For the special case where $r = n$, equation (11) becomes

$$_nP_n = n(n - 1)(n - 2) \cdots 1 = N! \tag{12}$$

This is referred to as n-factorial. Equation (11) can be expressed in terms of factorials as

$$_nP_r = \frac{n}{(n - r)!} \tag{13}$$

Note that equations (12) and (13) agree only when $0! = 1$. This is adopted as the definition of 0!

A permutation is important when we are interested in the order of arrangement of objects or actions. An example of this is in rubber compounding when mixing operations for masterbatches must be optimized (e.g., a straight mix consists of adding rubber, then carbon black and oil, whereas an upside-down mix might involve adding oil and carbon black first to make a paste, and then adding rubber). A more general example of 123 is a different permutation from 231. There are many problems, however, one is interested only in choosing objects or actions without

Definitions in Probability and Statistics

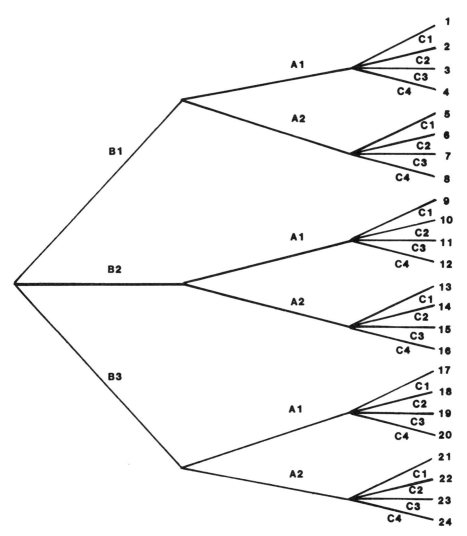

Figure 7.6 Tree diagram for masterbatch formulation problem.

regard to order. These selections are referred to as *combinations*. In the generalized example, 123 and 231 are the same combination.

If there are n objects, the total number of combinations of r objects selected can be denoted by n (r or $\binom{n}{r}$), and written in the following forms:

$$\binom{n}{r} = {}_nC_r = \frac{n!}{r!(n-r)!} \tag{14a}$$

$$\binom{n}{r} = \frac{N(n-1)\cdots(n-r+1)}{r!} = \frac{{}_nP_r}{r!} \tag{14b}$$

$$\binom{n}{r} = \binom{n}{n-r} \quad \text{or} \quad {}_nC_r = {}_nC_{n-r} \tag{14c}$$

So, for example, in an elastomer finishing line, the number of ways in which three samples can be chosen or selected from a total of eight different locations is

$$_8C_3 = \binom{8}{3} = \frac{3 \cdot 7 \cdot 6}{3!} = 56$$

The numbers of equation (14a) are referred to as binomial coefficients because they appear in the binomial expansion.

$$(x+y)^n = x^n \binom{n}{1} = x^{n-1}y + \binom{n}{2} = x^{n-2}y^2 + \cdots + \binom{n}{n}y^n \tag{15}$$

This will be useful later, particularly in discussions pertaining to experimental program planning and data regression.

When n is very large, a direct evaluation of $n!$ is not practical. In this case, Stirling's approximation to $n!$ is applied:

$$n! \sim \sqrt{2\pi n}\, n^n e^{-n} \tag{16}$$

The symbol "~" refers to the ratio of the left-hand side (LHS) to the right-hand side (RHS) approaches unity as $n \to \infty$. Thus the RHS of equation (16) is an asymptotic expansion of the LHS.

DEFINITIONS OF RANDOM VARIABLES AND PROBABILITY DISTRIBUTIONS

Let us assign a number to each point of a sample space. This enables us to define a *function* on the sample space. In this case the function is referred to as a *random variable*, denoted by X or Y. A random variable or function has some special physical, geometrical, or other outstanding property. As an example, suppose that a coin is tossed twice so that the sample space is $\mathscr{S} = \{HH, HT, TH, TT\}$. If X represents the number of heads that can come up, then with each sample point we can associate a number X as shown below:

Sample Point	HH	HT	TH	TT
X	2	1	1	0

In the case of two heads in a row (i.e., HH), $X = 2$, while for TH (one head) $X = 1$. It follows that X is a random variable. Many other random variables can also be defined on this sample space (e.g., the square of the number of heads, the number of heads plus the number of tails, etc.)

Any random variable that takes on a finite or countably infinite number of values is referred to as a *discrete random variable*. In contrast, one that takes on a noncountably infinite number of values is called a *nondiscrete* or *continuous random variable*.

Consider X to be a discrete random variable whose possible values are $X_1, X_2, X_3, \ldots$, arranged in an increasing order of magnitude. These values have probabilities given by

$$P(X = x_k) = f(x_k) \quad \text{where } k = 1, 2, \ldots \tag{17}$$

The probability function or *probability distribution* is given by

$$P(X = x_k) = f(x) \tag{18}$$

When $x = x_k$, equation (18) reduces to equation (17), while for other values of x, $f(x) = 0$.

In general, $f(x)$ is a probability function if the following criteria are met:

$$f(x) \geqslant 0 \tag{19a}$$

$$\sum_x f(x) = 1 \tag{19b}$$

The sum in Equation (19b) is taken over all possible values of x. A plot of $f(x)$ is referred to as the *probability graph*.

A probability graph can be constructed from a bar chart or histogram as illustrated in Figure 7.7. In the bar chart the sum of the ordinates is unity, whereas in the histogram the sum of the rectangular areas is 1. With the histogram, the random variable X is made continuous (e.g, $X = 1$ means that it lies between 0.5 and 1.5).

The cumulative distribution function for a random variable X is defined as

$$P(X \leq x) \, F(x) \tag{20}$$

where $-\infty < X < \infty$ (i.e., X is any real number). The distribution function can be derived from the probability function by noting that

$$F(x) = P(X \leq x) = \sum_{u \leq x} f(u) \tag{21}$$

Note that the sum on the RHS is taken over all values of u for which $u \leq x$. The probability function can be obtained from the distribution function.

Suppose that X has a finite number of values (i.e, $x_1, x_2, \ldots, x_n$). The distribution function is then given by

$$F(x) = \begin{cases} 0 & -\infty < x \, x_1 \\ f(x_1) & x_1 \leq x \leq x_2 \\ f(x_1) + f(x_2) & x_2 \leq x \leq x_3 \\ \vdots & \vdots \\ f(x_1) + \cdots + f(x_n) & x_n \leq x < \infty \end{cases} \tag{22}$$

To review some of these principles, consider the example of the double coin toss described earlier. The probability function corresponding to the random variable x is as follows: $P(\text{HH}) = \frac{1}{4}$, $P(\text{HT}) = \frac{1}{4}$, $P(\text{TH}) = \frac{1}{4}$, $P(\text{TT}) = \frac{1}{4}$. We may then state

$$P(X = 0) \, P(\text{TT}) = \tfrac{1}{4}$$

$$P(X = 1) = P(\text{WT} \cup \text{TH}) = P(\text{HT}) + P(\text{TH}) = \tfrac{1}{4} + \tfrac{1}{4} = \tfrac{1}{2}$$

$$P(X = 2) = P(\text{HH}) = \tfrac{1}{4}$$

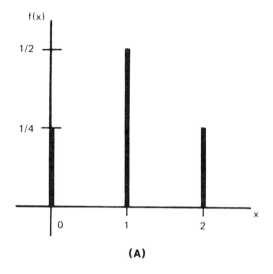

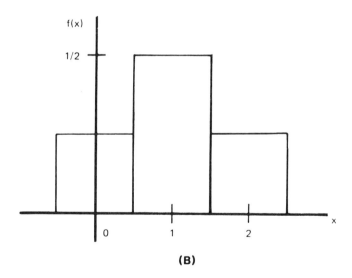

Figure 7.7 (A) Bar chart; (B) histogram.

We can summarize the probability function as

x	0	1	2
$f(x)$	¼	½	¼

Now we will determine the distribution function for the random variable X. The distribution function is

$$F(x) = \begin{cases} 0 & -\infty < x < 0 \\ \tfrac{1}{4} & 0 \leq x < 1 \\ \tfrac{3}{4} & 1 \leq x < 2 \\ 1 & 2 \leq x < \infty \end{cases}$$

A plot of $F(x)$ is shown in Figure 7.8. Because of the appearance of the

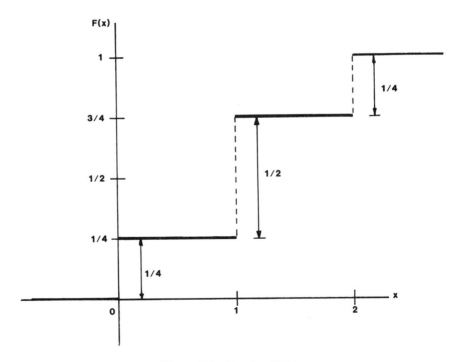

Figure 7.8 Graph of $F(x)$.

Definitions in Probability and Statistics

plot, the graph is referred to as a *step function*. The value of the function at an integer is obtained from the higher step. So, for example, the value at 1 is ¾, not ¼. This is expressed by stating that the distribution function is continuous from the right at 0, 1, 2.

Note that if X is a continuous random variable, the probability that X takes on any one particular value is generally zero. This means that we cannot define a probability function in the same manner as for a discrete random variable. To arrive at a probability distribution for a continuous random variable, we must realize that the probability that X lies between two different values is meaningful. We postulate the existence of a function $f(x)$ such that

$$f(x) \geqslant 0 \tag{23a}$$

$$\int_{-\infty}^{\infty} f(x)dx = 1 \tag{23b}$$

We define the probability that X lies between the limits of a and b as follows:

$$P(a < X < b) = \int_{a}^{b} f(x)\, dx \tag{24}$$

Any function $f(x)$ that satisfies requirement 24 is called the probability function or probability distribution for a continuous random variable. It is also referred to as the *probability density function*. When $f(x)$ is continuous, we can assume that the probability that X is equal to any particular value is zero.

The distribution function $F(x)$ for a continuous random variable is

$$F(x) = P(X \leqslant x) = P(-\infty < X \leqslant x) \int_{-\infty}^{x} f(u)du \tag{25}$$

Note that at points of continuity of $f(x)$, the symbol $\leqslant$ can simply be replaced by $<$.

The probability that X lies between x and $x + \Delta x$ is

$$P(x \leqslant X \leqslant x + \Delta x) = \int_{x}^{x+\Delta x} f(u)\, du \tag{26}$$

If Δx is small, we can state the following approximation:

$$P(x \leq X \leq x + \Delta x) = f(x)\, \Delta x \tag{27}$$

Differentiating both sides of this expression gives

$$\frac{dF(x)}{dx} = f(x) \tag{28}$$

Equation (28) applies at all points where $f(x)$ is continuous. In other words, the derivative of the distribution function is the density function.

A useful law applied to obtaining equation (28) is Leibniz's rule:

$$\frac{d}{dx}\int_a^x f(u)\, du = f(x) \tag{29}$$

The most general form of Leibniz's rule for differntiation of an intergral is

$$\frac{d}{dx}\int_{a_1(x)}^{a_2(x)} F(u,x)\, du \quad \int_{a_1(x)}^{a_2(x)} \frac{dF}{dx}\, du + F(a_2(x),x)\frac{da_2}{dx} - F(a_1(x),x)\frac{da_2}{dx} \tag{30}$$

Note that a_1, a_2, and F are assumed to be differentiable with respect to x.

Since $f(x)$ is the density function of a random variable X, $y = f(x)$ can be represented graphically by a curve as shown in Figure 7.9A. Since $f(x) \geq 0$, the curve cannot fall below the x-axis. The area bounded by the curve and the X-axis must be unity. Geometrically, the probability that X lies between a and b is shown by the shaded area in Figure 7.9A. The distribution function $F(x) = P(X \geq x)$ is a monotonically increasing function which increases from 0 to 1. It is represented by the curve shown in Figure 7.9B.

Let's now consider two random variables which can be either both discrete or both continuous. If X and Y are discrete random variables, the *joint probability function* is defined by

$$P(X = x,\ Y = y) = f(x,y) \tag{31}$$

Definitions in Probability and Statistics

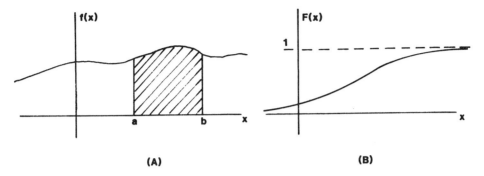

Figure 7.9 (A) Density function, $y = f(x)$; (B) Distribution function, $F(x) = P(X \leq x)$.

where $f(x,y) \leq 0$ and $\Sigma_x \Sigma_y f(x,y) = 1$. If X can assume any one of m values, $x_1, x_2, \ldots, x_m$ and Y can assume any one of n values $y_1, y_2, \ldots, y_n$, the probability of the event that $X = x_j$ and $Y = y_k$ is

$$P(X = x_j, Y = y_k) = f(x_j, y_k) \tag{32}$$

A joint probability function for X and Y can be represented in tabular form as shown in Table 7.1.

The joint probability that $X = x_j$ is determined by summing all entries in the row corresponding to x_j, given by

$$P(X = x_j) = f_1(x_j) = \sum_{k=1}^{n} f(x_j, y_k) \tag{33}$$

Note that $j = 1, 2, \ldots, m$ are indicated by the entry totals in the rightmost column of Table 7.1.

The probability that $Y = y_k$ is determined by adding all entries in the column corresponding to y_k,

$$P(Y = y_k) = f_2(y_k) = \sum_{j=1}^{m} f(x_j, y_k) \tag{34}$$

Where $k = 1, q, \ldots, n$, indicated by the entry totals in the bottom row of Table 7.1.

Table 7.1 Joint Probability Table

x	y y_1	y_2	...	y_n	Totals
x_1	$f(x_1,y_1)$	$f(x_1,y_2)$	...	$f(x_1,y_n)$	$f_1(x_1)$
x_2	$f(x_2,y_1)$	$f(x_2,y_2)$	...	$f(x_2,y_n)$	$f_1(x_2)$
⋮	⋮	⋮		⋮	⋮
x_m	$f(x_m,y_1)$	$f(x_m,y_2)$	...	$f(x_m,y_n)$	$f_1(x_m)$
Totals	$f_2(y_1)$	$f_2(y_2)$	...	$f_2(y_n)$	1

It is important to note that

$$\sum_{j=1}^{m}\sum_{k=1}^{n} f(x_j, y_k) = 1 \tag{35}$$

In other words, the total probability of all entries is unity. The joint distribution function of X and Y is given by

$$F(x, y) = P(X \leq x, Y \leq y) = \sum_{u \leq x}\sum_{v \leq y} f(u, v) \tag{36}$$

In Table 7.1, $F(x,y)$ is the sum of all entries for which $x_j \leq x$ and $y_k \leq y$.

Now consider the case where both X and Y are continuous. The joint probability function for the random variables is called the *joint density function*, given by

$$f(x,y) \geq 0 \tag{37a}$$

$$\int_{-\infty}^{\infty}\int_{-\infty}^{\infty} f(x,y)\, dx\, dy = 1 \tag{37b}$$

Note that the function $Z = f(x,y)$ represents a surface, called the *probability surface*. Figure 7.10 illustrates the function, where the total volume bounded by the surface and the x-y plane is equal to unity. The probability that X lies between a and b while Y lies between the limits of C and D is shown graphically by the shaded volume in Figure 7.10. Mathematically, this is expressed by

Definitions in Probability and Statistics

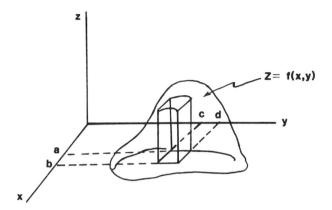

Figure 7.10 Joint probability function.

$$P(a < X < b, c < Y < d) = \int_{x=a}^{b} \int_{y=c}^{d} f(x,y) \, dx \, dy \tag{38}$$

If A represents any event, there will be a region $\mathcal{R}_A$ of the xy plane corresponding to it. The probability of A is

$$P(A) = \int_{\mathcal{R}_A} \int f(x,y) \, dx \, dy \tag{39}$$

The joint distribution function is

$$F(x,y) = P(X \leq x, Y \leq y) = \int_{u=-\infty}^{x} \int_{v=-\infty}^{y} f(u,v) \, du \, dv \tag{40}$$

The density function is obtained by differentiating the distribution function with respect to x and y:

$$\frac{\partial^2 F}{\partial x \, \partial y} = f(x,y) \tag{41}$$

Thus the distribution functions of X and Y are

$$P(X \leq x) = F_1(x) = \int_{u=-\infty}^{x} \int_{v=-\infty}^{\infty} f(u,v) \, du \, dv \tag{42a}$$

$$P(Y \leq y) = F_2(y) = \int_{u=-\infty}^{\infty} \int_{v=-\infty}^{y} f(u,v) \, du \, dv \qquad (42b)$$

The derivatives of these expressions are the density functions

$$f_1(x) = \int_{v=-\infty}^{\infty} f(x,v) \, dv \qquad (43a)$$

$$f_2(y) = \int_{u=-\infty}^{\infty} f(u,y) \, du \qquad (43b)$$

Let's now consider the case where X and Y are discrete random variables. When events $X = x$ and $Y = y$ are independent for all x and y, the X and Y are *independent random variables*.

$$P(X = x, Y = y) = P(X = x)P(Y = y) \qquad (44)$$

Another way of stating this is

$$f(x,y) = f_1(x)f_2(y) \qquad (45)$$

IMPORTANT THEOREMS OF RANDOM VARIABLES AND PROBABILITY DISTRIBUTIONS

We shall summarize important theorems for handling random variables and probability distributions. When the probability distributions of one or more variables are given, it is possible to determine distributions of other random variables which depend on them in some way. The following theorems outline procedures for obtaining them for the case of discrete and continuous variables.

Theorem(s). Define X as a discrete random variable whose probability function is $f(x)$. A discrete random variable U is further defined in terms of X by $U = \phi(X)$. To each value of X there is a corresponding U-value, so that $X = \psi(U)$. In this case the probability function for U is given by

$$g(u) = f[\psi(u)] \qquad (46)$$

Definitions in Probability and Statistics

Theorem (t). We define X and Y to be discrete random variables having joint probability function $f(x,y)$. Furthermore, there are two discrete random variables U and V, defined in terms of X and Y by $U = \phi_1(X,Y)$ and $V = \phi_2(X,Y)$. To each pair of X and Y values there is only one pair of values of U and V and, conversely, so that $X = \psi_1(U,V)$ and $Y = \psi_2(U,V)$. The joint probability function of U and V is given by

$$g(u,v) = f[\psi_1(u,v), \psi_2(u,v)] \tag{47}$$

Theorem (u). If X is a continuous random variable with probability density $f(x)$, we may define $U = \phi(X)$, where $X = \psi(U)$. The probability density of U is given by

$$g(u) = f(x) \frac{dx}{du} = f|\psi(u)||\psi'(u)| \tag{48}$$

Theorem (v). We define X and Y to be continuous random variables having joint density function $f(x,y)$. Further, we define $U = \phi_1(X,Y)$, $V = \phi_2(X,Y)$, where $X = \psi_1(U,V)$ and $Y = \psi_2(U,V)$. The joint density function of U and V is given by

$$g(u,v) = f(x,y) \frac{\partial(x,y)}{\partial(u,v)} \quad f[\psi_1(u,v), \psi_2(u,v)] |J| \tag{49}$$

where

$$J = \frac{\partial(x,y)}{\partial(u,v)} = \begin{vmatrix} \frac{\partial x}{\partial u} & \frac{\partial x}{\partial v} \\ \frac{\partial y}{\partial u} & \frac{\partial y}{\partial v} \end{vmatrix} \equiv \text{Jacobian determinant}$$

The following theorems are useful for determining the probability distribution of some specified function of several random variables:

Theorem (w). X and Y are continuous random variables, and $U = \phi_1(X,Y)$, $V = X$. The density function for U is the marginal density obtained from the joint density of U and V as stated in Theorem (v).

Theorem (x). $f(x,y)$ is the joint density function of X and Y. The density function $g(u)$ of the random variables $U = \phi_1(X,Y)$ is obtained by

$$G(u) = P[\phi_1(X,Y) \leq u] = \int \int_{\mathcal{R}} f(x,y)\, dx\, dy \tag{50}$$

where $\mathcal{R}$ is the region for which $\phi_1(x,y) \leq u$.

PRINCIPLES OF GEOMETRIC PROBABILITY

We will introduce this subject by way of an example. In the production of rubbers via the solution-based process (e.g., EPDM) the final product goes through a finishing operation where it is dried and formed into a bale. Since moisture is a contaminant that can affect cure and compounding properties, the rubber must meet a maximum permissible water content level before being shipped to the compounder. Samples of rubber from bales are randomly taken for moisture checks. Let's assume for the sake of simplicity that the bale is circular and that the sample is retrieved manually and at random. In this way, the sampling procedure is much like a game of darts, whereby the sample may be retrieved from any point along the radius of a circle). Figure 7.11 shows the geometric configuration of the system under consideration. We can show that the probability of a sample being taken between r and $r + dr$ is

$$P(r \leq R \leq r + dr) = c[1 - (r/a)^2]\, dr$$

This could be important in assessing the uniformity of drying for rubber bales. Note that R is the distance of the sample point from the center of

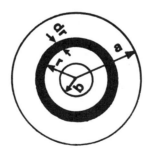

Figure 7.11 Circular bale problem.

the bale, c is a constant, and a is the radius of the bale. Let's find the probability that a sample will be retrieved from the center of the bale, which is assumed to have radius b. The density function is

$$f(r) = c[1 - (r/a)^2]$$

and hence

$$c \int_0^a [1 - (r/a)^2] \, dr = 1$$

from which we determine that $c = 3/(2a)$. The probability that a sample will be taken from the center of the bale is

$$\int_0^b f(r) \, dr = \frac{3}{2a} \int_0^b [1 - (r/a)^2] \, dr = \frac{b(3a^2 - b^2)}{2a^3}$$

There a variety of problems that arise from geometric considerations. The one above is an example. Another example is that we have a target in the form of a plane region of area K and a defined portion of it with area K_1. It is reasonable to suppose that the probability of hitting the region of area K_1 is proportional to K_1. Thus we define

$$P \text{ (hitting region of area } K_1) = \frac{K_1}{K} \tag{51}$$

It is assumed, as in the first example, that the probability of hitting the target is 1. We can, of course, assume some other probability value, where appropriate.

SUGGESTED READINGS

Cheremisinoff, N. P., *Practical Statistics for Engineers and Scientists,* Technomic Publishing, Lancaster, Pa., 1987.
Debing, L. M., ed., *Quality Control for Plastics Engineers,* Reinhold, New York, 1957.
Hayslett, H. T., jr., *Statistics Made Simple,* Doubleday, New York, 1968.
Snedecor, G. W., and W. G. Cochran, *Statistical Methods,* Iowa State University Press, Ames, Iowa, 1980.
Spiegel, M. R., *Theory and Problems of Probability and Statistics,* Schaum's Outline Series, McGraw-Hill, New York, 1975.

8
Application of Quality Control Techniques

FREQUENCY DISTRIBUTIONS

Every process has some variability. Automatic temperature controls are activated only when there is a difference between the actual and the desired temperature, and temperature fluctuations are inevitable. The preforming and preheating steps in manufacturing rubber and plastic articles are subject to variations. As an example, despite extraordinary precision in die sinking, two or more presumably identical dies will never be exactly alike. Measuring out charges to a die, whether by weight or volume, introduces other sources of variability. Also, lot-to-lot, drum-to-drum variations in the physical and chemical properties of raw materials are inevitable.

To illustrate this variability, take 100 molded articles, measure them, and set down the readings. If measured to the nearest thousandths of an inch or less, the variation will soon become apparent. This variation can also be followed by weighing pieces and plotting variations by weight instead of dimensions. This is easier and frequently just as sensitive as measuring dimensions.

All these sources of product viability, when reduced to their practical minimum, are grouped together and called "inherent variability." This is about as good as can be hoped for without radically changing the operation.

Application of Quality Control Techniques

Obviously, one should not be satisfied until the observed variability is close to this minimum. Statistical theory provides a picture of what this minimum variability should be. If the variability does not resemble an approximate distribution, the variability is more than should be condoned.

It is never possible in any production operation to manufacture two things exactly alike; it is the realization of this fact that makes possible the development of the science of statistical quality control. By making use of the variation patterns obtained over a period of time it is possible to compare the behavior of a process to its expected performance, and hence to determine what, if any, corrective action is required for better quality.

A study of a large majority of frequency histograms reveals that they have certain characteristics in common; they are relatively symmetrical, with the greatest number of observations to be found in the center of the scale, and fewer and fewer data points occur as the distance from the center increases.

It was Gauss, back in 1803, who first decided that since so many collections of data from many varied sources followed the curve that we have been discussing, there was something wrong with any data that did not behave in this manner. Therefore, he called the frequency curve the "normal curve of errors," or the "normal frequency distribution."

By analyzing a process that has been found to fit the normal distribution curve, it is possible to predict with a good degree of certainty what percentage of the measurements will fall between certain values on the scale. The mathematics have been worked out in great detail, so that with relatively simple calculations it can be determined whether a process is behaving as it should.

As described in Chapter 7 to define the curve completely, we need some idea of (1) its level (where is it located on the scale of values?) and (2), its dispersion (how far out along this scale does it spread?)

The most effective way of measuring or defining the level of a distribution is the average, or arithmetic mean. This is obtained by adding up all the values of the individual measurements and dividing the total by the number of measurements taken.

The most effective way of expressing the dispersion of a distribution is by using the standard deviation. This is defined as "the square root of the mean of the squares of the individual deviations from the average." To obtain the standard deviation, the following procedures may be used.

1. Determine the average of all the data observed.
2. Determine the difference between each observation and the average.
3. Square each of these differences.

4. Add all these squares together.
5. Divide the sum of these squared differences by the total number of data points.
6. Take the square root of this result.

An interesting point using the average and standard deviation in describing a collection of data is that for any given average and standard deviation, one and only one normal frequency distribution curve can be drawn. The standard deviation is usually expressed by σ and the arithmetic mean by the symbol $\bar{X}$ (X-bar).

It may well be asked at this point: How can we make use of this mathematical curve in controlling product quality? To answer this, we should first consider the relationship between the data collected from a process and the normal frequency distribution curve calculated from the data.

1. The mean or average of the frequency curve on the scale of values is equal to the average of all the collected data obtained.
2. The height of the curve above the baseline represents the relative frequency with which the corresponding measured value would occur in a large number of measurements.
3. The area under the curve between any two given measurements that percentage of the measurements that will fall between the two stated values in a large number of observations. The mathematics have been worked out and tables have been published to assist in using this information.
4. The standard deviation measures the expected spread of measured values, and the relationship between the standard deviation and the area under the curve is shown in Figure 8-1.

In controlling the quality of a production operation, we make use of the values already given. In Figure 8-1:

Between ±1 standard deviation from the mean we expect to find 68% of all observed values.
Between ±2 standard deviations from the mean we expect to find 95% of all observed values.
Between ±3 standard deviations from the mean we expect to find 99.7% of all observed values.

In general, as long as the observed values conform to the expected proportions, it can be assumed that the process is operating as it did when the original calculations were made. Any drastic shift in these percentages indicates a change in the process; it is to our advantage to investigate and

Application of Quality Control Techniques

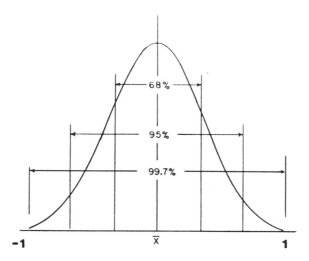

Figure 8-1. Details of a normal distribution curve.

maintain this shift if it represents better quality and to eliminate the causes if it represents poorer quality.

Since the probability of measurements falling between ±3 standard deviations is practically 100%, provided that no changes (intentional or otherwise) are made in the process, the values corresponding to ±3σ are called the natural tolerance of the process. Hence when we observe that a measured dimension on an extruded article average 0.015 in. when the molding operation is in a good state of control, and calculate the standard deviation at 0.0001 in., we call 0.015 ± 3 (0.001 in.) the natural tolerance of the dimension. We expect 97.73% of all pieces to fall between 0.012 and 0.018 in. Another term often used for these limits is the 3σ or control limits, discussed later.

When a dimensional tolerance is specified on a molded or extruded article it is advisable to determine the natural tolerance of the molding operation. If this natural tolerance is narrower than the specifications, everything is in good shape; if the natural tolerance is wider than the specifications, a certain amount of scrap will be produced. Figure 8-2 shows several operations in which the normal frequency distributions are compared with the tolerance and indicates the type of action that would be required for optimum quality control.

In discussing the properties of the normal frequency distribution we have covered the underlying mathematical basis for most production operations where the data represent a measured property. Properties that

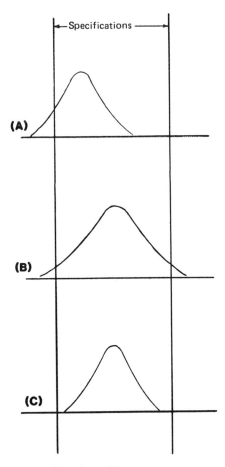

Figure 8-2. (A) Process level too low; (B) spec too narrow and process variation too great; (C) ideal case.

can be measured are called variables and include such things as dimension, tensile strength, chemical properties, electrical and chemical resistance, compression set, and so on.

In some processes there are certain quality characteristics that are not measured. For example, we may be interested in whether a particular property is between certain limits or outside these limits, or we may be interested in whether a certain item is considered good or bad as regards a certain visual standard. The quality decision that we have to make is

Application of Quality Control Techniques

merely a yes or no decision; is the item good or bad, acceptable or defective? These quality characteristics are referred to as *attributes*.

It is difficult to see at a glance how the properties of the normal frequency distribution can be applied to this type of quality characteristic. If we consider a fairly large number of individual items as a unit of production and determine by a go/no-go type of inspection how many of these individual items are good and how many are bad, the fraction defective is thus obtained and can be used as a quality measurement. A series of such observations will eventually build up into a particular type of frequency distribution. This type of distribution is called the "binominal frequency distribution."

Under certain conditions the binomial distribution conforms closely to the normal frequency distribution. For fairly high values of percent defective (20 to 80%), or where the number of items inspected is quite large, the normal distribution can reasonably be substituted for the binomial frequency distribution.

A special case of attribute inspection is that in which the number of defects in a certain unit of production is counted. For example, the total number of discolored specks on the surface of bales of rubber might be considered molded piece or extruded article might be others. There is a frequency distribution somewhat similar to the binomial which can be used to describe the type of variation occurring in counted data of this type. This distribution is called the *Poisson distribution*. Mathematical tables showing the percentages to be expected within certain limits are available for the Poisson distribution.

Much of the growth of mathematical interpretation of quality data (and other operating data) has resulted from the use of the normal and binomial distribution for representing the expected variation of a process and the expected chance error in the testing methods. Tests have been developed to show how confident we may be in making management decisions on available data; these are at the fingertips of quality control and management personnel.

Above, we showed that the use of frequency distributions for quality analysis is valuable for many reasons. The process history can be compared with specifications to show whether the product quality coincides with the specifications and if the variability lies within specification limits. The shape of the frequency distribution quantifies the type of variation encountered and gives clues as to the kind of corrective action required. Also, the frequency distribution provides a means of presenting quality information to all personnel—to operators, setup men, toolroom personnel, production and inspection supervisors, tool design engineers, and so on.

Although a histogram is valuable in analyzing past performance, it is not particularly suitable for evaluating real-time performance. For example, the order in which the data are obtained is lost when the frequency distribution is plotted. We are not able to determine from the distribution whether trends or cycles are occurring in the process.

Another shortcoming of this method is the time that it takes to present the information. The actual quantity of production during the period necessary to obtain, plot, and analyze the data may be such that a considerable amount of effort could be lost in defective material before anyone ever knew about it. The frequency distribution may well be an excellent tool for obtaining information about processes; it is most effective in achieving long-range quality improvements, but for the day-to-day quality decisions, something faster is needed. Key questions to ask are: How are we to make quality decisions more rapidly than by use of frequency distribution? How are we to establish the existence of trends, cycles, or excessive sample-to-sample variation in the data?

The answers lie in making use of the previous quality history of the process as a "frame of reference," against which the relative quality of each sample test value can be judged. Suppose that through the use of frequency histograms we were to determine the average and standard deviation of past quality performance. By observing the pattern of variations we were assured that only chance or normal variation had been present. We can now construct a chart with the scale of measured values as the vertical axis. By drawing a horizontal line at the value corresponding to the average, and two parallel lines above and below the average at a distance of three standard deviations, the "frame of reference" mentioned above can be derived. The horizontal axis of such a chart gives us something we did not have in our frequency histogram. It can be used as a time scale. This is, we can divide it into units representing the time at which the sample was taken, or merely record the order in which the various data values were obtained. Figure 8-3A shows the relationship between this chart and the frequency distribution, showing the past history of the process.

For each sample, its test value is plotted on the chart so that it represents both the value obtained and the order in which the sample was taken. Figure 8-3B shows the following values plotted against the frame of reference: sample 1, 14.2; sample 2, 15.7; sample 3, 15.2; sample 4, 14.5; sample 5, 17.0; sample 6, 13.7, and so on. Note the cryptic abbreviations LCL, AVE, and UCL on the chart. These stand for lower control limit, average and upper control limit. Traditionally, the control limits are three standard deviations above and below the average line.

Application of Quality Control Techniques

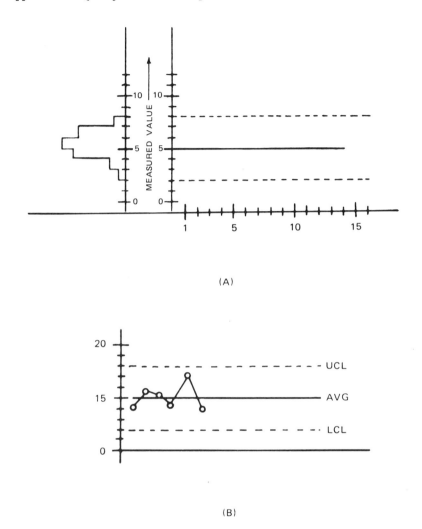

Figure 8-3. (A) Start of a control chart, illustrating relationship between control chart framework and frequency distribution; (B) chart showing distribution of group averages about total average.

The chart, with its upper and lower control limits, and individual measurements plotted in the order of production is known as a *control chart*. As long as the data points plotted remain within the limits and are fairly uniformly dispersed on both sides of the average, it can be assumed that the same chance variation that we had when we calculated the limits continues to exist.

The control chart is useful for separating the "only to be expected" variation from that which indicates a change in the process. In fact, this is the purpose of statistical control charts. There is a catch, however. For control charts to work properly, we must be able to make a reasonably good assumption that the distribution pattern conforms to the normal frequency distribution. This assumption is however not always valid. A dilemma often arises on how far a distribution may deviate from normal before the control chart system breaks down.

There is a way out of this problem through the "central limit theory." Briefly, it may be stated as follows: "No matter what the distribution of individual values in a population, the distribution of the averages of several items taken at random from the population approaches the normal distribution as the sample size increases." Shewhart (1939) demonstrated that for most processes, taking averages of three to seven values is sufficient to normalize the distribution.

Starting off then with the averages of small random samples being considered as normally distributed, it is possible, as in Figure 8-3, to construct a control chart of averages that is related to the distribution of averages as the chart was related to the distribution of individual measurements. If the process is operating with only the expected variation present, a group of individual successive values can be considered as a random sample for the purpose of obtaining average readings.

A shortcoming in the suggested method can immediately be seen. The average values show process variation on a group-to-group basis, but there is no way of estimating the spread of the variation within the group. A sample with an average of 20 could be made up of four individual readings of 19, 20, 20, and 21, or 15, 20, 27, and 18, or even a wider spread of values. There is no way of telling from the chart what the situation actually is.

The original suggestion by Shewhart was that another chart, showing the variation within the group by means of the standard deviation of the individuals, could be used along with the averages. The two charts taken together would then show the within- and between-groups variation. Figure 8-4 shows how such a chart would be set up.

Application of Quality Control Techniques

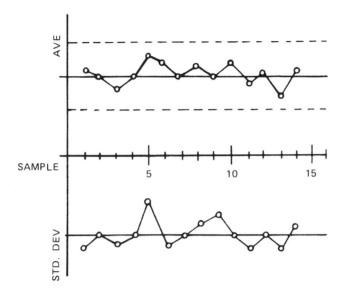

Figure 8-4. Variation of group averages (top chart) and the standard deviation within each of the groups (bottom chart).

We can further reason that variation within small samples provided a good estimate of the chance or expected variation of the process, and develop tables by which control limits for both averages and standard deviations can be estimated from the average standard deviation of all samples.

For small samples (2 to 15 individuals) the range is closely related to the standard deviation and is more readily determined. The range is the difference between the largest and smallest individual value in the group. Shewhart (1939) proposed the use of average and range control charts as an effective means of controlling product quality (known as the Shewhart control chart).

The advantages of average and range charts over control charts are:

- The use of the normal distribution probabilities is more valid.
- Trends or shifts in level are more readily detected.
- Changes in variability can easily be seen on the range chart.

- Control limits are easy to calculate.
- Out-of-control points can be systematically eliminated prior to calculation of the control limits.
- Averages are less sensitive to experimental error.
- More data can be plotted in a given space, making recognition of long-term variation clearer.

The control limits for averages can be calculated by the same procedure for calculating limits for individuals. For example, if each group average is considered as one data point, the mean and standard deviation for these averages can be calculated, and 3σ (3 standard deviation) limits set up around the mean.

The following terminology should be noted.

Subgroup: individuals that are grouped together and averaged for plotting on the control chart for averages.

Subgroup size: indicated by the symbol n; number of individuals in subgroup.

Subgroup average: indicated by the symbol $\bar{X}$; average of the individual values in the subgroup. Averages of the subgroup averages is denoted by the symbol $\bar{\bar{X}}$.

Range: indicated by the symbol R; difference between the highest and lowest values in the subgroup.

Average range: indicated by the symbol $\bar{\bar{R}}$; average of all the subgroup ranges.

The factors for calculating the control limits for average and range charts are given in Table 8-1. The following procedure outlines the calculations.

Table 8-1 Factors for Calculating Control Limits for Average and Range Charts

n	A_2	D_3	D_4
2	1.880	0	3.268
3	1.023	0	2.574
4	0.729	0	2.282
5	0.577	0	2.114
10	0.308	0.223	1.777
15	0.223	0.348	1.652

Application of Quality Control Techniques

1. Select the desired group size and plot the averages and ranges.
2. Obtain the average of all the ranges ($\bar{R}$).
3. Find the value of D_4 in the table corresponding to the group size selected.
4. Multiply $\bar{R}$ by D_4. This gives the upper control limit for the ranges.
5. Multiply $\bar{R}$ by D_4. This gives the upper control limit for the ranges.
6. Calculate the average of the averages ($\bar{\bar{X}}$).
7. Find the value of A_2 corresponding to the sample group size.
8. Multiply this factor by the average range.
9. Add the product ($A^2\bar{R}$) obtained to the average of the averages ($\bar{\bar{X}}$). This is the upper control limit for the averages.
10. Subtract the product $A_2\bar{R}$ from $\bar{\bar{X}}$. This gives the lower control limit for the averages.

More complete tables may be found in many standard statistics texts. We illustrate the procedures with an example. Table 8-2 lists 100 successive wall thickness measurements taken on an extruded vulcanized tube for a radiator hose application. Table 8-3 shows the same data broken down into 20 successive groups of 5 with the calculations conducted according to the steps outlined above.

$\bar{X}$ = sum of the averages = 12.765
$\bar{\bar{X}}$ = average of the averages = 0.638
R = sum of the ranges = 0.430
$\bar{R}$ = average range = 0.022

Control limits for averages
= $\bar{\bar{X}} \pm A_2\bar{R}$ = 0.638 ± (0.577)(0.022)

Upper control limit for averages
= 0.638 + 0.013 = 0.651

Lower control limit for averages
= 0.638 − 0.013 = 0.625

Upper control limit for ranges
= $D_4\bar{R}$ = (2.114)(0.022) = 0.047

Lower control limit for ranges
= $D_3\bar{R}$ = (0)(0.022) = 0

Note: Values of A_2, D_3 and D_4 are given in Table 8-1.

In an operation like the one just discussed, a certain variation is expected in the measured dimension. According to our calculation, 99.73% of the averages of four measurements should lie between 0.661 and 0.709 mm, provided that the process operates essentially as it did when the data were being collected. The control chart is shown in Figure 8-5, which

Table 8-2 Example Problem: Wall Thickness Measurements

Sample no.	Thickness (mm)	Sample no.	Thickness (mm)	Sample no.	Thickness (mm)	Sample no.	Thickness (mm)
1	0.525	26	0.667	51	0.714	76	0.695
2	0.528	27	0.682	52	0.691	77	0.696
3	0.560	28	0.698	53	0.668	78	0.697
4	0.592	29	0.697	54	0.686	79	0.697
5	0.624	30	0.697	55	0.704	80	0.696
6	0.623	31	0.697	56	0.722	81	0.696
7	0.621	32	0.697	57	0.740	82	0.695
8	0.620	33	0.696	58	0.758	83	0.695
9	0.618	34	0.696	59	0.751	84	0.694
10	0.617	35	0.696	60	0.744	85	0.694
11	0.630	36	0.696	61	0.737	86	0.693
12	0.643	37	0.695	62	0.730	87	0.693
13	0.656	38	0.695	63	0.723	88	0.692
14	0.669	39	0.695	64	0.716	89	0.692
15	0.665	40	0.705	65	0.709	90	0.691
16	0.662	41	0.682	66	0.702	91	0.691
17	0.659	42	0.693	67	0.695	92	0.693
18	0.656	43	0.704	68	0.688	93	0.696
19	0.652	44	0.705	69	0.689	94	0.696
20	0.649	45	0.716	70	0.690	95	0.699
21	0.646	46	0.727	71	0.691	96	0.704
22	0.642	47	0.738	72	0.692	97	0.707
23	0.639	48	0.749	73	0.693	98	0.709
24	0.636	49	0.760	74	0.693	99	0.712
25	0.651	50	0.737	75	0.694	100	0.715

shows the normal state of control for this process, as calculated from the data.

Suppose that in the same process, the pattern observed in the previous chart (Figure 8-5) continued as in Figure 8-6A, with point A indicating the spot at which the new data plot starts. A survey of the control chart shows that at the two points marked B the process actually went outside the limits of expected variability. We can now make the statement, with the 99.73% probability being correct, that there has been a change in the process. These are action points and indicate the presence of an "assignable cause" of variation, that is, a cause of variation other than those which are normal. A further look at the chart shows that at point C there are more than

Application of Quality Control Techniques

Table 8-3 Subclassification of Data From Table 8-2

Sample X (mm)	Average X	Max X	Min Range	numbers
1–5	0.637	0.653	0.628	0.025
6–10	0.631	0.654	0.623	0.031
11–15	0.642	0.676	0.615	0.061
16–20	0.627	0.629	0.625	0.004
21–25	0.645	0.653	0.625	0.028
26–30	0.642	0.632	0.628	0.015
31–35	0.637	0.638	0.623	0.015
36–40	0.638	0.646	0.615	0.031
41–45	0.643	0.660	0.625	0.035
46–50	0.650	0.658	0.625	0.033
51–55	0.640	0.648	0.636	0.012
56–60	0.635	0.645	0.625	0.020
61–65	0.635	0.648	0.625	0.023
66–70	0.639	0.642	0.636	0.006
71–75	0.635	0.645	0.623	0.022
76–80	0.635	0.642	0.625	0.017
81–85	0.636	0.645	0.623	0.022
86–90	0.637	0.639	0.632	0.007
91–95	0.635	0.652	0.625	0.027
96–100	0.646	0.650	0.642	0.008
Sum =	12.765			0.430
Avg. of averages =	0.638			0.022

seven consecutive values on one side of the mean. At this point it can be concluded that the process average has shifted and that if nothing is done of a corrective nature, that data can soon be expected to fall outside the control limits. In reviewing the process history to find the new cause of variation, the chart tells us to go back to some point prior to C rather than to look for the new source point B.

Interpretation of the foregoing data has made use of two control chart rules, the "rule of 7" and the existence of points outside the control limits. There are many other indications of nonrandom variation that can be recognized through the use of control charts.

One case is where points are found to lie above or below control limits. This is the most obvious indication of an out-of-control process. The

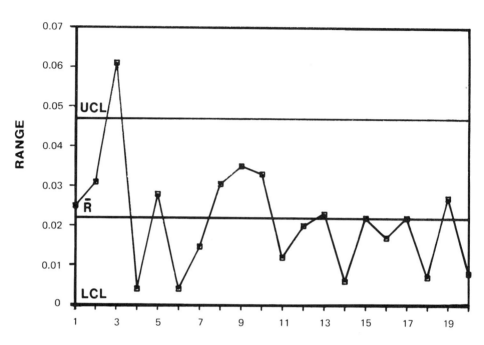
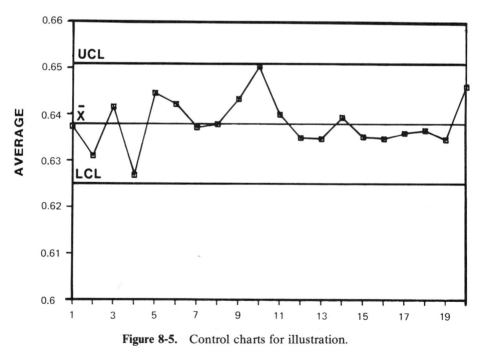

Figure 8-5. Control charts for illustration.

chances of getting the two points circled at B from the process as it was run at A are about 1 in 500,000. It can be concluded, therefore, that some factor under control at A was not controlled at B in Figure 8-6A. An assignable cause of variation has crept in, which unless located and corrected will continue to cause more points out of control. It may be difficult to assign causes for isolated points out of control, but as long as this condition exists, we are not getting the quality the process and equipment is capable of producing.

Examining Figure 8-6B, over section A-B the process control was good. Near B a process variable changed and the process after B is definitely operating at a new level. This new level may be desirable or undesirable; in either case, now is the time to diagnose, correct or adopt the cause, because sooner or later some points will run outside the control limits. This is one of the most common inferences from the statistical method since it detects assignable process changes before serious trouble has occurred.

Another common situation is the appearance of too many points near a control limit (i.e., a nonnormal distribution). With 5 points out of 14 in the "2 to 3σ zone" it would be wrong to say that an assignable cause of variation existed in the process not more than once in a thousand times. Obscure as this interpretation might be, it is extremely valuable in a carefully controlled process because it asks for corrective action long before substandard material is produced.

A third common situation is a nonnormal distribution that is bimodal (Figure 8-6C). A number of things could result in this type of chart, one of which is process "overcontrol." A process is overcontrolled when corrections are applied on the basis of extreme values when those extreme values are actually perfectly normal to the process. Normal variations can be determined by maintaining everything under as close control at a constant level as possible. To correct for normal variations does nothing but superpose an artificial variation on one that already exists. It is in fact unwise to change operating levels until several successive points are out of control; if the cause cannot be found, one should make an arbitrary process change. Even then, one can get into trouble if the unknown cause suddenly disappears.

Figure 8-6D shows yet another commonly encountered situation, that of cylical trends. It is often said that product uniformity is at least three times poorer than the process is capable of maintaining. Slow changes in blended raw materials, due to pumping variant tank car loads, could be one possible explanation. So could cyclical supervisory pressure to produce higher quality as well as the slow unrecognized effect of time on

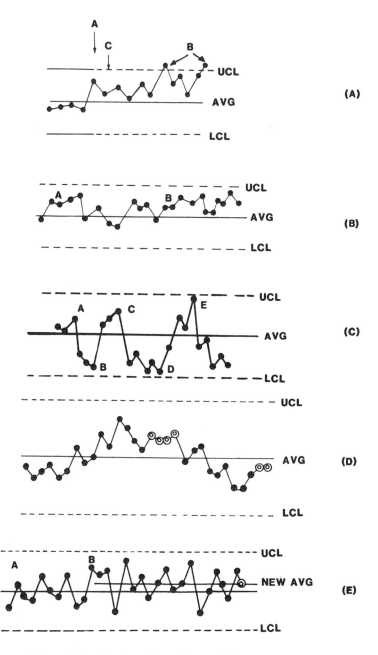

Figure 8-6. Continuation of control chart example.

Application of Quality Control Techniques

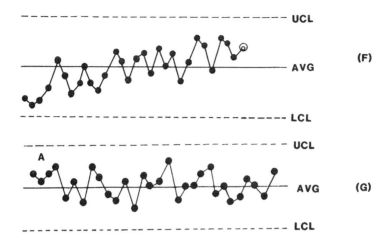

operating procedures where these have not been pinpointed to precise values.

In addition to showing gross variations in quality, the control chart is a sensitive indicator of minute variations that could hardly be detected by any other method. In Figure 8-6E an assignable cause of variation entered at point B, shifting the average by a barely perceptible but truly significant amount. Perhaps someone changed the catalyst feed to the polymerization reactor at B. In any case, a small but significant changes of level, and identification and adoption of causes leading to better quality, are major constructive functions of control charts.

Figure 8-6F illustrates a trending chart. In this case all points should be circled because the process is not in statistical control unless it is producing at a stable level of quality. Although it is obvious enough that a trend exists and that an assignable cause is producing it, it is not so obvious that if that cause were removed the finished quality would be more uniform. Example causes for this type of process response are gradually changing raw materials, transient production standards or procedures, catalyst poisoning, or laboratory standard solutions that are off standard.

Finally, Figure 8-6G illustrates the key features of a control chart indicative of a process that is under good control. These features are:

All process variables controlled prior to A were as adequately controlled through B.
Quality variation is completely random, centered about the average, free of runs (trend), and points are more dense near the average (normal distribution).

This does not mean that the process cannot be improved. More accurate instrumentation can always be installed. What the chart tells us is that we are doing a fine job with the present process and equipment, and that it is unlikely that a better job can be done unless we have a new process, new raw materials, or new equipment.

We now direct attention to a variation on the average and range chart—the so-called "modified control chart," which is useful when the process specifications are fairly wide compared with the process control limits. In using this type of chart we are primarily interested in those changes that will result in the production of out-of-spec product. In using the modified control chart, the control limits are measured inward from the specification limits instead of outward from the process average, as is the case with the average and range chart. Figure 8-7A shows the amount of process variation permitted with the modified control chart compared with the conventional type of chart. If the cost of controlling the process within narrow limits at a given average is negligible, the conventional control chart approach will result in a much more uniform product and greater customer satisfaction. If this type of control is costly and the customer can permit some shifting of process level, then the modified chart is useful.

The so-called central limit theorem, on which the average and range chart, or Shewhart control chart, is based is noteworthy. The medium of the group of data is an estimate of the central value and is the middle value in the order of magnitude of the data. Thus for 25 data points, that value which has 12 observations above it and 12 below it is considered the median. The median of the five values 2, 7, 3, 4, and 6 is 4, while the average of these values is 22/5 or 4.4 If the group contains an even number of data points, the median is the average of the two middle values. Thus the median of the values 5, 6, 3, and 8 is the average of 5 and 6 or 5.5. In this case the average of the four values is also 5.5. To make a comparison between the average and range technique and the median range technique, the same data given as an illustrative example in Table 8-2 is reproduced in Table 8-4. The symbol X is used to indicate the median.

When these data are plotted on a median and range control chart, the data pattern is as indicated in Figure 8-8. To calculate the control limits, the following method is used.

1. Determine the median of all the subgroup ranges. Since the tenth and eleventh values in the order of magnitude are both 0.022, this is the median value of the 20 groups. Draw this median line on the chart and label it R.

Application of Quality Control Techniques

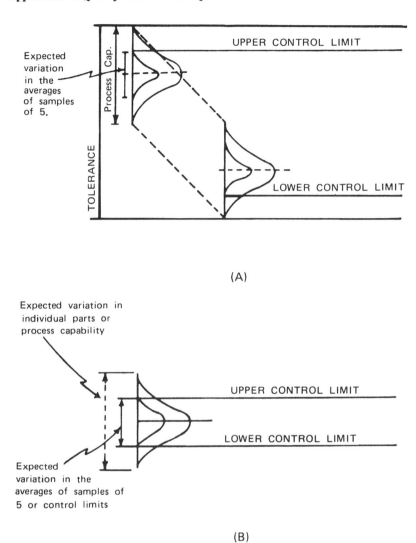

Figure 8-7. (A) Modified and (B) conventional control limits.

Table 8-4 Subclassification of Data from Table 8-2

Sample numbers	Average X (mm)	Range	Median
1–5	0.637	0.025	0.635
6–10	0.631	0.031	0.627
11–15	0.642	0.061	0.636
16–20	0.627	0.004	0.627
21–25	0.645	0.028	0.651
26–30	0.642	0.004	0.646
31–35	0.637	0.015	0.637
36–40	0.638	0.031	0.636
41–45	0.643	0.035	0.645
46–50	0.650	0.033	0.652
51–55	0.640	0.012	0.639
56–60	0.635	0.020	0.636
61–65	0.635	0.023	0.633
66–70	0.639	0.006	0.639
71–75	0.635	0.017	0.636
76–80	0.635	0.017	0.636
81–85	0.636	0.022	0.641
86–90	0.637	0.007	0.638
91–95	0.635	0.027	0.629
96–100	0.646	0.008	0.645
Sum =	12.765	0.430	
Avg. of averages =	0.638	0.022	
Median values =	0.637	0.022	0.637

2. Determine the median of the subgroup medians. The median of the medians in this example is 0.637.
3. To calculate the control limits for the ranges, multiply the median range by the appropriate $\tilde{D}_3$ and $\tilde{D}_4$ factors in Table 8-5. Thus the lower control limit for ranges of subgroup size of 5 becomes

$$D_3 R = (0)(0.022) = 0$$

The upper control limit for the ranges becomes

$$D_4 R = 2.179(0.022) = 0.048$$

Application of Quality Control Techniques

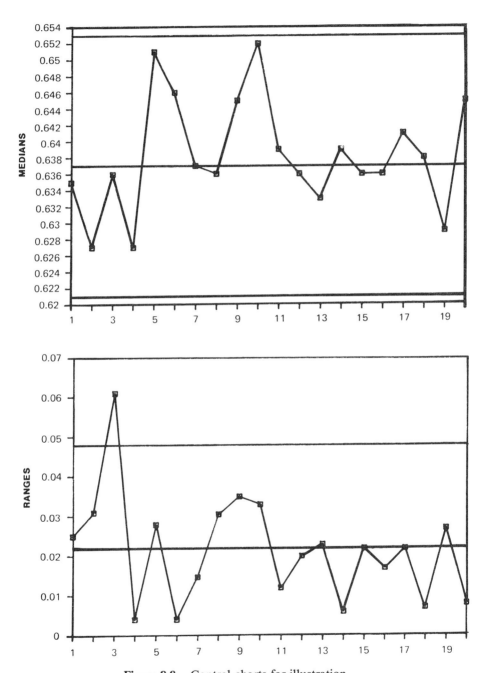

Figure 8.8. Control charts for illustration.

Table 8-5 Normal Factors for Control Charts Using Medians and Ranges

Subgroups Size	$\tilde{A}_2$	d_3	$\tilde{D}_2$	$\tilde{D}_4$
2	2.232	0.954	0	3.865
3	1.264	1.588	0	2.745
4	0.828	1.978	0	2.375
5	0.712	2.257	0	2.179
6	0.562	2.472	0	2.055
7	0.519	2.645	0.078	1.967
8	0.442	2.791	0.139	1.901
9	0.419	2.916	0.187	1.860
10	0.368	3.024	0.227	1.809
11	0.356	3.12	0.260	1.773
12	0.321	3.21	0.288	1.742
13	0.311	3.29	0.312	1.716
14	0.281	3.36	0.333	1.694
15	0.281	3.42	0.352	1.675

99.73% limits for medians = $\tilde{\tilde{X}} \pm \tilde{A}_2 \tilde{R}$
99.73% limits for ranges = $\tilde{D}_4 \tilde{R}$ upper limit
$\tilde{D}_4 \tilde{R}$ lower limit

$$\frac{\tilde{R}}{d_2} = \sigma \text{ for individuals}$$

4. To calculate the control limits for the medians, multiply the median range by the appropriate $\tilde{A}_2$ factor in Table 8-5. The value thus obtained is added to the median of the medians to give the upper control limit, and subtracted from the median of the medians to give the lower control limit. Thus

$$\text{UCL} = \tilde{\tilde{X}} + \tilde{A}_2 \tilde{R} = 0.637 + (0.712)(0.022) = 0.653$$

$$\text{LCL} = \tilde{\tilde{X}} - \tilde{A}_2 \tilde{R} = 0.637 - (0.712)(0.022) = 0.621$$

5. Draw these limits in on the control charts to complete Figure 8-8.

The values given in the column headed d_2 in Table 8-5 are of value in estimating the standard deviation of a given process by making use of the median range R. The standard deviation of the universe of individual values is estimated by dividing R by the appropriate $\bar{d}_2$ value in the table. In the example given,

$$\sigma = \frac{R}{d_2} = \frac{0.022}{2.257} = 0.0097$$

An advantage of this technique is that individual stray values of range or median do not inordinately effect the calculations of the limits. Thus, for a process that is out of control, the basic process capabilities are more directly determined by median and range. For processes in control, the two methods give comparable results. The interpretation of the variation pattern on a median and range chart is the same as that described for average and range charts. The techniques may be used interchangeably, and selection will depend on the needs of the individual.

The greatest application of control charts in industry has probably been by laboratory or inspection personnel, or by research chemists and engineers who are interested in basic product improvement and development of new products. The control chart, however, is designed primarily for process control, and as such is most useful to production personnel. The control chart supplies information in a form that is most readily available for immediate action. Charts should actually be followed by the operators, using data that they themselves, and giving them freedom of action based on the interpretation of the data. Control charts supply an excellent means of recording routine quality data and form a sound basis for performance review and quality improvement programs. The charts should be circulated freely within the organization, so that the groups responsible for various phases of the quality improvement function will be working in the same direction on the basis of the same information.

It is important to note that there are many production processes that are not readily adaptable to the control chart treatment discussed above. For example, if an item is produced in extremely large quantities, it may be too time consuming and/or costly to measure its quality characteristics. In such cases, the testing is often done by means of a go no-go gauge when a dimension is involved, or by comparison with a standard of acceptibility when another quality characteristic is being determined. In other situations we may be inspecting a complex item or assembly for the presence of more than one given defect, and as in the cases above, we are interested primarily in whether or not a defect is present. Quality parameters are not necessarily a precise measurement but rather a subjective rating of several characteristics, individually or collectively, which can qualify the product to be unsatisfactory for shipment. In the discussion above we emphasized quality characteristics that can be measured; these are called "variables." The problem of control of quality in an operation like the one just described concerns itself with what we call "attributes." Quality control of variables is concerned with the question "how much?" or "to what extent?" the quality characteristic exists. Quality control of attributes is concerned with the presence or absence of a given characteristic or defect. The acceptability or rejectability based on the surface characteristics of an extruded article can be considered an attribute parameter, for example.

If we had the luxury of considering each individual item alone, we could assess whether it is good or bad, and the only obvious improvement in quality would be complete elimination of all the defects. Such perfection is practically unattainable. How, then, can we measure improvement or know when the control of product quality changes?

By considering the number of defective items in a given number of items, a quality index can be defined. This index will enable the use of techniques for separating the expected variation of quality from the variation that indicates a change in the state of control. Earlier we observed that the distribution of averages of measurements could be presented by the normal frequency distribution curve, and made use of the mathematics of that curve to determine whether the process was behaving as expected. Under certain limited conditions, the number of defective items in a group of items can be considered as normally distributed quality, but in order to handle the majority of cases encountered in practice, it will be necessary to develop another distribution, which can be applied more generally. What is required is a frequency distribution that represents the amount of variation from sample to sample of the number of defective items or defects occurring in a process that is operating in a fixed state of control. This variation then represents the chance variation in a fixed state of control. This variation then represents the chance variation of the process, the differences to be expected from sample to sample due to chance alone.

As we learned from Chapter 7, as the sample size increases, the distribution becomes more symmetrical, and as the sample size increases indefinitely, the distribution approaches as a limit the normal frequency distribution. The mathematical expression for histograms derived in this manner is known as the "binomial expansion," and the distribution so obtained is called the "binomial distribution." The relative frequency of samples containing r defective items in a large number of samples is given by the expansion of the bionial

$$(p + q)^n$$

Application of Quality Control Techniques

where n is the number of items in the sample (sample size), p the expected proportion of defective items in the sample, and q is equal to $1 - p$. The value of r can vary from 0 to n, and the terms of the expanded binomial tell us how often to expect 0, 1, 2, ..., or n defective items in the sample.

The transition from the frequency distribution (i.e., bar charts as in Chapter 7) to a system of control charts is handled in a manner similar to the $\bar{X}$ and R charts. The basic distribution establishes the pattern of variation to be expected from the process, and the operation is considered to be "in control" as long as the variation does not exceed this pattern in degree.

If the value of p is close to 0.50, and n is sufficiently large, the distribution of the number of defects is reasonably close to the normal frequency distribution. In such cases we can calculate the average and 3σ limits for the distribution and use these factors on control charts as we did for the control charts for variables. The standard deviation may be calculated directly from the data, where X is the number of defectives in a given sample and N is the total number of samples taken.

The binomial distribution differs from the normal in that the variation is dependent on the average. The average number of defective items in the samples of n items is equal to np, where p is the probability of one item being defective. The standard deviation can therefore also be computed from

$$\sigma = \sqrt{np(1 - p)}$$

The use of the standard deviation as a measure of variability is only approximate unless the sample size is sufficiently large that the distribution is symmetrical. For small samples, or low values of p that give an appreciably skewed distribution, the probability limits of chance variation are not balanced and should be obtained from the binomial probability tables.

As an example, let's consider the mass production of weighted plastic floats to be used in a certain type of laboratory rotameter that we are marketing. These items are being checked on a go–no-go gauge for dimension and weight. We are manufacturing 100 units per hour. The number of articles rejected per hour is given in Table 8-6. From these data the mean or average number of defective articles per 100 units is

$$\frac{\Sigma \text{ (no. defective items)}}{\Sigma \text{ (no. samples taken)}} = \frac{158}{24} = 6.6$$

Table 8-6. Example Problem Data

Time (hrs)	Number of articles checked	Number of defective articles
0100	100	7
0200	100	8
0300	100	11
0400	100	6
0500	100	9
0600	100	9
0700	100	7
0800	100	6
0900	100	10
1000	100	8
1100	100	6
1200	100	7
1300	100	5
1400	100	7
1500	100	7
1600	100	7
1700	100	6
1800	100	4
1900	100	5
2000	100	6
2100	100	3
2200	100	4
2300	100	6
2400	100	5
Total	2400	158
Mean		6.6
Standard Deviation		2.48

The probability of an individual sample that is defective can be obtained by dividing the average number of defects per sample by the number of items in the sample:

$$p = \frac{6.6}{100} = 0.066$$

Application of Quality Control Techniques

To calculate the standard deviation for this set of data,

$$\sigma = \sqrt{np(1-P)}$$
$$= \sqrt{(100)(0.066)(0.934)}$$
$$= 2.48$$

The control limits for the number of defective items per sample of 100 in this case is given by $X \pm 3\sigma$ or $6.6 \pm 3(2.48)$ (i.e., the limits are 0 to 14.0). In other words, if this process continued to operate in the same state of control that existed when the data were collected, we would expect 99.73% of all the samples to have between 0 and 14 defective items.

The example above is, of course, not very practical because it is not cost-effective to check every article and it is not always possible to select a constant sample size. It is often more convenient to think in terms of the percent or fraction defective in a sample. In the example given above, we might consider the data collected in such terms, so that the fraction defective figures for the individual samples are 0.07, 0.08, 0.11, 0.06, and so on. The average fraction defective for the distribution is arrived at by dividing the total number of defective items obtained by the total number of items examined:

$$\text{average fraction defective} = \frac{158}{2400} = 0.66$$

The standard deviation for fraction defective is

$$\sigma = \sqrt{\frac{(p)(1-p)}{n}} = \sqrt{\frac{(0.066)(0.934)}{100}} = 0.0248$$

and the control limits are $0.066 \pm 3(0.0248)$. Note that the variation in sample fraction defective is inversely related to the sample size, so that for very small samples, the range of chance variability is quite large. In contrast, large samples give precise estimates of quality with very little inherent error.

In practical situations there are always variations in production rate, manpower requirements in the inspection department, and the state of control of the process itself that can affect possible variations in sample size. Suppose in the process just considered that we had to go through a period in which the sample size varied between 25 and 150 over a period of time. The data tallied are given Table 8-7. Shown also are the standard deviations and upper and lower control limits for each sample group. The control limits will obviously differ for each portion of the production process since the standard deviation depends on the sample size. Interpretation of these control charts follows the same general rules as discussed earlier. Runs, trends, shifts in level, and excessive variability are shown by the same symptoms as were indicative of such deviations from the expected variation for X and R charts.

Table 8-7. Example Problem Data for Varying Sample Sizes

Time (hrs)	Sample size	Number of defective articles	Fraction defective	Standard deviation	UCL	UCL
0100	25	2	0.080			
0200	25	3	0.120			
0300	25	2	0.080			
0400	25	1	0.040			
0500	25	3	0.120			
0600	25	1	0.40	0.0536	0.239	0
0700	50	3	0.60			
0800	50	4	0.080			
0900	50	3	0.060			
1000	50	6	0.120			
1100	50	5	0.100			
1200	50	4	0.080	0.038	0.192	0
1300	100	8	0.080			
1400	100	9	0.090			
1500	100	7	0.070			
1600	100	7	0.070			
1700	100	8	0.080			
1800	100	13	0.180	0.027	0.158	0
1900	150	9	0.060			
2000	150	11	0.073			
2100	150	8	0.053			
2200	150	15	0.100			
2300	150	9	0.060			
2400	150	12	0.080	0.022	0.144	0
Total	1950	153	1.927			

Average fraction defective = 153/1950 = 0.078

Application of Quality Control Techniques

There are often situations where we are interested in the total number of defects in a product. For example, a complicated molding may have several possible defects, any one of which might cause the rejection of the piece. With the *p* and *pn* charts we were concerned with defective items, but more precise control of quality and troubleshooting can often be assisted by talking in terms of the number of defects present. Also, we might be interested in counting contamination specks in bales of rubber, fisheyes in extruded plastic film, or surface defects in extruded articles. To separate the chance variation of an operation from the type of variation that indicates a new or "assignable cause," a control chart such as that shown in figure 8-9 should be constructed with control limits adjusted to fit the sample size.

The distribution underlying the variation in counted data of this type is the Poisson distribution. The Poisson distribution is a simplified approximation to any term of the binomial; as shown in Chapter 7. It is developed by the expansion of the transcendental number *e*:

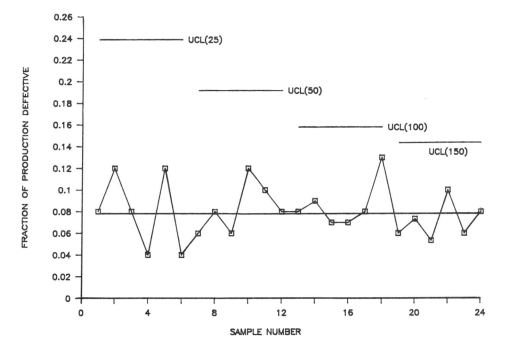

Figure 8-9. Example of a *C* chart.

$$1 = e^{-Z} \times e^{Z} = e^{-Z}\left(1 + Z + \frac{Z^2}{2!} + \frac{Z^3}{3!} + \cdots + \frac{Z^n}{n!}\right)$$

When plotted for appropriate values of z and n, a distribution similar to the binomial is obtained.

The probability limits of chance variation for this distribution are most appropriately obtained from the published tables, but as in the case of number of defectives, a workable approximation is available. This approximation is developed from the formula for the standard deviation of pn, by the following line of reasoning: If there is a large number of possible defects, and the number of defects is quite small compared to the number of defects that could have occurred, the formula for standard deviation of the pn chart can be simplified. Recall that

$$\sigma = \sqrt{np(1-p)}$$

where n is the sample size, p is the probability of an item being defective, and $1 - p$ is the probability of an item being nondefective. Substituting the value c for np as representing the number of defects likely to be found in a sample of a given size, and selecting the size of the sample so that the value of p is small, $1 - p \approx 1$, and the expression becomes

$$\sigma = \sqrt{c}$$

The control limits for the so-called c charts, shown in Figure 8-9 are $\bar{c} \pm \sqrt{\bar{c}}$, where $\bar{c}$, the average number per unit, is the best estimate of c. These charts used and interpreted in the same way as np and p charts.

There are several approaches to addressing the problem of varying sample size in attributes inspection. One commonly used method involves basing the control limits on an average sample size and recomputing for points that appear dangerously close to the limits (between 2 and 3σ for example). Another approach is the so-called stabilized p chart. On this chart, each point is plotted in relation to its probability of occurrence due to the chance and provides a means of detecting trends and shifts in level on fraction defective or percent defective charts.

We will close this subsection by making some general comments on the relation between specifications and quality control. It is important to note that specifications must be defined such that the requirements made of the product performance can be met properly within the constraints of the process. At the same time a specification states the sample size and whether zero, one, or some other number of tests can be beyond a limit. It

Application of Quality Control Techniques

must also require that every portion of a lot be as likely to be selected for the sampling as any other portion. This, in fact, embodies the statistical definition of randomness, for without a known risk of obtaining incorrect conclusions. Let's consider a very simple example: Suppose that the impact strength on a plastic bumper should not be less than 800 lb. We take 6 bumpers from a lot of 3000 and subject them to 800 lb. One out of the six fails the test. What is really meant by the specification in this case? What should we do with the remaining 2994 bumpers? Suppose we continue with the evaluation and select six other bumpers from the lot and subject them to the same test. This time the entire sample lot passes the test. If this had occurred during the testing of the first six samples, we would be asking no questions and the entire lot would be accepted. Would this decision, in fact, be correct? Obviously, the same lot can have samples that give different results, but from a practicality standpoint we can afford neither the time nor money to test each and every bumper in order to qualify the lot.

It should be clear from the example that any specification imposed on a product must be clearly defined as to what constitutes acceptable and nonacceptabe. Limits should be given that define products that will be successful in meeting real end-use requirements. Typical areas of coverage include dimensional fits and clearances, minimum tensile and impact strengths, crush strength and similar functional properties, all the way to weight loss resistance to solvents, water absorption, and electrical properties. If at the same time these limits result in products that can readily be made within the constraints of the process, we have a good specification. To transform the good specification into a workable one, make available practical standards against which the limits can be judged, and include rules for inspection. A specification that does all this represents a successful balance between the viewpoints of the producer and the consumer.

The natural tolerance of a given process refers to the numerical width of the band of the expected variation of individual results when the process is operating in statistical control. Earlier we showed that this special control condition can be expected to exist when the points for averages and those for ranges from successive samples fall within the control limits of the control chart. To get the natural tolerance or process capability, multiply the numerical distance between the control limits for averages by the square root of the sample size. As an example, if we have 0.10 mm between control limits for X for a sample size of 4 units, the process capability is $2 \times 0.10 = 0.20$ mm.

Later, the analysis of variance estimates from a factorial experiment are discussed, in which we can see how much of this total variability comes from the variability of each certain suspected factors and how much comes from the remaining factors as a group, which have not been selected for the analysis known as the precision of the test. As an introduction to analysis of variance, here is a brief description of the two techniques of balancing and randomizing, which are critical to this method.

An experiment is to consist of a series of test runs, each to be made with a certain combination of the factors in question. Say that three levels or values of each factor are to be studied. One combinations can be with all factors simultaneously at their lowest level, another the same but with one factor at its intermediate level, and so on. The design is laid out systematically in one of several possible statistical patterns with, of course, some selected reliability. But the combinations are so chosen that when the results common to one level of a factor are compared with the results common to another level of the same factor, the levels of all the other factors will have contributed equally to each of those two sets of results. In this manner, the effects of the other factors have been balanced out, and the comparison will estimate the differences in results attributed to moving from one level to another just one factor alone. Similar estimates can be individually obtained for all levels of all factors.

Randomizing prevents any one or more of the remaining, and nonselected, factors from having anything but a random effect on these estimates. It is usually the selection of test material and the order of the test run that is made to follow the random pattern in the application of randomization. From the standpoint of process economics, the relative cost evaluation approach to determining which requirements should be subjected to a statistical analysis can be used. Figure 8-10 shows an orderly plotting of process problems against percent of total cost of all problems with the items produced. Problem S was the most costly for the period. Its percentage of the total cost is shown by the crosshatched bar. Trouble C was the next most costly, and so on, as we move from right to left. The curve connecting the tops of the bars represents the accumulated cost.

This type of analysis reveals that a noticeable change of slope occurs in the accumulated curve. The place where the curve starts to approach some constant level asymptotically can be considered a point of diminishing returns. In the illustration we stopped showing the increments of cost per problem at this region. Notice that a reduction of 75% of the total cost would accompany a successful troubleshooting application based on the total cost would accompany a successful troubleshooting application based on only eight problem areas identified. The general experience is that a large cost is concentrated among relatively few items.

The concept of tolerance evolved from recognizing that certain specific and known things would make it "uneconomical" to require all product to come out identical. Recognized were such things as die dif-

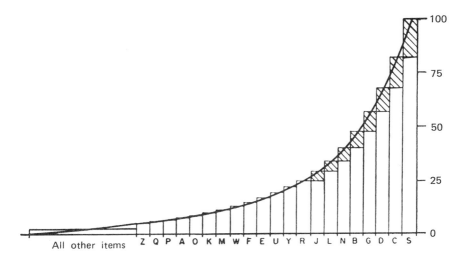

Figure 8-10. Plot of troubles versus total cost of troubles illustrating maximum cost reduction.

ferences; the complications involved if one charge to a die had to be just like the last; and that temperature changes and lot-to-lot and drum-to-drum variations in raw material properties might require "frequent" operator compensating adjustments. Management almost invariably believes that it is possible to make all product identical. In the author's own experiences dealing with customers, I have often heard the cry to hold a particular lot, for a specific reason, to the high limit or within an amount representing only a small fraction of the tolerance. If we were to construct a histogram with fine enough cell divisions, it would reveal the variation inherent in a process, and the demands of consistently identical product would no longer be feasible. One can employ a charting technique in which we maintain the measurements in their product order. Figure 8-11A shows such a charting example which can be converted to the process capability in the form of a control chart as shown in Figure 8-11B. This minimum spread of results will apply when the process runs as it did during the collection of the data. Also, this measure recognizes no assignable causes of extra variation as being present.

Process capability is essentially the natural tolerance of the process, or its 6σ width. If the tolerance is one-sided, having just a maximum or only a minimum limit, this point of view calls for at least 4σ between the limit and the desirable or expected average value of the process.

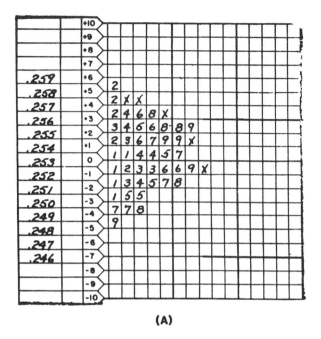

(A)

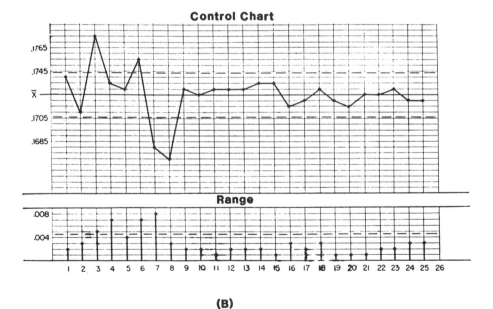

(B)

Figure 8-11. (A) Histogram of inherent variation in a process; (B) control chart for example.

Application of Quality Control Techniques

From the standpoint of specifications, then, a list of the process capabilities in a plant are needed. As an example, consider the production of parts that are assembled or mated. Distributions exist of each characteristic of mating ones, or of those that accumulate at assembly to give a possible large buildup. When the peaks or modes of these distributions lie near the center of the tolerance band, the relative frequency or number of parts that occur in the tails of the distributions become fewer as each tolerance limit is approached. This means that the probability of one of these near-limit parts being selected at random for a given assembly is rather small. But still more unlikely is the situation where two of these parts near conflicting limits are selected for a single assembly.

Statistical methods enable one to compute these probabilities and express "natural tolerance" of the assembly fit, clearance, or dimensional buildup. The assembly tolerance is made up of the "statistical" sum (as contrasted to the arithmetic sum) of the component natural tolerances. If there are two mating components controlled in production to remain within four units in one case and within three in the other, one will produce assemblies that will vary five units from minimum to maximum clearance, instead of the seven units of possible variation from an arithmetic addition. For three parts, say two, three, and six units, component natural tolerance, the statistical sum will be seven units, a saving of four units from the conventional sum of eleven. Each natural tolerance is squared, added, and the square root of the sum gives the assembly natural tolerance.

In an earlier chapter we echoed the cry that "quality cannot be inspected into a product; it must be built it." In the preparation of specifications this statement must not be overlooked. Inspection and production departments each have a specific role in the quality control procedure.

The customer of the product and the inspection department bear the same relationship to the producing group. Although they always have the task of considering a product for acceptance and sometimes specify tolerances and standards, they should never be given the authority to control quality. This emphasizes the unsatisfactory nature of the almost universal practice of having the quality control and inspection personnel in the organization closely allied. Such an arrangement inherently dulls the possible effectiveness of the quality control effort. Usually, one group reports directly to the other, and in certain cases there is no distribution between the two; just one department—inspection and quality control department—if it is not working under a still shorter name. The quality controll group should be part of production, and inspection can well be separate.

Tolerances exist as instructions to production. It is expected that they will strive, as a goal, to make all product conform completely. The concept of small or very moderate amounts of production permitted beyond tolerance limits is not a part of the control part of the specification. As a practical matter it would be rather difficult for a production group to produce just small or very moderate amounts of product beyond limits. It is much simpler to make it all within limits.

SAMPLING INSPECTION PHILOSOPHY

A successful sampling inspection program is based on obtaining a truly representative sample. It should be noted that there is no fixed relationship that has to be maintained between the lot size and the sample size. As long as the lot is homogeneous, the same size sample can be representative of the distribution of values (dimensions, strengths, visual defects, etc.) within a small lot or within a large lot. In this manner, the general impression that obtaining a sample from a lot is an unreliable way of assessing its disposition is correct only if the items are not homogeneously distributed within the lot. If we were selling jellybeans with a quality criteria that each lot of 1000 contain an equal number of four different flavors, a successful sampling program could be executed by placing the entire lot in a tumbler for some time period to ensure homogeneity. Also, the jellybeans would probably be fairly uniform in size and mass, and hence there would be little concern over segregation that could bias the lot's sampling.

Product forms are rarely as accommodating as jellybeans are to sampling programs. In such cases we might take the tack of assigning a number to each unit, from one on up through the lot size, and mix these numbers in a barrel (e.g., by writing the numbers down on disks): Then draw out a sample of disks from the mixture, and select those items of product for test that correspond to the numbers drawn. The results of such sample drawings have been tabulated in the form of tables of random numbers. The expression, "take a random sample" or "take a sample without regard to the quality differences among items of the lot" means that tables of random numbers should be used to guarantee a selection equivalent to that from a homogeneous lot. It must be emphasized that one cannot get a random sample simply by taking pieces "here and there" from a lot, or by "stirring" the lot first, or by assuming that handling between operations packing, and so on, has mixed the lot sufficiently. This approach can lead to samples that are not representative of the lot.

The obvious question to ask is: How representative is a random sample? Let's consider a sample of two items taken from a lot that was 5% defective. In this case the odds are rather high that the sample would be

Application of Quality Control Techniques

made up of two acceptable parts. In this case of random sampling, there is 1 chance in 20 (there being 5 defectives per 100) that the first piece drawn will be a defective one, and 19 chances in 20 that the second item will be a good one (there being 95 good items per 100). The probability of such a sample is 2 in 19/400 (1/20 times 19/20). The probability of the reverse sample, first good and second defective, is 19/400. There is 1 chance in 400 (the product of 1/20 times 1/20) that both items will be defective. The chance of either the first or the second being defective (19/400 + 19/400 = 38/400) is added to 1/400 to give 39/400, or 0.0975, as the probability of one or more defectives occurring in the sample of 2. This means that there is a remainder of 9.025, or 90.25% of the time that the sample will not indicate any defectiveness in the lot. In this example, a sample of 2 can only show 0, 50, or 100% defective. Sampling for this case would not be representative of a 5% defective lot.

In contrast, suppose that we have a very large sample taken at random from the same lot. This sample will almost always contain close to 5% defective items. We can conclude that while the size of the lot is not generally involved, the size of the sample does have a bearing on the reliability of the picture presented of the lot condition. Chance or probability calculations show just how often a lot of given percent defective will produce random samples that are some other percent defective.

Sampling programs are designed to be decision makers for the inspector. That is, we take a random sample of n items from the lot; if c or fewer items are found defective, we accept the lot. The computation results are then plotted as a graph or curve of the "probability of acceptance" against a scale of values of "lot percent defective."

To illustrate the approach, let's consider a lot size of 5 and a random sample of 2 with an acceptance number, c of zero. We represent the lot of 5 units by a column of circles in Figure 8-12, where acceptable units are open circles and defective ones are crosses. The lot is shown five times so that we can illustrate the lot percent defective. The possible combinations or pairs of units we can select for our sample are

1-2	1-4	2-3	2-5	3-5
1-3	1-5	2-4	3-4	4-5

In other words, there are 10 possible samples comprised of two each that can be picked. Now, the probability of acceptance for the first column is 1.00 because no possible sample can show more than zero defective pieces. Obviously when the acceptance number is never exceeded, the lot is accepted 100% of the time.

	1	O	X	X	X	X
PIECE NUMBER	2	O	O	X	X	X
	3	O	O	O	X	X
	4	O	O	O	O	X
	5	O	O	O	O	O
% DEFECTIVE		0	20	40	60	80
PROBABILITY OF ACCEPTANCE		100	60	30	10	0

Figure 8-12. Example of percent defective probability of acceptance.

When the lot is 20% defective, 4 of 10 possible samples have unit 1 represented (1-2, 1-3, 1-4, and 1-5). Hence the acceptance number will be exceeded 40% of the time. In other words, a 20% defective lot will be accepted 60% of the time. The reader can continue with the exercise moving from the left column to the right; the probability of acceptance for each case is summarized on the bottom of Figure 8-12. From the values in Figure 8-12 an operating characteristic curve can be constructed as shown in Figure 8-13.

There are obviously an infinite number of different sampling plans that can be devised, each with its own operating characteristic curve. These curves conain important information about the sampling plan; the upper end of the scale reports on the probability of acceptance at the lower values of lot percent defective; and the low end of the curve gives the probability of acceptance at the higher values of lot defectiveness. The former is the region where lots are considered acceptable; the latter is the region that should be rejected a high percentage of the time. In sampling plan design, increasing the acceptance number c keeps the curve up high as it moves to the right from "0" lot defectiveness. This has the effect of broadening the band within which material is considered acceptable. Increasing the sample size n steepens the slope of the drop-off portion of the curve, which increases the plan's ability to discriminate between acceptable and nonacceptable material. A small sample size results in a small

Application of Quality Control Techniques

slope, giving a large area of indefinite decision between acceptable and nonacceptable lots. In the illustration with an n of 2, this area runs from about 10 to 50% defective where the plan makes no consistent decision.

Two points can be selected on the curve to describe the acceptable region and the nonacceptable region. The first is taken at the 0.95 probability of acceptance, which is of course a 5% probability of rejection. The other is at 10% probability of acceptance.

The corresponding lot percent defective points (A and B in Figure 8.13) are referred to as the acceptable quality level (AQL) and the lot tolerance percent defection (LTPD). If material happens to come to inspection exactly at the AQL value of defectiveness, the producer stands a 5% chance of such a lot being rejected. If a lot happens to reach inspection at the LTPD value, consumers will, 1 time in 10, find such material passed on to them as acceptable even though it is in the rejectable region. A second case for a lot size of 10, sample size 2, and $c = 0$ is illustrated in Table

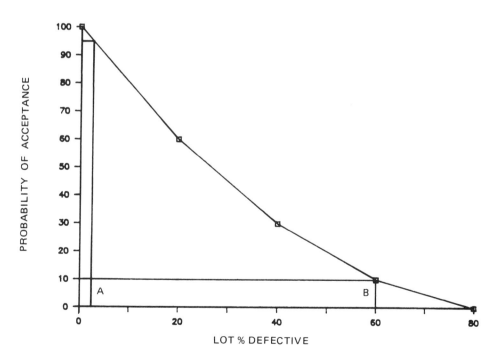

Figure 8-13. Probability of acceptance chart.

8-8. The total number of possible combinations can be computed from the formula

$$M = \frac{n(n-1)(n-2)\cdots(n-r+1)}{r!} = \frac{n!}{r!\,(n-r)!}$$

where M denotes the number of combinations of n distinct items taken r at a time. That is, r is the sample size (2 in our example 5) and n is the number of individual items in a lot (5 and 10 in the examples denoted by symbol N). The percent probability of acceptance is then

$$\% \text{ prob. acceptance} = 100\left[1 - \frac{\Sigma(N - n_i)}{M}\right]$$

At this point we might well ask what is to be done with a lot rejected by sampling. A thorough screening of the remaining parts in the lot might be prudent; this rule would make possible the so-called limiting risk concept. The operating characteristic curve portrays the long-run picture of acceptance probabilities without any limits. It is not possible to estimate a satisfactory maximum percent defective that may reach the customer. The curve shows that occasionally a "good" lot (low percent defective) will be rejected, and occasionally a "poor" lot will be sent to the consumer. In fact, if rejected lots are not screened but simply returned to the supplier, the supplier could resubmit them with the knowledge that if this was done often enough, the operating curve shows that eventually the lot will pass sampling (try to sell that one to your boss!).

If we invoke the screening rule, all lots rejected by sampling will have the defective pieces removed. In the example, half of the 2% defective lots would be made 0% defective. If these cleared lots are mixed with the other half of the lots which are still 2% defective, and if the lots are all the same size, the mixture will be only 1% defective. Hence, by effective screening of rejected lots, a reduction in the net percent defective is achieved by sampling.

In the analyses, we have counted the number of defective pieces. Figure 8-14 essentially summarizes the bounds of this approach. The solid curves show, for this type of sampling, how risk (expressed AOQL) varies with sample size and with lot size. The latter effect disappears when the lot size >500. The acceptance number c for this chart is always zero, and the screening of all rejected lots is assumed to be completely effective in removing all defective pieces. Also, removed pieces are assumed to have been replaced by good pieces—in practice, this is generally not done with

Application of Quality Control Techniques

Table 8-8. Tabulation of Operating Characteristic Curve for $n = 2$; Sample Size 10; $c = 0$

	1	2	3	4	5	6	7	8	9	10	Percent of defective	Possible number defects	Probability of acceptance
1	0	X	X	X	X	X	X	X	X	X	10	9	80.0
2	0	0	X	X	X	X	X	X	X	X	20	17	62.2
3	0	0	0	X	X	X	X	X	X	X	30	24	46.7
4	0	0	0	0	X	X	X	X	X	X	40	30	33.3
5	0	0	0	0	0	X	X	X	X	X	50	35	22.2
6	0	0	0	0	0	0	X	X	X	X	60	39	13.3
7	0	0	0	0	0	0	0	X	X	X	70	42	6.7
8	0	0	0	0	0	0	0	0	X	X	80	44	2.2
9	0	0	0	0	0	0	0	0	0	X	90	45	0.0
10	0	0	0	0	0	0	0	0	0	0			

AQL 95
LTPD 65

Possible Combinations of Samples to Select

1-2	1-7	2-4	2-9	3-7
1-3	1-8	2-5	2-10	3-8
1-4	1-9	2-6	3-4	3-9
1-5	1-10	2-7	3-5	3-10
1-6	2-3	3-6	4-5	4-10
4-6	5-6	6-7	7-9	
4-7	5-7	6-8	7-10	
4-8	5-8	6-9	8-9	
4-9	5-9	6-10	8-10	
5-10	7-8	9-10		

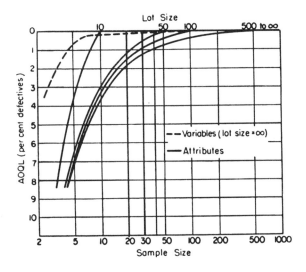

Figure 8-14. AOQL plot.

defects. On the curve marked "500 to infinity," the sample size for any desired AOQL can be read.

A general rule of thumb is that sampling becomes ineffective whenever the percent defective coming to inspection exceeds either the AOQL or one-half the LTPD of the plan being used. In this case, too much material is sorted, resulting in a mediocre job because of oversights. The best procedure is to forget sampling temporarily and concentrate on doing a slower, carefully organized 100% initial check (followed by a sample check of the material accepted to see if the 100% was effective).

NOTATION

c average number of defects per unit; acceptance number
M number of combinations of n distinct items taken r at a time
N total number of samples
n number of individuals in a subgroup
p probability
R range
$\underline{R}$ sample size
$\bar{X}$ mean or average value

Greek Symbol

σ   standard deviation

REFERENCES

Cheremisinoff, N.P., *Practical Statistics for Engineers and Scientists,* Technomic Publishing, Lancaster, Pa. 1987.

Cochran, W.G., and G.M. Cox, *Experimental Designs,* Wiley, New York 1957.

Duncan, A.J., *Quality Control and Statistics,* Richard D. Irwin, Homewood, Ill. 1965.

Shewhart, W.A., in *Statistical Methods from the Viewpoint of Quality Control,* W. Demming, Washington, D.C., Graduate School, Department of Agriculture. 1939.

Snedecor, G.W., and W.G. Cochren, *Statistical Methods,* 7th ed., Iowa State University Press, Ames, Iowa 1980.

9
Design of Experiments and Application to Product Design

BASIC DEFINITIONS AND TERMINOLOGY

When planning any experiment the first step is to define the "independent" and the "dependent" variables. All experiments should have a defined purpose. The purpose will heavily influence both the experimental procedure and the specification of independent and dependent variables. Often the precise purpose of the experiment emerges only during discussions that lead to the lists of independent and dependent variables.

Independent variables are those that will deliberately be controlled (i.e., set at or near predetermined target values at each "run" of the experiment). Examples of independent variables (x's) are factors (in experimental design literature), predictor variables (in regression and analysis literature), process variables (in manufacturing plants), and so on in other fields. The measured variables are the "dependent" variables whose values are dependent on the settings of the independent variables. Examples of dependent (Y's) variables are responses (in the statistics literature), product properties (in manufacturing plants), and so on in other fields.

The variable's name does not necessarily define whether it is an independent variable or a dependent variable. For example, temperature, concentration, and time are typical independent variables that are important in chemical reaction, while product yield, conversion, and purity are typically dependent variables. However, temperature may become a dependent variable, depending on the experimental procedure. For example, in an exothermic reaction the maximum temperature achieved may be the dependent variable.

It is important to realize that experimental error is unavoidable. We must therefore agree that our experimental strategy should always be one that will give good results in the presence of experimental error. It is useful to distinguish between two broad types of experimental error: "bias" error and "random" error.

Bias error is experimental error whose numerical value tends to remain constant or to follow a consistent pattern over a number of experimental runs. For example, a source of experimental error may remain constant for all experiments involving a single batch of raw material, a single operator, a single machine, and so on. Similarly, a source of experimental error may follow a consistent pattern, or cycle, depending on the hour within a day, season of the year, catalyst age, age of equipment, and so on.

For example, the yield of a chemical plant process may depend on the temperature of the cooling water taken from an adjacent river. The plant yield may then follow a seasonal pattern. If an unknowing experimenter made plant runs using say, high ingredient concentration in the summer and low ingredient concentration in the winter, one might attribute the difference in yield to ingredient concentration when, in fact, the difference is due to river water temperature. Analogous extraneous effects abound in all experimental situations. Another example is the density of rubber bales sitting in a warehouse. During summer months some of the bales may experience cold flow and compress into denser bales, causing mixing problems with the customer. Better mixing/dispersion achieved from lots of the same grade during colder months might erroneously be attributed to differences in production rather than mean ambient storage temperatures.

In designing an experiment it is important to come as close to the ideal case as possible by careful choice of experimental apparatus, good experimental controls, and so on. No experimenter is ever completely successful in this quest. At the National Bureau of Standards some physical properties are measured to six significant figures, but the experimenters are vitally concerned about bias errors that affect the last significant figure. Many manufacturing plants are concerned with bias in the second or third significant figure. At whichever level of precision, the statistical tools for dealing with bias error are "randomization" and "blocking," described earlier. It is important that these terms be understood and properly applied. Randomization simply means running the experiments in random order so that the bias errors do not persistently confuse the average effects of the (independent) variables deliberately varied in the experiments.

Random error is experimental error whose numerical value changes from one run to the next. The average value of random error is zero. Small positive or negative random errors are more common, usually, than larger or Gaussian, error distribution. The principal statistical tool for dealing with random error is "replication." By replication we mean that the average of a number of observations is more precise than a single observation. If y is a dependent variable whose standard error for a single observation is σ, and n independent observations of Y are made with identical settings of the x's in the experiment, the standard error of the average Y is given by

$$\sigma_{\bar{Y}} = \frac{\sigma}{\sqrt{n}} \tag{1}$$

Let's see how the function $1/\sqrt{n}$ behaves:

n	$1/\sqrt{n}$
1	1.0
4	0.5
9	0.33
16	0.25
100	0.10

Since the values of $1/\sqrt{n}$ versus n decrease rapidly at first, but only slowly as n becomes large, the general principle is clear that a small amount of replication is helpful, even essential. Let's keep this in perspective, however, because large amounts of replication are grossly wasteful of experimental time and money. Experimental programs should therefore be planned for moderate amounts of effective replication, never for large amounts.

The preceding considerations are part of the justification of another general principle. Experiments should have an appropriate bite size. In other words, they should always be large enough to have a high probability of learning something but never so late that the cost and elapsed time to achieve an answer is impractical. Against these criteria, the proper size of experiment is almost always between 8 and 80 experimental runs, but most often, the proper size is between 12 and 60 runs.

The "best" experiment depends on the experimental environment. Some characteristics of the environment should have little influence on the choice of experimental strategy, while others have a major influence. There are several important characteristics that should be kept in mind when planning an experiment. First, consider the number of independent variables in the experiment. If there are only three x's, a quite complete exploration of the effects of these variables should be possible in moderate number of experimental runs. On the other hand, to explore the effects of 40 x's with comparable thoroughness would require an inordinate number of experimental runs.

A continuous factor (independent variable) is one that can take on any value over some numerical range. There will be a lower limit and an upper limit, in practice, but the factor can be set at any value in the range. Temperature, pressure, and concentration are typical examples of continuous independent variables. In contrast, a discrete factor is one that can take on only specific (usually qualitative) levels. Examples would be catalyst type materials of construction (stainless steel versus ceramic), operator, batch of raw material, machine type, and so on. Discrete factors usually are of prime interest only during the early stages of an experimental program. Later, the catalyst type, materials of construction, operator techniques, machine design, and so on, will have been quantified and attention will be focused on the quantitative effects of the continuous factors, for the configuration of discrete factors selected.

The next important characteristic is the quality of prediction. This will vary with the number and type of factors and the size of the experiment. However, every experiment should lead to a model that will give valid predictions about future behavior of the system. Experimenters who merely fill notebooks with records of observation but do not produce useful predictive models have not earned their keep, nor have they functioned as scientists and/or engineers.

Let us not forget theoretical considerations. If there is adequate theory available, one should use the theoretical model and derive an experimental strategy attuned specifically to the mathematical form of the model. Most often there are two impediments to exclusive dependence on a theoretical model:

> The adequacy of the model may not yet have been thoroughly established (i.e., empirically verified).

Even when a theoretical model can predict accurately *some* of the critical responses (say, compound green strength or yield of a product from a polymerization), there usually are other important responses for which no theoretical model exists (e.g., responses such as color, hardness, impurity, aesthetic properties, etc.).

Given these considerations, most experimental programs must be designed on the basis of an appropriate empirical model. Good data for an empirical model will also be useful to develop a theoretical model if necessary.

Every experimental program has a defined life span; that is, each program has a beginning and an end. During its lifetime, every program evolves through a characteristic sequence of stages. The "bests" experimental strategy changes greatly from stage to stage. Therefore, experimenters must learn to recognize where their experimental programs really are in the natural evolutionary process. Figure 9-1 summarizes these evolutionary stages.

The initial phase of a program is referred to as the "screening" stage. At this point in time, the experimenter does not yet know which variables really are important. Through discussion with others, reviewing reports and through his or her own analysis, the experimenter can list a number of physical mechanisms that may affect the response of interest. Each potentially important mechanism carries with it some corresponding variables. For example, if heat transfer may be important, some candidate variables may be temperatures, agitation rates, materials of construction, and so on. The more diligent the experimenter is in identifying the potentially important mechanisms, the longer is the list of candidate variables. By traditional methods of experimentation this might appear to make the experimental problem utterly beyond all hope for a solution with finite time and money. However, screening designs can effectively examine 10, 20, even 30 factors to determine which are important and should be carried into future stages of experimentation, while the remaining factors are dropped. The term "dropped" refers to maintaining the factor constant at a convenient level or a level that is economically favorable.

Its quite possible that an experimental program may never pass the screening stage, even though it may have been under way for some time. In such a situation the experimenter will have to contend with a body of "folklore" that has built up as to the effects (or lack of effects) of the candidate variables. Good experimenters will allow the screening experiment to help them to sort out the facts from the folklore.

Following the screening stage, we identify a stage called a "limited response surface," characterized by the use of two-level factorial ex-

Experiment and Product Design

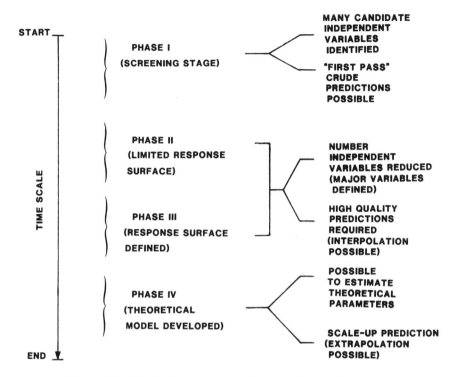

Figure 9-1. Evolutionary states of an experimental program.

periments, then a "response surface" stage characterized by the use of response surface designs. As the program proceeds, the number of independent variables decreases, the model complexity increases, and the quality of prediction improves. The sequence of events is summarized in Table 9-1.

History can often teach us much on what mistakes to avoid in our own programs. Looking back one can identify two broad classes of experimental strategy that have been employed by scientist. For lack of better names, we may call these the "classical" viewpoint and the "statistical" viewpoint. The classical viewpoint is characterized by one-factor-at-a-time experimentation, holding all other factors constant. As already discussed, this strategy rarely has an appealing simplicity, both in concept and in the ease with which experiments may be carried out. However, those who follow this strategy rarely give explicit consideration to the

Table 9.1 Experimental Environments

Experimental phase	Typical number of independent variables	Typical information sought	Typical statistics design
Screening	6 to 30 (continuous and/or discrete)	Identify which ones of independent variables are important; show approximate overall effect of each	Plackett-Burman
Limited response surface	3 to 8 (continuous and/or discrete)	Prediction over experimental region, with estimates of linear effects and interactions only	Two-level factorial, possibly with center point
Response surface	2 to 6 (continuous)	Higher-quality prediction over region where linear, interaction, and curvature effects are needed	Box-Behnken
Theoretical model	1 to 5	Estimate parameters in theoretical model to enable outside range of experiments	Special-purpose computer-rived design
Sampling experiment		Quantify contribution of various sources to the overall product variability under standard operating conditions	Nested or cross-classification

assumptions they are making, assumptions about the "true" character of the problem. The shortcoming of this methodology is that it tends to ignore what happens in the real world; namely, interactions between variables.

The inherent assumptions of the classical strategy are:

The response is likely to be a complicated function of any one x that will not result in a smooth relationship. This requires many experimental levels for each variable.

Experiment and Product Design

The effect of any other x is simply to raise or lower the function but not to change its shape or slope. That is, "interactions" are negligible.

The experimental error is negligible relative to the effects of the x's.

If these assumptions hold, one factor at a time is a good strategy. In contrast, the statistical viewpoint assumes that:

Over the experimental range most response functions are smooth, with perhaps an upward or a downward curvature.

The slope of this function or its shape may change significantly when the levels of other x's are changed. Such "interactions" are considered typical.

The experimental error is not negligible relative to the effects of the x's and the experimental error may contain both bias and random components.

If these assumptions hold, the statistical approach is a good strategy, but the classical approach will do poorly. The statistical approach is characterized by experimental designs employing "factorial" structures.

It is important to note that no magic formula exists that enables one to be correct 100% of the time. History is filled with examples where the classical approach has been effectively applied. Both judgment and experience, along with a clear understanding of the problem, establishes which approach should be used. On the whole, however, a statistical strategy has a better chance of resulting in a successful experimental program.

STATISTICS BASICS FOR EXPERIMENTATION STRATEGY

Much of the mathematics needed to apply statistical analysis methodology was presented in earlier chapters. We will review some of the more crucial aspects first, and then press on to the subject of interest, experimental design methodology.

In scientific and/or engineering studies we are frequently interested in describing quantitatively the changes in a response, or dependent variable, arising out of deliberate changes in the controllable, or independent variables. Generally, the investigation consists of a series of experimental runs, each at different levels of the independent variables. Let's consider a set of experimental runs involving a single independent variable (x) and a dependent variable (Y). The results are tabulated below and graphed in Figure 9-2. There are eight data points—two each at four values of x. Note

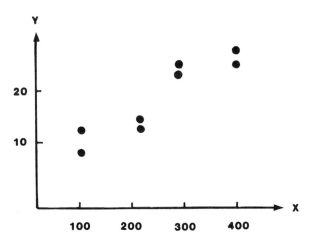

Figure 9-2. Sample problem

that the data point sequence does not imply that the observations were obtained in that order.

Data Point	x	Y
1	100	11
2	100	9
3	200	14
4	200	15
5	300	23
6	300	24
7	400	26
8	400	24

A typical analysis may conclude that the change in Y per unit change in x is about 0.5. A statistical analysis would proceed from the assumption that measurements are not precisely reproducible—they fluctuate. Here we assume that the x-values can be set with precision and that the experimental error is (predominantly) in Y. We would assume that we have four *populations* of Y-values. Each population is of infinite size. Each population corresponds to a level of x and has its own central value (the population mean, μ) and dispersion (the population variance, σ^2). Recall that these are population parameters. Since it is impossible to determine every measurement belonging to a specific population, we obtain a *sample,* and

Experiment and Product Design

we estimate the population parameters from sample values. For any specific x-level, say x_0, the estimate of the central Y-value, the mean, is the average of the Y_i observations taken at $x = x_0$.

$$\bar{Y}_{x=x_0} = \frac{Y_1 + Y_2 + \cdots + Y_n}{n} = \frac{\sum_{i=1}^{n} Y_i}{n} \tag{2}$$

Variance is the estimate of the dispersion:

$$s^2_{x=x_0} = \frac{(Y_1 - \bar{Y})^2 + (Y_2 - \bar{Y})^2 + \cdots + (Y_b - \bar{Y})^2}{n - 1}$$

$$= \frac{\sum_{i=1}^{n}(Y_i - \bar{Y})^2}{n - 1} \tag{3}$$

The square root of the variance is the standard deviation:

$$s = \sqrt{s^2} \tag{4}$$

We seldom have the luxury of repeating experimental conditions and infinitum to obtain precise estimates. The obvious question is: How many measurements do we need to estimate population parameters from sample values? In many cases, two or three measurements provide a minimum adequate estimate of the mean. However, for estimating the variance with any precision, we would need more measurements than the minimum adequate for estimating the mean. If we may assume that the population variance, $\sigma^2_{x=x_0}$, is about the same for all levels of x in the experiment, we can pool our $s^2_{x=x_0}$ estimates into a single number, s^2_{pooled}, which is

$$s^2_{\text{pooled}} = \frac{(n_1 - 1)s_1^2 + (n_2 - 1)s_2^2 + \cdots + (n_k - 1)s_k^2}{(n_1 - 1) + (n_2 - 1) + \cdots + (n_k - 1)} \tag{5}$$

$$s_{\text{pooled}} = \sqrt{s^2_{\text{pooled}}}$$

where s_j^2 is the variance of Y for the jth x-level, $j = 1, \ldots, k$, and, n_j is the number of repeat measurements at the jth x-level.

The pooled variance is a weighted average of the individual variance estimates. The advantage of pooling is that the pooled estimate is a more precise estimator of the population variance than any of the estimates from which the pooled value was computed. The denominators of equations (3) and (5) are the degrees of freedom for the respective estimates. The pooled estimate has degrees of freedom equal to the sum of the degrees of freedom of the individual variance estimates, and gains precision accordingly.

A "degree of freedom" corresponds to one linear combination among the observations. For example, the mean is one linear combination of the observations:

$$\bar{Y} = \frac{1}{n} Y_1 + \cdots + \frac{1}{n} Y_n$$

Therefore, the mean carries one degree of freedom. If the mean $\bar{Y}$ of n observations is specified, then only $(n - 1)$ of the observations can take on arbitrary values; the nth observation is completely determined.

In the *special case* where we have only two observations at each of k-levels of x, equations (3) and (5) simplify to

$$s^2_{pooled} = \frac{\sum_{i=1}^{n} (Y_{j1} - Y_{j2})^2}{2k} \qquad (6)$$

(For the sample data given earlier,

$$s^2_{pooled} = \frac{(11 - 9)^2 + (14 - 15)^2 + (23 - 24)^2 + (26 - 24)^2}{2(4)}$$

$$= \frac{(2)^2 + (-1)^2 + (-1)^2 (2)^2}{4} = 2.5$$

$$s_{pooled} = 1.58$$

The standard deviation, whether based on a variance computed by formula (3), (5), or (6), has the following interpretation. If the measurements in a population are approximately normally distributed, about two-thirds of the measurements will lie within one standard deviation of the mean, [i.e., within $(\mu - \sigma)$ and $(\mu + \sigma)$]. In practice, the normal distribution is a good model for most situations.

$$f(y) = \frac{1}{\sigma\sqrt{2\pi}} \exp\left[-\left(\frac{1}{2}\right) \frac{(y - \mu)^2}{\sigma}\right] \qquad (7)$$

Experiment and Product Design

For a function to be statistical frequency function, or density function, the area under the $f(y)$-y curve must equal unity. Furthermore, the density function must be positive for all values of the variable. The area under any portions of the density function is the probability that the random variable will fall in the corresponding intervals of the absissa.

In applied statistics, probability statements concerning average must be made. If for given x-setting there are multiple samples, each of size n, then regardless of the distribution of the individual measurements, the sample averages will tend to be approximately normally distributed (for large n). The variance of each of the sample average is

$$\sigma_{\bar{Y}}^2 = \frac{\sigma_Y^2}{n} \qquad (8)$$

For each of the x-settings we have x settings we have $\bar{Y}$ as an estimate of the population mean (μ_j) corresponding to the x-level. A question that may then be asked is: If we do not know σ but have a valid estimate of the population standard deviation, how close is $\bar{Y}_j$ to μ_j? To answer this question we must make use of the *t-distribution*. If $Y, \ldots, Y_n$ are a random sample from a normal population with mean μ and variance σ^2, the quantity

$$t = \frac{\bar{Y} - \mu}{s/\sqrt{n}} \qquad (9)$$

follows the *t*-distribution. The *t*-distribution is bell-shaped like the normal distribution, but it has longer tails, depending on the number of "degrees of freedoms" associated with the estimated standard deviation s. In the limit, for infinite degrees of freedom the *t*-distribution becomes identical with the normal distribution.

Table 9-2 tabulates the fraction of t for varying probability levels and degrees of freedom. For our purposes we need only "double-sided" *t*-tests where we wish to detect the occurrence of significant deviation, whether positive or negative. The probabilities in Table 9-2 apply for this situation. If the standard deviation of the population is estimated from a sample, the population mean is expected to fall within

$$\bar{Y} \pm t_{(1-\alpha)}, \, df \frac{s}{\sqrt{n}}$$

Table 9.2 Probability Points of the t-Distribution: Double-Sided Test

df	\multicolumn{4}{c}{P}			
	0.995	0.99	0.95	0.90
1	127	63.7	12.7	6.31
2	14.1	9.92	4.30	2.92
3	7.45	5.84	3.18	2.35
4	5.60	4.60	2.78	2.13
5	4.77	4.03	2.57	2.01
6	4.32	3.71	2.45	1.94
7	4.03	3.50	2.36	1.89
8	3.83	3.36	2.31	1.86
9	3.69	3.25	2.26	1.83
10	3.58	3.17	2.23	1.81
11	3.50	3.11	2.20	1.80
12	3.43	3.05	2.18	1.78
13	3.33	2.98	2.14	1.76
15	3.29	2.95	2.13	1.75
16	3.25	2.92	2.12	1.75
17	3.22	2.90	2.11	1.74
18	3.20	2.88	2.10	1.73
19	3.17	2.86	2.09	1.73
20	3.15	2.85	2.09	1.72
21	3.14	2.83	2.08	1.72
22	3.12	2.82	2.07	1.72
23	3.10	2.81	2.07	1.71
24	3.09	2.80	2.06	1.71
25	3.08	2.79	2.06	1.71
26	3.07	2.78	2.06	1.71
27	3.06	2.77	2.05	1.70
28	3.05	2.76	2.05	1.70
29	3.04	2.76	2.05	1.70
30	3.03	2.75	2.04	1.70
40	2.97	2.70	2.02	1.68
60	2.91	2.66	2.02	1.67
120	2.86	2.62	1.98	1.66
∞	2.81	2.58	1.96	1.64

Experiment and Product Design

where n is the number of observations used to calculate $\bar{Y}$ and $(1 - \alpha)$ = probability level. Continuing with the same example, at $X = 200$, $\bar{Y} = 14.5$. The pooled standard deviation is 1.58 with 4 degrees of freedom. Here, $n = 2$. From the t- table we have

$$t_{0.95,4} = 2.78$$

$$t_{0.99,4} = 4.60$$

Hence the population mean for $x = 20$ is expected to fall, with a probability of 0.95, within the range

$$14.5 \pm 2.78 \frac{1.58}{\sqrt{2}}$$

or

$$11.39 < \mu < 17.61$$

and with a probability of 0.99 in the range

$$14.5 - 4.60 \frac{1.58}{\sqrt{2}} < \mu < 14.5 + 4.60 \frac{1.58}{\sqrt{2}}$$

or

$$9.36 < \mu < 19.64$$

These ranges are known, respectively, as the 95% *and* 99% *confidence intervals* for the mean. The importance of the t-distribution is that it allows us to put "confidence bands" around population estimates. Later, we will use the t-distribution for making quantitative statements about the effect of variable on a response.

In statistical analysis it is often necessary to test whether two sample variances, s_1^2 and s_2^2, could reasonable have come from populations with the same variance, $\sigma_1^2 = \sigma_2^2$, against the alternative $\sigma_1^2 > \sigma_2^2$. The statistic used for this purpose is the F-statistic, defined as follows:

$$F = \frac{s_1^2}{s_2^2} \tag{10}$$

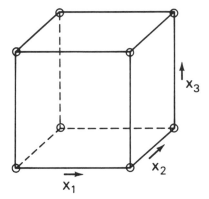

Figure 9-3. A 2^3 factorial design.

where s_1^2 and s_2^2 are estimates of σ_1^2 and σ_2^2. The estimates are assumed independently obtained with df_1 and df_2, respectively. Furthermore, the population values are assumed to obey the relation $\sigma_1^2 \geqslant \sigma_2^2$. Table 9-3 provides probability points of F. It must be entered with both df_1 and df_2 as well as the confidence probability. The F-statistic will be used in later discussion.

With the review above, we may now discuss in detail some of the concepts introduced in Chapter 6. Specifically, we wish to concentrate on the principles of the factorial design.

Figure 9-3 illustrates a three-dimensional cube as a perspective drawing on the two-dimensional paper. Each cube dimension represents one factor in the experiment. Since this is a three-factor problem (i.e., x_1, x_2, x_3), this perspective drawing permits us to visualize the factor space quite adequately. With more factors, the cube becomes a hypercube. Our ability to draw meaningful pictures is more limited in higher dimensions, but the mathematical properties generalize rather easily. The design pattern in Figure 9-3 is referred to as a 2^3 factorial—three factors at two levels (high and low)—eight design points on the corner points of the cube. Each design point represents a set of experimental conditions at which one or more experimental runs will be made. For example, as indicated by the arrows on Figure 9-3, factor x_1 has its low level at the left side of the cube (four points) and its high level at the right side of the cube (four points)—similarly for x_2 and x_3. The left, front, lower point represents the experimental condition where x_1, x_2, and x_3 are all at their low values. Consequently, the cube and its interior represent the available factor space in

Experiment and Product Design

which experiments can be run. The important feature of the 2^3 design is that it incorporates the extreme points of the factor space. This "space-filling" concept is characteristic of all well-designed experiments.

The two-level factorial can determine the "main effect" of each factor plus the "interactions" of the factors in combination. Here we want to show how these effects are related to the geometry of the cube. Later the corresponding computational methods are developed.

The main effect of x_1 is based on a comparison of the response values at the left and right planes of the cube, as shown in Figure 9-4A. Consider

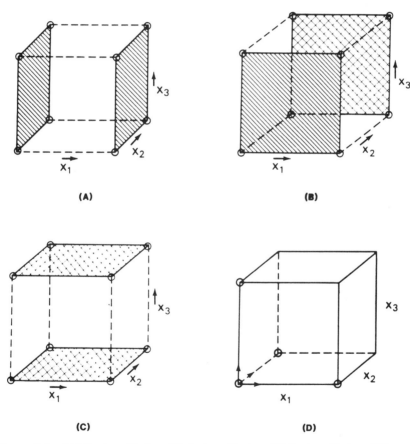

Figure 9-4. (A) Factor effect x_1; (B) factor effect x_2; (C) factor effect x_3; (D) one-factor-at-a-time.

Table 9-3. Fractiles of F

P	df_2	\multicolumn{14}{c}{df_1}
		1	2	3	4	5	6	7	8	9	10	12	15	20	30	60	120	∞
0.90	1	39.9	49.5	53.6	55.8	57.2	58.2	58.9	59.4	59.9	60.2	60.7	61.2	61.7	62.3	62.8	63.1	63.3
0.95		161	200	216	225	230	234	237	239	241	242	244	246	248	250	252	253	254
0.99		4050	5000	5400	5620	5760	5860	5930	5980	6020	6060	6110	6160	6210	6260	6310	6340	6370
0.90	2	8.53	9.00	9.16	9.24	9.29	9.33	9.35	9.37	9.38	9.39	9.41	9.42	9.44	9.46	9.47	9.48	9.49
0.95		18.5	19.0	19.2	19.2	19.3	19.3	19.4	19.4	19.4	19.4	19.4	19.4	19.5	19.5	19.5	19.5	19.5
0.99		98.5	99.0	99.2	99.2	99.3	99.3	99.4	99.4	99.4	99.4	99.4	99.4	99.4	99.5	99.5	99.5	99.5
0.90	3	5.54	5.46	5.39	5.34	5.31	5.28	5.27	5.25	5.24	5.23	5.22	5.20	5.18	5.17	5.15	5.14	5.13
0.95		10.1	9.55	9.28	9.12	9.01	8.94	8.89	8.85	8.81	8.79	8.74	8.70	8.66	8.62	8.57	8.55	8.53
0.99		34.1	30.8	29.5	28.7	28.2	27.9	27.7	27.5	27.3	27.3	27.4	26.9	26.9	26.5	26.3	26.2	26.1
0.90	4	4.54	4.32	4.19	4.11	4.05	4.01	3.98	3.95	3.93	3.92	3.90	3.87	3.84	3.82	3.79	3.78	3.76
0.95		7.71	6.94	6.59	6.39	6.26	6.16	6.09	6.04	6.00	5.96	5.91	5.86	5.80	5.75	5.69	5.66	5.63
0.99		21.2	18.0	16.7	16.0	15.5	15.2	15.0	14.8	14.7	14.5	14.4	14.2	14.0	13.8	13.7	13.6	13.5
0.90	5	4.06	3.78	3.62	3.52	3.45	3.40	3.37	3.34	3.32	3.30	3.27	3.24	3.21	3.17	3.14	3.12	3.11
0.95		6.61	5.79	5.41	5.19	5.05	4.95	4.88	4.82	4.77	4.74	4.68	4.62	4.56	4.50	4.43	4.40	4.37
0.99		16.3	13.3	12.1	11.4	11.0	10.7	10.5	10.3	10.2	10.1	98.9	9.72	9.55	9.38	9.20	9.11	9.02
0.90	6	3.78	3.46	3.29	3.18	3.11	3.05	3.01	2.98	2.96	2.94	2.90	2.87	2.84	2.80	2.76	2.74	2.72
0.95		5.99	5.14	4.76	4.53	4.39	4.28	4.21	4.15	4.10	4.06	4.00	3.94	3.87	3.81	3.74	3.70	3.67
0.99		13.7	10.9	9.78	9.15	8.75	8.47	8.26	8.10	7.98	7.87	7.72	7.56	7.40	7.23	7.06	6.97	6.83
0.90	7	3.59	3.26	3.07	2.96	2.88	2.83	2.78	2.75	2.72	2.70	2.67	2.63	2.59	2.56	2.51	2.49	2.49
0.95		5.59	4.74	4.35	4.12	3.97	3.87	3.79	3.73	3.68	3.64	3.57	3.51	3.44	3.38	3.30	3.27	3.23
0.99		12.2	9.55	8.45	7.85	7.46	7.19	6.99	6.84	6.72	6.62	6.47	6.31	6.16	5.99	5.82	5.74	5.65
0.90	8	3.46	3.11	2.92	2.81	2.73	2.67	2.62	2.59	2.56	2.54	2.50	2.46	2.42	2.38	2.34	2.31	2.29
0.95		5.32	4.46	4.07	3.84	3.69	3.58	3.50	3.44	3.39	3.35	3.28	3.22	3.15	3.08	3.01	2.97	2.93

Experiment and Product Design

0.99		11.3	8.65	7.59	7.01	6.63	6.37	6.18	6.03	5.91	5.81	5.67	5.52	5.36	5.20	5.03	4.95	4.86
0.90		3.36	3.01	2.81	2.69	2.61	2.55	2.51	2.47	2.44	2.42	2.38	2.34	2.30	2.25	2.21	2.18	2.16
0.95	9	5.12	4.26	3.86	3.63	3.48	3.37	3.29	3.23	3.18	3.14	3.07	3.01	2.94	2.86	2.79	2.75	2.71
0.99		10.6	8.02	6.99	6.42	6.06	5.08	5.61	5.47	5.35	5.26	5.11	4.96	4.81	4.65	4.48	4.40	4.31
0.90		3.29	2.92	2.73	2.61	2.52	2.46	2.41	2.38	2.35	2.32	2.28	2.24	2.20	2.15	2.11	2.08	2.06
0.95	10	4.96	4.10	3.71	3.48	3.33	3.22	3.14	3.07	3.02	2.98	2.91	2.84	2.77	2.70	2.62	2.58	2.54
0.99		10.0	7.56	6.55	5.99	5.64	5.39	5.20	5.06	4.94	4.85	4.71	4.56	4.41	4.25	4.08	4.00	3.91
0.90		3.18	2.81	2.61	2.48	2.39	2.33	2.28	2.24	2.21	2.19	2.15	2.10	2.06	2.01	1.96	1.93	1.90
0.95	12	4.75	3.89	3.49	3.26	3.11	3.00	2.91	2.85	2.80	2.75	2.69	2.62	2.54	2.47	2.38	2.34	2.30
0.99		9.33	6.93	5.95	5.41	5.06	4.82	4.64	4.50	4.39	4.30	4.16	4.01	3.86	3.70	3.54	3.45	3.36
0.90		3.07	2.70	2.49	2.36	2.27	2.21	2.16	2.12	2.09	2.06	2.02	1.97	1.92	1.87	1.79	1.76	0.95
0.95	15	4.54	3.68	3.29	3.06	2.90	2.79	2.71	2.64	2.59	2.54	2.48	2.40	2.33	2.25	2.16	2.11	2.07
0.99		8.68	6.36	5.42	4.89	4.56	4.32	4.14	4.00	3.89	3.80	3.67	3.52	3.37	3.21	3.05	2.96	2.87
0.90		2.97	2.59	2.38	2.25	2.16	2.09	2.04	2.00	1.96	1.94	1.89	1.84	1.79	1.74	1.68	1.64	1.61
0.95	20	4.35	3.49	3.10	2.87	2.71	2.60	2.51	2.45	2.39	2.35	2.28	2.20	2.12	2.04	1.95	1.90	1.84
0.99		8.10	5.85	4.94	4.43	4.10	3.87	3.70	3.56	3.46	3.37	3.23	3.09	2.94	2.78	2.61	2.52	2.42
0.90		2.88	2.49	2.28	2.14	2.05	1.98	1.93	1.88	1.85	1.82	1.77	1.72	1.67	1.61	1.54	1.50	1.46
0.95	30	4.17	3.32	2.92	2.69	2.53	2.42	2.33	2.27	2.21	2.16	2.09	2.01	1.93	1.84	1.74	1.68	1.62
0.99		7.56	5.39	4.51	4.02	3.70	3.47	3.30	3.17	3.07	2.98	2.84	2.70	2.55	2.39	2.21	2.11	2.01
0.90		2.79	2.39	2.18	2.04	1.95	1.87	1.82	1.77	1.74	1.71	1.66	1.60	1.54	1.48	1.40	1.35	1.29
0.95	60	4.00	3.15	2.76	2.53	2.37	2.25	2.17	2.10	2.04	1.99	1.92	1.84	1.75	1.65	1.53	1.47	1.39
0.99		7.08	4.98	4.13	3.65	3.34	3.12	2.95	2.82	2.72	2.63	2.50	2.35	2.20	2.03	1.84	1.73	1.60
0.90		2.75	2.35	2.13	1.99	1.90	1.82	1.77	1.72	1.68	1.65	1.60	1.54	1.48	1.41	1.32	1.26	1.19
0.95	120	3.92	3.07	2.68	2.45	2.29	2.18	2.09	2.02	1.96	1.91	1.83	1.75	1.66	1.55	1.43	1.35	1.25
0.99		6.85	4.79	3.95	3.48	3.17	2.96	2.79	2.66	2.56	2.47	2.34	2.19	2.03	1.86	1.66	1.53	1.38
0.90		2.30	2.08	1.94	1.85	1.77	1.72	1.67	1.63	1.60	1.55	1.49	1.42	1.34	1.24	1.17	1.00	0.95
0.95	2.71	3.00	2.60	2.37	2.21	2.10	2.01	1.94	1.88	1.83	1.75	1.67	1.57	1.46	1.32	1.22	1.00	0.99
∞	3.84 6.63	4.61	3.78	3.32	3.02	2.80	2.64	2.51	2.41	2.32	2.18	2.04	1.88	1.70	1.47	1.32	1.00	

the lower front edge of the cube. Along this edge the values of x_2 and x_3 are consistent (i.e., both are at their values). At the ends of this edge are two experimental points that differ only with respect to x_1, holding x_2 and x_3 constant (a one-factor-at-a-time estimate). There are, however, three other edges of the cube along each of which x_2, and x_3 are constant. Each of the four edges gives an estimate of the effect of x_1. The "main effect" of x_1 is defined as the average of the estimates from these four edges. Equivalently, we are looking at the differences in average response between the right-hand plane and the left- hand plane. Since all combinations of x_2 and x_3 are represented in this comparison of planes, any main effect that is found will be known to apply over these ranges of x_2 and x_3, not just at a particular combination of x_2 and x_3. In the same manner, the effect of x_2 (Figure 9-4B) is determined from the difference in average response between the front and backplanes. The effect of x_3 involves the top and bottom planes, as shown in Figure 9-4C.

Notice that all eight points in the experiment are used to determine the effect of each x. This is called the "hidden replication" in a factorial experiment, because each factor effect and each interactoin effect is based on all the data points. Each effect is a difference between two averages, hence the "hidden replication." Effects determined in this way are mathematically independent ("orthogonal"), so they can be interpreted as truly separate estimates of separate characteristics of the physical system being investigated. Recall that the x's may be quite different physical quantities such as x_1 = temperature, x_2 pressure, and so on.

It is instructive to compare the geometry of these factorial factor effects with the geometry of (a simplified) one-factor-at-a-time. Figure 9-4D assumes that the base point from which one-factor-at-a-time explores is the (x_1, x_2, x_3) = (low, low, low) point at the bottom, front left. One-factor-at-a-time determines the effect of x_1 by comparing the base point only with the (high, low, low) point at the other end of the lower front edge, as shown by the arrow. Similarly, the effects of x_2 and x_3 are determined by comparing the base point to only one other point. There is no hidden replication, not is there any exploration of the effects of any factor at more than one combination of other factors from which interactions could be found.

By definition, two factors, say x_1 and x_2 are said to "interact" if the effect of x_1 is different at different levels of x_2. Figure 9-5 illustrates two interactive examples and one noninteractive example. As shown, the absence of interaction requires that the y versus x_1 plot have the same slope at both high and low values of x_2, creating a constant difference in y between the two plots.

Experiment and Product Design

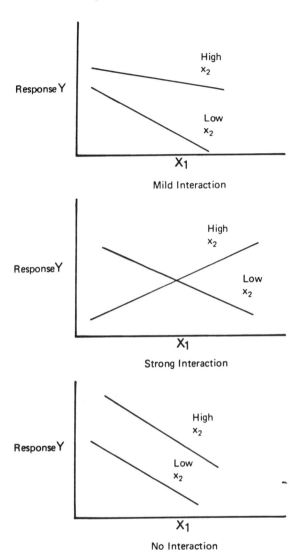

Figure 9-5. Interaction of x_1 and x_2.

Taking the simplest possible case, consider a 2^2 experiment in x_1 and x_2, as shown in Figure 9-6.

The interaction is defined in terms of differences of the response values:

$$\text{interaction} = \tfrac{1}{2}[(Y_D - Y_c) - (Y_B - Y_A)]$$

$$= \frac{Y_A + Y_D}{2} - \frac{Y_C + Y_B}{2} \tag{11}$$

Note that the result is a comparison of diagonally opposite pairs of points. In the three-factor case the interaction geometry involves diagonally opposite planes, as in Figure 9-7A.

Factorial designs permit estimation of the effects of several factors simultaneously. Main effects and interactions can be estimated by making experimental runs at all combinations of the p-factors, with l levels per factor. The number of experimental runs required is $n = l^p$. In particular, for $l = 2$ the number of runs is $n = 2^p$. These two-level factorial designs are highly useful for a wide variety of problems, are easy to plan and analyze, are readily adaptable to both continuous and discrete factors, and provide adequate prediction models for responses that have no strong curvature over the experimental range.

The discussions that follow describe procedures for up to five factors. Generalization to larger number of factors is straightforward, but the

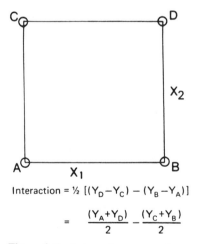

Figure 9.6 Interaction geometry.

Experiment and Product Design

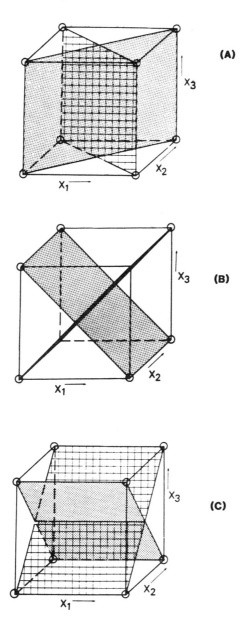

Figure 9-7. (A) Interaction x_1x_2; (B) interaction x_1x_3; (C) interaction x_2x_3.

number of runs required becomes large (e.g., $2^6 = 64$). For six, seven, or i factors, a fraction of the full factorial may be appropriate. If there are many factors, a Placket-Burman screening design would be an appropriate first step (discussed later).

It should be remembered that any number of responses can be measured at each experimental point. For best results the responsive variables in a factorial experiment should be continuous and have uniform, independent errors. A "continuous" response is one that gives a numerical value on some continuous scale for each experimental point. A counterexample would be a go/no-go response. In the latter case one can use as the response the fraction of runs that "go" at each experimental point. To do so, at least 10 independent repeat runs must be made at each point in order that the fraction may be nearly a "continuous" variable.

The requirement of "uniform error" means that the experimental error has approximately the same magnitude at all experimental points. The most common departure from this requirement is that the error may be a constant percent of the response. In the latter case it is appropriate to take logarithms of the individual response readings and to analyze the logarithms, which will then have approximately the same magnitude of error at all experimental points. The requirement of "independent" experimental errors means that the size and sign of error at any one experimental point is not affected by the sizes and signs of errors that occur at other points.

Factor coding is a technique used with two-level designs for discrete factors (such as two kinds of polymer additives) or for continuous variables (such as two levels of temperature). The two levels are coded "+" and "−".

In this method, arbitrarily assign either condition + and the other − for discrete variables. In dealing with continuous variables the object is to code the higher value + and the lower −. The levels of x should be chosen consistent with the guide shown in Figure 9-8. The (+) and (−) levels should far enough apart so that the "effect" of factor x is clearly larger than the experimental error. This guarantees a large signal-to-noise ratio. Coding of the factors into (+) and (−) levels (actually +1 and −1 in mathematical terms) is simply a convenient linear tranformation that converts the various natural scales of the factors into the same numerical range (−1 to +1). Computational details are described shortly.

Table 9-4 provides design schematics or patterns. The factors are identified across the top of the table. The trials are denoted by a serial number at the left of the table. Note that each design is contained in the next-larger design. Note also the simple pattern of alternating − and + signs, which

Experiment and Product Design

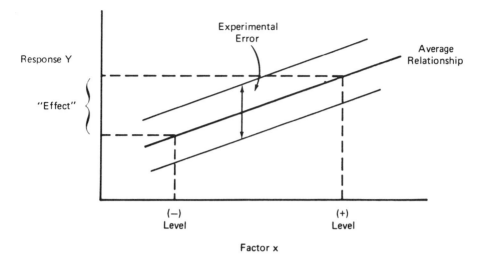

Figure 9-8. Principles of factor coding.

can easily be memorized. The general method involves copying the design required in coded form, then rewriting the table, assigning the actual (decoded) factor levels for each + and − in the coded design table.

The order of the trials in Table 9-4 is not good for running the tests. It is important that the trials be randomized before running. Table 9-5 lists several random orders for 16 and 32 runs. Select a new one for each test. (The random schedules can be used for designs with less than the listed number of runs by dropping the unnecessary run numbers.) Copy the design over in the specfic random order to be used.

The use of these designs enables the estimation of factor effects more precisely than one-factor-at-a-time testing, due to the hidden replication included. The improved precision of the factor effects due to the hidden replication is given by the so-called precision ratio

$$\frac{\sigma_{FE}}{\sigma} = \frac{2}{\sqrt{n}} \tag{12}$$

where σ_{FE} is the standard deviation of a factor effect, σ the standard deviation of a single observation, and n = total number of observations in the

Table 9-4. Two-Level Factorial Design Patterns

Trial	Factors				
	x_1	x_2	x_3	x_4	x_5
1	−	−	−	−	−
2	+	−	−	−	−
3	−	+	−	−	−
$4 = 2^2$	+	+	−	−	−
5	−	−	+	−	−
6	+	−	+	−	−
7	−	+	+	−	−
$8 = 2^3$	+	+	+	−	−
9	−	−	−	+	−
10	+	−	−	+	−
11	−	+	−	+	−
12	+	+	−	+	−
13	−	−	+	+	−
14	+	−	+	+	−
15	−	+	+	+	−
$16 = 2^4$	+	+	+	+	−
17	−	−	−	−	+
18	+	−	−	−	+
19	−	+	−	−	+
20	+	+	−	−	+
21	−	−	+	−	+
22	+	−	+	−	+
23	−	+	+	−	+
24	+	+	+	−	+
25	−	−	−	+	+
26	+	−	−	+	+
28	+	+	−	+	+
29	−	−	+	+	+
30	+	−	+	+	+
31	−	+	+	+	+
$32 = 2^5$	+	+	+	+	+

Experiment and Product Design

Table 9-5. Random Orders for Test Trials

16 trials					32 trials				
3	9	11	10	3	21	12	22	12	28
1	15	1	7	5	18	20	5	24	10
11	16	12	11	10	3	2	29	13	31
5	12	4	5	1	11	16	18	4	1
8	10	14	8	14	26	7	11	15	3
4	5	3	4	16	17	32	17	25	5
12	13	15	15	8	31	3	8	14	8
15	1	10	14	13	1	5	23	28	15
10	11	13	3	9	13	15	15	26	11
2	4	6	2	11	23	22	14	22	26
7	6	8	13	12	7	25	10	16	23
9	8	7	1	6	20	9	16	27	19
6	14	5	9	7	12	24	13	20	12
14	3	2	12	4	16	31	30	5	27
16	7	9	16	2	8	26	3	7	14
13	2	16	6	15	25	30	32	6	17
					6	10	25	23	2
					14	17	7	21	7
					4	13	4	11	13
					19	18	27	17	29
					32	4	9	18	9
					5	8	12	10	32
					30	29	28	2	16
					15	19	19	31	18
					10	1	20	3	21
					24	6	31	29	24
					27	14	6	1	6
					2	11	1	19	22
					29	21	21	32	20
					9	28	26	8	4
					28	23	2	9	25
					22	27	24	30	30

design ($= k2^p$, where k = number of replicates). This expression for the precision ratio applies to all balanced two-level factorial experiments: both multiple replicates (k = 1, 2, 3...) and balanced fractional replicates (k = ½, ¼, ⅛, etc.).

The precision ratio depends only on n. It does not matter whether the value of n is achieved by replicating a design with a given number of factors p or by increasing the number of factors. The precision ratio required in a given experiment depends on three parameters that can be specified by the experimenter.

The parameter α specifies the confidence level $(1 - \alpha)$ that will be used for the t-statistic in testing the data for minimum significant factor effects

The parameter Δ is the size of the factor effect it is desired to detect.

The parameter $(1 - \beta)$ is the desired probability of concluding that a factor has a significant effect when it does have a true effect of size Δ. The probability $(1 - \beta)$ is known as the "power of the test."

A common specification is $(1 - \alpha) \geqslant 0.95$, $(1 - \beta) \geqslant 0.090$. These will be satisfied if we require that $\sigma_{FE} = \Delta/4$. Substituting this into equation (12) and solving for n, we have a reliable guide to the required number of runs:

$$n = \left(\frac{8\sigma}{\Delta}\right)^2 \qquad (13)$$

The factor of 8 guarantees that $(1 - \alpha)$ is *at least* 0.95 and $(1 - \beta)$ is *at least* 0.90; however, a factor of 7 will be adequate in many cases. This range of factors in equation (13) leads to an important rule of thumb: To detect an effect twice as large as the experimental error ($\Delta = 2\sigma$), the required number of runs is about 12 to 16. To detect an effect the same size as the experimental error ($\Delta = \sigma$), the required number of runs is four times as large, (i.e., 48 to 64). Choose the number of replicates accordingly.

These designs do not give any estimate of curvature of the response in the experimental region. It is always desirable for a design to provide a check for "lack of fit" of the assumed model. To estimate the curvature of each factor separately, a full response surface design would be needed. The number of runs in the design would be increased 50 to 100%. However, a relatively economical estimate of the overall curvature can be obtained by running points at a middle value of all factors. These center points can be run only for continuous factors. The severity of the curvature is estimated by the difference between the average of the design points

Experiment and Product Design 465

and the average of the center points as illustrated in Figure 9-9. *If the curvature is severe, the prediction model will not give accurate predictions except near the cube points.*

Let's consider a hypothetical case where we wish to optimize the heat-aging properties (elongations) of an EPDM. From experience we know that elongations will depend on molecular weight (reflected by Mooney viscosity), the level of crystallinity (ethylene content of the EPDM), and the efficiency of curve (reflected by the level of diene in the polymer). This problem lends itself to a three-factorial design. To establish an adequate number of degrees of freedom for error, a reasonable sensitivity for a test of overall curvature and an ability to detect with high probability any effects more than two times the experimental error, we select two replicates of 2^3 design, plus four center points. This gives a total of 20 data points.

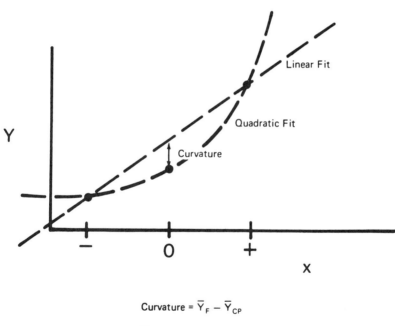

Curvature = $\bar{Y}_F - \bar{Y}_{CP}$

$\bar{Y}_F$ = Average of Factorial Points
$\bar{Y}_{CP}$ = Average of Center Points

Figure 9-9. Two-level factorials. The center points are used as a test for curvature. Note curvature = $\bar{Y}_F - \bar{Y}_{cp}$, where $\bar{Y}_F$ = average of factorial points and $\bar{Y}_{cp}$ = average of center points.

The range of dependent and independent variables of interest in this exercise is as follows:

			Range
Independent Variables	x_1	Mooney viscosity (1 + 4) at 12 SC	40–70
	x_2	Ethylene content (wt %)	65–75
	x_3	Diene content (wt %)	3–6
		Factor Coding	(−) (+)
Dependent	Y_1	Elongation after 500 hrs at 100°C(%)	60–500

The two-level factorial design is outlined in Table 9-6. The design is shown in both coded and actual independent variables, including the centerpoint, trial 9. Table 9-7 shows the 20-point design in randomized order. Center points were deliberately positioned at run numbers 1, 7, 14, 20, to spread them throughout the design. The factorial points were, however, randomized into the remaining 16 runs using one of the permutations of 16 in Table 9-5.

Table 9-7 also shows the values of the responses Y (i.e., elongations) measured at each of the 20 runs. Table 9-8 shows the data points tabulated in nonrandom trial order, with averages at each trial.

Small factorial experiments are more readily suitable to graphical interpretation. Figure 9-10A shows the data averages on the three-dimensional cube and center point. Comparing planes in the manner described

Table 9-6. Two-Level Factorial Design

Trial	Coded			Actual		
	X_1	X_2	X_3	X_1	X_2	X_3
1	−	−	−	40	65	3
2	+	−	−	70	65	3
3	−	+	−	40	75	3
4	+	+	−	70	75	3
5	−	−	+	40	65	6
6	+	−	+	70	65	6
7	−	+	+	40	75	6
8	+	+	+	70	75	6
9	0	0	0	55	70	4.5

Experiment and Product Design

Table 9-7. Two-Level Factorials: Randomized Design

Order	Trial	Mooney	Ethylene	Diene	Elongations
1	9	55	70	4.5	380
2	1	40	65	3	110
3	5	40	65	6	75
4	1	40	65	3	123
5	3	40	75	3	175
6	2	0	65	3	300
7	9	55	70	4.5	370
8	7	40	75	6	145
9	6	70	65	6	165
10	3	40	75	3	197
11	6	70	65	6	159
12	5	40	65	6	87
13	2	70	65	3	287
14	9	55	70	4.5	395
15	4	70	75	3	450
16	8	79	75	6	215
17	7	40	75	6	157
18	4	70	75	3	410
19	8	70	75	6	209
20	9	55	70	4.5	377

Table 9-8. Data Summary for Two-Level Factorials

Trial	Y observations	$\bar{Y}$	Range	Variance
1	110, 123	116.5	13	84.5
2	300, 287	243.5	13	84.5
3	175, 197	186	22	242
4	450, 410	430	40	800
5	75, 87	81	12	72
6	159, 165	162	6	18
7	145, 157	151	12	72
8	215, 209	212	6	18
9	377, 395, 370, 380	380.5	25	111

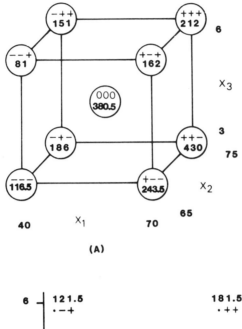

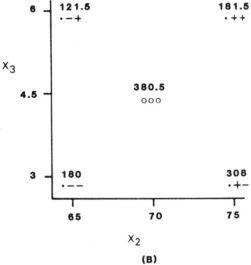

Figure 9-10. (A) Two-level factorial and center points; (B) two-way plot showing interaction; (C) two-way plot showing no interaction.

Experiment and Product Design

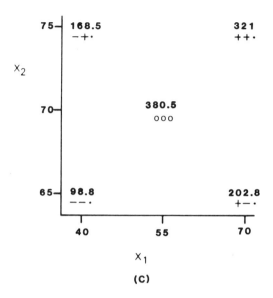

(C)

earlier indicates that all three factors—x_1, x_2, x_3—undoubtedly have positive main effects. One can average over a variable, say x_1, to produce a lower-dimension plot, as in Figure 9-10B. A dot in place of a (+) or (−) indicates a variable averaged over. This is an example where the effect of x_2 is quite different at the high and low levels of x_3, suggesting that the x_2, x_3 interaction is significant. Figure 9-10C shows the x_1, x_2 plan, where the data are averaged over x_3. This plot does not show evidence of any strong interaction. These plots give a quick insight into the "structure" of the system under study. They also provide an opportunity to recognize any anomalous data points. This analysis can be performed computationally following the procedure outlined below.

1. Copy the computation table for the design from Table 9-9. (e.g., this could be constructed on a Lotus spreadsheet).
2. Write the average response for each run to the right of the table. (*Note:* Trial order, not random run order.)
   Total the values of the response on the lines with a + sign for each column. Write the total on the line called Sum +.
4. Do the same for the responses on the lines with − signs. Write their totals on the line called Sum −.

Table 9.9. Computing Table for Analysis of 2^p Factorial Designs

		2^2			2^3				2^4								2^5																
Trial	Mean	x_1	x_2	x_1x_2	x_3	x_1x_3	x_2x_3	$x_1x_2x_3$	x_4	x_1x_4	x_2x_4	$x_1x_2x_4$	x_3x_4	$x_1x_3x_4$	$x_2x_3x_4$	$x_1x_2x_3x_4$	x_5	x_1x_5	x_2x_5	$x_1x_2x_5$	x_3x_5	$x_1x_3x_5$	$x_2x_3x_5$	$x_1x_2x_3x_5$	x_4x_5	$x_1x_4x_5$	$x_2x_4x_5$	$x_1x_2x_4x_5$	$x_3x_4x_5$	$x_1x_3x_4x_5$	$x_2x_3x_4x_5$	$x_1x_2x_3x_4x_5$
1	+	−	−	+	−	+	+	−	−	+	+	−	+	−	−	+	−	+	+	−	+	−	−	+	+	−	−	+	−	+	+	−
2	+	+	−	−	−	−	+	+	−	−	+	+	+	+	−	−	−	−	+	+	+	+	−	−	+	+	−	−	−	−	+	+
3	+	−	+	−	−	+	−	+	−	+	−	+	+	−	+	−	−	+	−	+	+	−	+	−	+	−	+	−	−	+	−	+
4	+	+	+	+	−	−	−	−	−	−	−	−	+	+	+	+	−	−	−	−	+	+	+	+	+	+	+	+	−	−	−	−
5	+	−	−	+	+	−	−	+	−	+	+	−	−	+	+	−	−	+	+	−	−	+	+	−	+	−	−	+	+	−	−	+
6	+	+	−	−	+	+	−	−	−	−	+	+	−	−	+	+	−	−	+	+	−	−	+	+	+	+	−	−	+	+	−	−
7	+	−	+	−	+	−	+	−	−	+	−	+	−	+	−	+	−	+	−	+	−	+	−	+	+	−	+	−	+	−	+	−
8	+	+	+	+	+	+	+	+	−	−	−	−	−	−	−	−	−	−	−	−	−	−	−	−	+	+	+	+	+	+	+	+
9	+	−	−	+	−	+	+	−	+	−	−	+	−	+	+	−	−	+	+	−	+	−	−	+	−	+	+	−	+	−	−	+
10	+	+	−	−	−	−	+	+	+	+	−	−	−	−	+	+	−	−	+	+	+	+	−	−	−	−	+	+	+	+	−	−
11	+	−	+	−	−	+	−	+	+	−	+	−	−	+	−	+	−	+	−	+	+	−	+	−	−	+	−	+	+	−	+	−
12	+	+	+	+	−	−	−	−	+	+	+	+	−	−	−	−	−	−	−	−	+	+	+	+	−	−	−	−	+	+	+	+
13	+	−	−	+	+	−	−	+	+	−	−	+	+	−	−	+	−	+	+	−	−	+	+	−	−	+	+	−	−	+	+	−
14	+	+	−	−	+	+	−	−	+	+	−	−	+	+	−	−	−	−	+	+	−	−	+	+	−	−	+	+	−	−	+	+
15	+	−	+	−	+	−	+	−	+	−	+	−	+	−	+	−	−	+	−	+	−	+	−	+	−	+	−	+	−	+	−	+
16	+	+	+	+	+	+	+	+	+	+	+	+	+	+	+	+	−	−	−	−	−	−	−	−	−	−	−	−	−	−	−	−

5. Add Sum + and Sum −. Write the total on the line called Overall Sum. This step is a check. It should be the same for all columns.
6. Subtract Sum − from Sum +. Write the difference on the line called Difference.
Finally, divide the Difference by the number of + signs in that column. This number is the factor effect for any main effect or interaction (second-through-last columns). For the first (mean) column the result in the row labeled "effect" is the mean of all the cube points.

To assess the degree of curvature in the analysis, first compute the average of the center points, and then calculate the difference between the average of the factorial points computed above the average of the center points.

We can now direct attention to the significance of factor effects and curvature. If a computed factor effect is larger (in absolute value) than the "minimum significant factor," the experimenter can safely conclude that the true effect Δ is nonzero. Similarly, if the curvature effect is larger than the minimum significant curvature effect, we conclude that at least one variable has nonzero curvature associated with it. The minimum significant factor effect [MIN] and the minimum significant curvature [MIN C] are derived from an appropriate t-test significance. The formulas are

$$[\text{MIN}] = ts\sqrt{\frac{2}{mk}} \qquad (14a)$$

$$[\text{MIN C}] = ts\sqrt{\frac{1}{mk} + \frac{1}{c}} \qquad (14b)$$

where

[MIN] = minimum significant factor effect
[MIN C] = minimum significant curvature
t = value of Student's t at the desired probability level for the number of degrees of freedom in the estimate s
m = number of + signs in the column ($= 2^{p-1}$) for factor effect columns and 2^p for the mean column)
k = number of replicates or each trial
c = number of center points
s = pooled standard deviation of a single response observation

Experiment and Product Design

Examination of the formula for MIN C shows that it is helpful to replicate the center point more heavily than any one cube point.

In the previous example there are three positive main effects (x_1, x_2, and x_3), with x_3 being the strongest. Also, there is one significant interaction (x_2, x_3), also positive, meaning that the positive effect of x_2 is increased in magnitude at the high level of x_3. Finally, there is significant curvature, the average observed response at the center point being significantly lower than predicted on the basis of the cube points. All other effects are zero or too small to be declared significant. These significant effects are useful knowledge per se. Coefficients of the model equation follow directly from them, as described later. Main effects and interactions are illustrated in Figure 9-11. Note that in all cases, including the interaction case, a vertical

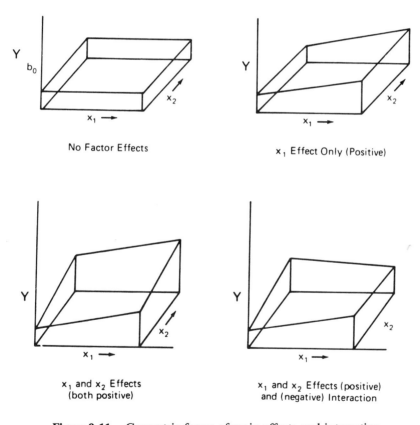

Figure 9-11. Geometric forms of main effects and interaction.

slice through the figure parallel to either the x_1- or x_2-axis would always result in a y-curve that is a straight line. Even though the interaction case is a twisted surface to look at, in three dimensions any such cross section is a straight line. If curvature is present, in addition to main effects and interactions, the surface is bent upward or downward giving cross sections that are not straight lines.

The factor effects represent the difference between the response at the high and low levels of the factors. Thus if the factor effect is divided by the difference between the high and low levels of the factors, the result will be the change in the response for a coded-unit change in the factor.

The model underlying the two-level factorial is written in terms of coded factors x_j. The model is

$$y = b_0 + b_1 x_1 + b_2 x_2 + \cdots + b_p x + b_{12} x_1 x_2 + b_{13} x_1 x_3 + \cdots + b_{p-1,p} x_{p-1} x_p + \text{higher-order interactions} \tag{15}$$

where

y = predicted response
b_j = ½(factor effect for x_j)
b_{ij} = ½(interaction effect for $x_j x_j$)
$b_0 = \bar{Y}$

Note that the coefficients of the model are half of the corresponding factor effects since the coded levels ($+1$) and (-1) differ by two units. It is common practice to omit the terms whose effects are not significant.

SCREENING DESIGNS

The application of screening designs is normally used first in any experimental program. The main use of a screening design is to screen out the few really important variables from a larger number of possible variables with a minimum amount of testing. Most screening designs, in particular the Plackett-Burman designs, are obtained by using a fraction of the 2^p factorial design. A Plackett-Burman design is a specific fraction that has properties that allow efficient estimation of the effects of the variables under study (Plackettt and Burman, 1946).

Table 9.10 illustrates the number of runs required to study 6 to 18 variables by several alternative design strategies. Shown are the recommended screening designs. There is a tremendous reduction in ex-

Table 9-10. Required Number of Runs

Number of factors	2^p factorial	Screening Designs[b]			
		Fraction of 2^p		Recommended Plackett-Burman	
		(1)	(2)	(3)	(4)
6	64	<u>16</u>	<u>32</u>	12	24
7	128	<u>16</u>	64	20	40
8	256	<u>16</u>	64	20	40
9	512		128	<u>20</u>	<u>40</u>
10	1024		128	<u>20</u>	<u>40</u>
11	2048		128	<u>20</u>	<u>40</u>
12	4096		256	<u>20</u>	<u>40</u>
13–18			256	<u>28</u>	<u>56</u>

[a] Designs that are especially attractive for screening ($n \geqslant 40$) are highlighted by a underscore.

[b] (1) Estimates main effects clear of two-factor interactions (reflected design in case of P-B).

(2) Estimates main effects and two-factor interactions clear of each other.

(3) Estimates main effects clear of each other.

periments by using a Plackett-Burman design. The Plackett-Burman designs, however, do not provide estimates of the interactions between variables. It is important to note that the purpose of screening designs is to select a few variables from many so that these few can be investigated in more detail. This can be done by either a more complete factorial or a response surface design, discussed later. For the specific cases of 6, 7, or 8 factors, there are 16- and 32-run fractional factorials that are attractive alternatives to the Plackett-Burman designs. Those are underlined in Table 9-10.

The screen designs can be used for discrete factors (such as two kinds of polymer additive) or for continuous variables (such as two levels of carbon black in compounding). The two levels have been coded + and −. Call either condition + and the other − for discrete variables. In dealing with continuous variables, code the higher value + and the lower −.

Although Plackett-Burman designs are available for nearly every multiple of four trials from 4 to 100, the most useful ones are for 12, 20, or

28 trials. They nominally handle up to 11, 19, and 27 factors, respectively. Tables 9-11 to 9-13 show these designs. The factors are identified across the top of the tables. The trials are denoted by a serial number at the left of the table. As a rule of thumb, a *maximum* of 7 factors for the 12-trial designs, 15 for the 20-trial design, and 23 for the 28-trial design are recommended. This makes it possible to estimate the experimental error from the design data. The p independent variables of the experiment should be assigned to p different columns of the design (most conveniently, the first p columns). The remaining "unassigned" columns provide the error estimate during analysis of the data. As in the 2^p factorial design, the trials should be randomized before running. Random schedules discussed earlier can be used.

The precision ratio may be calculated as described earlier. Plackett-Burman designs usually are not replicated directly. Additional precision with the Plackett-Burman designs is best achieved by one or both of the following routes.

1. "Partial" replication is obtained by using the next larger design. For example, with six factors, the recommended design has 12 runs. This leaves 5 degrees of freedom to estimate experimental error. Also each factor effect is the difference of two averages—six runs in each average. Alternatively, the six factors can be run in 20 runs, leaving 13 degrees of freedom to estimate experimental error. Further, each factor effect becomes the difference of two averages of 10 runs each.

Table 9-11. Twelve-Run Plackett-Burman Design

Trial	Mean	x_1	x_2	x_3	x_4	x_5	x_6	x_7	x_8	x_9	x_{10}	x_{11}
1	+	+	+	−	+	+	+	−	−	−	+	−
2	+	+	−	+	+	+	−	−	−	+	−	+
3	+	−	+	+	+	−	−	−	+	−	+	+
4	+	+	+	+	−	−	−	+	−	+	+	−
5	+	+	+	−	−	−	+	−	+	+	−	+
6	+	+	−	−	−	+	−	+	+	−	+	+
7	+	−	−	−	+	−	+	+	−	+	+	+
8	+	−	−	+	−	+	+	−	+	+	+	−
9	+	−	+	−	+	+	−	+	+	+	−	−
10	+	+	−	+	+	−	+	+	+	−	−	−
11	+	−	+	+	−	+	+	+	−	−	−	+
12	+	−	−	−	−	−	−	−	−	−	−	−

Table 9-12 Twenty-Run Plackett-Burman Design

Trial	Mean	x_1	x_2	x_3	x_4	x_5	x_6	x_7	x_8	x_9	x_{10}	x_{11}	x_{12}	x_{13}	x_{14}	x_{15}	x_{16}	x_{17}	x_{18}	x_{19}
1	+	+	+	−	−	+	+	+	+	−	+	−	+	−	−	−	−	+	+	−
2	+	+	−	+	−	−	+	+	+	+	−	+	−	+	−	−	−	−	+	+
3	+	−	+	−	+	−	−	+	+	+	+	−	+	−	+	−	−	−	−	+
4	+	+	−	+	−	+	−	−	+	+	+	+	−	+	−	+	−	−	−	−
5	+	+	+	−	+	−	+	−	−	+	+	+	+	−	+	−	+	−	−	−
6	+	+	+	+	−	+	−	+	−	−	+	+	+	+	−	+	−	+	−	−
7	+	+	+	+	+	−	+	−	+	−	−	+	+	+	+	−	+	−	+	−
8	+	−	+	+	+	+	−	+	−	+	−	−	+	+	+	+	−	+	−	+
9	+	+	−	+	+	+	+	−	+	−	+	−	−	+	+	+	+	−	+	−
10	+	−	+	−	+	+	+	+	−	+	−	+	−	−	+	+	+	+	−	+
11	+	+	−	+	−	+	+	+	+	−	+	−	+	−	−	+	+	+	+	−
12	+	−	+	−	+	−	+	+	+	+	−	+	−	+	−	−	+	+	+	+
13	+	+	−	+	−	+	−	+	+	+	+	−	+	−	+	−	−	+	+	+
14	+	+	+	−	+	−	+	−	+	+	+	+	−	+	−	+	−	−	+	+
15	+	+	+	+	−	+	−	+	−	+	+	+	+	−	+	−	+	−	−	+
16	+	+	+	+	+	−	+	−	+	−	+	+	+	+	−	+	−	+	−	−
17	+	−	+	+	+	+	−	+	−	+	−	+	+	+	+	−	+	−	+	−
18	+	−	−	+	+	+	+	−	+	−	+	−	+	+	+	+	−	+	−	+
19	+	+	−	−	+	+	+	+	−	+	−	+	−	+	+	+	+	−	+	−
20	+	−	−	−	−	−	−	−	−	−	−	−	−	−	−	−	−	−	−	−

Table 9.13 Twenty-Eight-Run Plackett-Burman Design

Trial	Mean	x_1	x_2	x_3	x_4	x_5	x_6	x_7	x_8	x_9	x_{10}	x_{11}	x_{12}	x_{13}	x_{14}	x_{15}	x_{16}	x_{17}	x_{18}	x_{19}	x_{20}	x_{21}	x_{22}	x_{23}	x_{24}	x_{25}	x_{26}	x_{27}
1	+	+	−	+	+	+	+	−	−	−	−	+	−	−	−	+	−	−	+	+	+	−	+	−	+	+	−	+
2	+	+	+	−	+	+	+	−	−	−	−	+	+	−	−	−	+	−	−	+	+	+	+	+	+	+	+	−
3	+	−	+	+	+	+	−	+	+	−	+	−	−	−	+	−	−	+	−	+	−	+	+	+	−	−	+	+
4	+	+	+	−	−	−	+	−	+	+	+	+	−	−	+	−	−	+	+	+	−	−	−	+	+	+	+	−
5	+	+	−	+	−	−	−	+	+	+	+	−	+	+	−	−	+	+	+	+	+	+	−	+	−	+	+	−
6	+	+	−	+	+	−	−	+	−	+	−	−	−	+	+	+	−	+	−	+	+	−	+	−	+	−	+	+
7	+	−	+	+	−	+	+	+	+	−	+	−	+	−	+	+	+	+	+	+	−	−	−	−	+	−	+	−
8	+	+	−	+	+	+	−	−	+	−	+	+	+	−	+	−	+	+	+	−	−	+	+	+	+	−	−	−
9	+	+	+	+	+	+	−	−	−	−	+	−	−	−	+	−	−	+	+	+	−	+	+	+	+	+	+	−
10	+	−	+	−	−	−	+	−	+	+	+	+	−	+	+	+	+	+	+	−	+	−	−	+	+	+	−	+
11	+	+	−	+	−	+	+	+	−	+	+	+	+	−	−	−	+	+	+	−	+	+	+	−	−	−	+	+
12	+	+	−	+	+	+	−	−	+	−	+	+	+	+	+	+	+	+	−	+	+	+	+	+	+	+	+	+
13	+	+	−	−	−	+	−	+	−	−	+	−	+	+	+	+	+	+	−	+	−	+	+	+	+	−	−	−
14	+	+	+	+	+	+	−	+	+	+	+	−	+	+	+	+	+	−	+	+	+	+	−	+	+	+	+	+
15	+	−	−	−	−	−	+	−	−	−	+	+	−	+	+	−	−	+	+	+	+	+	+	−	+	+	+	+
16	+	−	+	−	−	−	−	+	−	+	+	−	+	+	+	−	+	+	+	−	+	+	−	−	+	+	+	+
17	+	+	−	−	+	+	−	−	+	−	+	+	+	+	−	+	+	+	+	−	+	+	+	+	+	+	−	+
18	+	+	−	+	+	+	+	+	+	+	−	+	−	+	−	+	+	−	+	−	−	+	−	+	+	+	+	−
19	+	+	+	−	−	−	−	−	+	−	−	−	−	+	+	+	−	−	+	+	−	+	−	−	+	+	+	−
20	+	+	−	+	−	+	+	+	+	+	+	+	−	+	−	+	+	+	+	−	+	−	−	+	−	+	+	−
21	+	+	+	−	+	−	+	−	−	−	−	+	+	+	−	+	+	+	−	−	+	−	+	−	+	+	−	−
22	+	−	−	+	+	−	+	+	+	+	+	−	+	+	−	+	+	+	+	+	−	−	+	−	−	+	+	−
23	+	+	+	+	−	+	−	+	+	−	−	+	−	+	+	−	−	+	−	+	−	+	−	+	−	+	+	−
24	+	−	−	−	+	−	+	−	+	+	−	+	−	−	−	+	+	+	+	−	+	−	+	−	+	+	+	−
25	+	+	+	+	+	+	+	−	+	+	+	+	−	−	−	+	+	+	+	−	−	−	−	−	+	−	−	−
26	+	−	+	−	+	+	−	+	+	+	+	+	+	+	+	+	−	+	−	−	−	−	+	−	−	+	+	−
27	+	−	+	+	−	+	−	+	−	−	−	+	+	−	−	−	+	−	−	−	+	−	+	+	−	−	−	−
28	+	−	−	+	−	−	−	+	−	−	−	+	−	−	−	−	−	−	−	−	−	−	−	−	−	−	−	−

Experiment and Product Design

2. If a full second replicate is economically feasible, it is preferable to run the "reflected" design for the second replicate. The reflected design is obtained by reversing every + and every − sign in the original design. This gives a completely new set of runs for the second replicate. The reflected design can be analyzed separately, just like the original design. The advantage of the reflection procedure stems from the improved mathematical properties of the two designs, analyzed together. In either the original design or its reflection, estimates of all main effects are uncorrelated with each other, but main effects are slightly correlated with groups of two-factor interactions.

Computation of factor effects is performed as follows:

1. Copy the computation table for the design from the appropriate table (Tables 9-11 to 9-13).
2. Record the average response for each run to the right of the table. (*Note:* Trial order, not random run order.)
3. Total the values of the response on the lines with a plus sign for each column. Write the total on the line called Sum+. Repeat this for all columns.
4. Do the same for the responses on the lines with minus signs. Write the total on the line called Sum −.
5. For each column add Sum+ and Sum−. This step should be the same for all columns and is essentially a check.
6. For each column substract Sum − from Sum + and tabulate the results on the line called Difference.
7. For each column divide the Difference value by the number of plus signs in that column. This number is the effect for that column. For those columns with assigned factors this number is the factor effect. For extra columns for which no factor is assigned, the value is an estimate of the experimental error.

A ranking of the assigned factor effects from largest to smallest (in absence value) establishes the best estimate of the relative importance of the factors. Future work should concentrate on the three to five highest-ranked factors.

The size of the experimental error must be determined. This can be used to compute the minimum significant factor effect. Any factor effect whose absolute value is larger than (MIN) is established as statistically significant. The value of (MIN) is computed following the procedure given below.

1. Square each of the unassigned factor effects.

2. Add them together and divide by the number of unassigned factor effects.
3. Compute the square root as follows:

$$S_{FE} = \sqrt{\frac{1}{q}(UFE_1^2 + UFE_2^2 + \cdots + UFE_q^2)} \qquad (16)$$

4. Compute [MIN] = tS_{FE}. (t has degrees of freedom equal to the number of unassigned factor effects q); q is the number of unassigned factor effects = $n - p - 1$.)

When there are only a small number of degrees of freedom, as in the unassigned columns of a typical Plackett-Burman experiment, it is preferable to pick a significance level lower than 0.95. By so doing, the "power of test," that is, the likelihood of detecting a significant effect if it exists, is greater. The following table can be used as a rule of thumb.

Degrees of freedom	Significance level
5	0.90
5–30	0.95
30	0.99

In cases where there are a small number of degrees of freedom for error it also is important to avoid experimental mistakes. One bad data point can have a significanct effect on the calculations.

When two-factor interactions are present, the Plackett-Burman designs have the property that the interaction effects are mathematically separated into many parts. These parts behave as an inflation of the experimental error. Thus, when a factor effect is found to be significant, it means that the factor stands out over a pool of background noise containing both experimental error and interactions, if present.

Once the important variables are identified and we know how they effect the response, we can take action. A more detailed investigation of these factors by either a 2^p factorial or a response surface design can be executed. We also could apply the results directly, and probably would do so if a substantial improvement in the response is obvious at some experimental condition in the screening experiment.

RESPONSE SURFACE CONCEPTS AND CONSTRUCTION OF POLYNOMIAL MODELS

Interest often lies in describing the response of one or more dependent variables to several independent variables. This relationship can be expressed by one or more mathematical equations or models. Models can be theoretical or empirical. The exact theoretical model is rarely known; therefore, empirical models must be used to approximate the response. Polynomial approximations are widely used and have proven very valuable in modeling process data or in scale-up of product performances.

A process can be a lab- scale test, a pilot plant, or even a full-scale plant trial. All processes operate on inputs and produce one or more outputs. Some of these outputs may be color, impurity level, yield, green strength, modulus, and so on. We want to predict how the outputs (y's or dependent variables) responsd to changes in the inputs (x's or independent variables such as temperature, pressure, catalyst concentration, ethylene content, molecular weight, etc.). Mathematically, these relations can be as follows:

$$y_1 = f_1(x_1, x_2, \ldots, x_p)$$

$$y_2 = f_2(x_1, x_2,, x_p)$$
$$\vdots \tag{17}$$
$$y_m = f_m(x_1, x_2,, x_p)$$

These mathematical expressions state that the ouptputs of the process (y_1 to y_m) are functions of the inputs (x_1 to x_p). In other words, each is a different function of the x's.

A model, is a counterpart, a likeness, or a representation of some property or event. In our case, we consider a mathematical model as a mathematical representation of the press or property response of a material. Exact theoretical relationships are very rarely known for a physical system. Even in cases where some Y's can be predicted from known theory, other Y's of commercial importance cannot be predicted from theory.

Statisticians have found the polynomial models are widely useful for empirical models. A polynomial has the same form as a Taylor series that has been truncated after a specified number of terms. Any function, continuous and having derivatives of all orders in the interval being considered, can be approximated by a Taylor series. This is a true statement regardless of how complex the function may be. If the function is a very complex, many terms may be needed in the Taylor series.

The first-order Taylor series is a linear approximation. In the operable region of most processes a second- order Taylor series is an adequate approximation.

If a known function is expanded into a Taylor series about some point, the error is zero at the point of expansion. If only second-order terms are retained, the error is not likely to be small at points some distance from the point of expansion. An obvious question is: How can a second-order polynomial be a widely useful approximation? The answer is that the coefficients in the polynomial are not these of the Taylor series but are adjusted (by the method of least squares) to give a small prediction error throughout the region of interest. The region of interest is here assumed to be the region in which data have been obtained.

A general polynomial model (second order) with one independent variables can be expressed mathematically as follows:

$$y = b_0 + b_1 x_1 + b_{11} x_1^2 \tag{18}$$

This expressive contains a constant term, a linear term, and a quadratic term. The subscripts identify the terms.

For two independent variables, the expression has the form

$$y = b_0 + b_1 x_1 + b_2 x_2 \, b_{12} x_1 x_2 + b_{11} x_1^2 + b_{22} x_2^2 \tag{19}$$

This expression contains linear, cross-product, and quadratic terms.

We can generalized the model equation for p independent variables as follows:

$$y = b_0 + \sum_{j=1}^{p} b_j x_j + \sum_{\substack{j=1 \\ j'=1 \\ j'>j}}^{p} b_{jj'} x_j x_{j'} + \sum_{j=1}^{p} b_{jj} x_j \tag{20}$$

Plackett-Burman designs essentially construct "very limited" response surface models becuase they estimate the constant and linear terms only. Similarly, the 2^p designs support "limited" response surface models comprised of main effects and interactions. A "response surface" design refers to a design that supports a full quadratic model.

The polynomial approximation for a particular process is obtained by running an experiment in the independent variables (x's) and observing the dependent variable (y's). These observations on the y's are then used to obtain estimates of the coefficient (b) for the general model. The b's we obtain are those that will result in an equation that will fit the data observed in the least-squares scheme.

Theoretical models, such as physical laws, can be of use in predicting the behavior of a system quite far from the area in which their parameters were estimated. A simple polynomial approximation interpolates very well but extrapolates not so well. The farther away you get from the region in which the data were collected, the poorer the results of extrapolation, even to the extent of being nonsense and completely useless. Consequently, the data should cover the entire region in which predictions will be made.

In cases where scale-up or extrapolation is ultimately required, it is necessary to develop a theoretical model for the purpose. The response surface data from a good basis for developing and fitting the theoretical model. In particular, adequacy of the proposed theoretical model requires, at the least, that it fit the response surface data as precisely as the empirical model.

The concept of response surface experimentation has been around for nearly 50 years. In very simple terms, response surface methodology is a means of seeking a maximum or a minimum in a response surface. The optimum is approached in a series of experiments that climb up (or down) a hill in states until the optimum (peak, minimum, or ridge) is eventually pinpointed and predicted by a quadratic response surface model. This is, however, an over simplification of the technique, as there are a number of subtleties to the methodology that need to be understood.

First, as noted by Bacon (1970), experiments dealing with real processes rarely have the luxury of more than one or two opportunities in getting process data during any one period of study of the process. Also, its important to bear in mind that processes have limited operating ranges, and the unconstrained "optimum" for any one response almost always is outside the available operating region. Thus we can have, at best, a "constrained optimum" at a boundary of the region. To make our task even more complicated, there is almost never only one response, and the optima of several responses rarely coincide. Indeed, the multiple response often "trade off" with each other. In research applications of response surface methodology, delineation of these trade-offs is the most important feature, leading to the identification of an attainable combination of properties.

Consider the point at which a process is introduced to the plant stage. Here the product as described to the customer has a specified quantitative attainable balance of properties. It is much more important to maintain this combination of properties than to optimize any one of them. As pointed out in earlier chapters, consistency is more important than optimality to the customer. A final point to bear in mind is that the response surface in a real process is affected by extraneous variables not in the response surface model. The primary usefulness of this methodology in a plant is as a tool determine how the plant responds to the independent variables. This provides a guide for process supervision on how to compensate for changes in the response surface due to effects of extraneous variables. In many cases these extraneous variables, such as raw material variations, environmental drifts, and others, do not change the shape of the response surface significantly but have only small additive effects. In terms of the model equation, the coefficient b_0 is affected only. This makes it easy to use contour plots as tools for process control. The response surface model can be thought of as a multivariate "process calibration" equation, to be used in the same manner as the calibration equation for any instrument.

With the foregoing background information, we can now lay down the principles behind response surface design experimentation. When contemplating a response surface type of experimental program, the investigation will have narrowed its attention to relatively (typically two to six) continuous independent variables. The objective of the experiment will be to provide information regarding linear, interaction, and curvature effects with respect to all or most of these continuance independent variables. The response surface is likely to be described by a full quadratic polynomial. The designs discussed here are highly efficient for this common system.

Curvature effects must be estimated; hence the experimental designs must have at least three levels for each independent variables. It would be feasible to use full three-level factorials that would provide orthogonal estimates of the linear, quadratic, and interaction effects. A disadvantage of the full three-level factorials is the large number of experimental runs required. This is shown below.

Number of independent variables (factors)	Number of trials in full three-level factorial	Number of coefficients in full quadratic
2	9	6
3	27	10
4	81	15
5	243	21
6	729	28

The last column of values provides the number of coefficients in the full quadratic polynomial model that is to be determined from the data. This is the absolute minimum number of data points required. In reality, the number of data points must be greater than the minimum to provide degrees of freedom from which an estimate of the error variance can be obtained. If only the minimum number of points is used, the fitted polynomial will pass through every observed point exactly, and will therefore exactly "fit" the response surface plus the random errors. For every experimental point beyond the minimum number, an additional degree of freedom for error is obtained, and additional capacity is provided for the fitted equation to smooth out the random errors when fitting the response surface. Ideally, we desire designs providing reasonable but no excessive numbers of degrees of freedom for error.

In a response surface setting, the concept of the prediction-variance profile is a relevant criterion in selecting a design. If we imagine ourselves standing at the center of the experimental region (in coded-scaled units of the x's), to predict the response at a point some specified distance from the center, it would be desirable that the precision (i.e., variance) of the prediction to be independent of the direction of that point from the center. Designs having this property are referred to as "rotable" designs. The prediction variance should also be relatively constant at varying distances from the center of the design out to the boundary of the experimental region. These two properties produce a relatively constant prediction variance throughout the experimental region.

In the case where a large number of runs is involved, a desirable feature to have in the design is the possibility of running it in separate blocks of points, with the property of being able to subtract out the effect of a shift of response between blocks without this block effect causing a bias in the estimates of any of the coefficients in the polynomial. These designs allow "orthogonal blocking." The block effect is estimated by adding to the model one additional "dummy" term for each block beyond the first block. The value of x for the additional term(s) is zero or 1—zero except for observations in that block. The coefficient(s) of the added term(s) are the amounts by which the responses differ in each block compared to the first block. Note that three levels is the minimum number for each factor; however, it is possible to develop designs using more than three levels. For example, one popular class of response surface designs are the "central composite" or "Box-Wilson (1951)" designs. These employ five levels of each factor. In general, it is experimentally convenient to use only the minimum number of levels necessary.

Box-Behnken (1960) designs enable use of the minimum number of levels. This system uses a subset of the points in the corresponding full three-level factorial. For example, the three-factor design uses 13 of the 27 points from the full factorial, with two extra replicates the center point, for a total of 15 points. This is five more than the minimum number and provides five degrees of freedom for error. The Box-Behnken designs are rotable and from 4 up to 10 factors they provide for orthogonal blocking into a least two blocks. Except for the center points, all points are at the midpoints of the edges (or faces) of a hypercube whose dimension is the number of factors. All these points lie on a single sphere and are therefore equally distant from the center. This is a geometric property associated with rotatability. The replicated center points provide a measure of inherent experimental error. In addition, they are sufficient in number to give relatively constant prediction variance as a function of distance from the center, within the design region. At the edge of the region and beyond, the prediction variance increases rapidly with additional extrapolation distance from the center. Figure 9-12 illustrates the geometry of the Box-Behnken design for three variables. Three designs are summarized in Tables 9-14 to 9-18 for three to seven independent variables, respectively. Figure 9-12 shows the three-variable design. In Table 9-14 it is seen that the 15 points are made up of three 2^2 designs holding one variable at its middle value for each 2^2, plus the center points. Tables 9-15 and 9-16 shows similar structures for the larger designs.

To construct the Box-Behnken response surface design, first list the

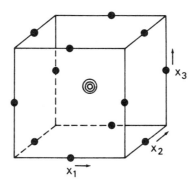

Figure 9-12. Geometry of Box-Behnken design for three variables.

Experiment and Product Design

Table 9-14 Three-Variable Box-Behnken Design (No Orthogonal Blocking Possible)

	x_1	x_2	x_3	
1	+1	+1	0	
2	+1	−1	0	
3	−1	+1	0	
4	−1	−1	0	
5	+1	0	+1	
6	+1	0	−1	
7	−1	0	+1	
8	−1	0	−1	
9	0	+1	+1	
10	0	+1	−1	
11	0	−1	+1	
12	0	−1	−1	
13	0	0	0	Center
14	0	0	0	point
15	0	0	0	

important independent variables and their levels. The levels should be equally spaced. As an example, consider the following:

Symbol	Name	Low (−1)	Middle (0)	High (+)	Unit
x_1	Temperature	80	100	120	°C
x_2	Extrusion speed	30	50	70	rpm
x_3	Green strength	200	400	600	kPa

Next, list the dependent variables to be studied and their units; for example:

Symbol	Name	Unit
Y_1	Extension rate	Grams per minute
Y_2	Swell	Percent

Table 9-15 Four-Variable Box-Behnken Design (Three Blocks of Nine Points)

x_1	x_2	x_3	x_4	
+1	+1	0	0	
+1	−1	0	0	
−1	+1	0	0	
−1	−1	0	0	
−1	−1	0	0	Block 1
0	0	+1	+1	
0	0	+1	−	
0	0	−1	+1	
0	0	−1	−1	
0	0	0	0	
+1	0	0	+1	
+1	0	0	−1	
−1	0	0	+1	
−1	0	0	−1	
0	+1	+1	0	Block 2
0	+1	−1	0	
0	−1	+1	0	
0	−1	−1	0	
0	0	0	0	
+1	0	+1	0	
+1	0	−1	0	
−1	0	+1	0	
−1	0	−1	0	
0	+1	0	+1	Block 3
0	+1	0	−1	
0	−	0	+1	
0	−1	0	−1	
0	0	0	0	

Table 9-16 Five-Variable Box-Behnken Design (Two Blocks of 23 Points)

x_1	x_2	x_3	x_4	x_5
+1	+1	0	0	0
+1	−1	0	0	0
−1	+1	0	0	0
−1	−1	0	0	0
0	0	+1	+1	0
0	0	+1	−1	0
0	0	−1	+1	0
0	0	−1	−1	0
0	+1	0	0	+1
0	+1	0	0	−1
0	−1	0	0	+1
0	−1	0	0	−1
+1	0	+1	0	0
+1	0	−1	0	0
−1	0	+1	0	0
−1	0	−1	0	0
0	0	0	+1	+1
0	0	0	+1	−1
0	0	0	−1	+1
0	0	0	−1	−1
0	0	0	0	0
0	0	0	0	0
0	0	0	0	0
0	+1	+1	0	0
0	+1	−1	0	0
0	−1	+1	0	0
0	−1	−	0	0
+1	0	0	+1	0
+1	0	0	−1	0
−1	0	0	+1	0
−1	0	0	−1	0
0	0	+q	0	+1
0	0	+1	0	−1
0	0	−1	0	+1
0	0	−1	0	−1
+1	0	0	0	+1
+1	0	0	0	−1
−1	0	0	0	+1
−1	0	0	0	−1
0	+1	0	+1	0
0	+1	0	−1	0
0	−1	0	+1	0
0	−1	0	−1	0
0	0	0	0	0
0	0	0	0	0
0	0	0	0	0

Table 9-17 Six-Variable Box-Behnken Design (Two Blocks of 27 Points)

x_1	x_2	x_3	x_4	x_5	x_6
+	+	0	−	0	0
+	−	0	+	0	0
−	+	0	+	0	0
−	−	0	−	0	0
0	+	+	0	−	0
0	+	−	0	+	0
0	−	+	0	+	0
0	−	−	0	−	0
0	0	+	+	0	−
0	0	+	−	0	+
0	0	−	+	0	+
0	0	−	−	0	−
+	0	0	+	−	0
+	0	0	−	+	0
−	0	0	+	+	0
−	0	0	−	−	0
0	+	0	0	+	−
0	+	0	0	−	+
0	−	0	0	+	+
0	−	0	0	−	−
+	0	+	0	0	+
+	0	−	0	0	−
−	0	+	0	0	−
−	0	−	0	0	+
0	0	0	0	0	0
0	0	0	0	0	0
0	0	0	0	0	0
+	+	0	+	0	0
+	−	0	−	0	0
−	+	0	−	0	0
−	−	0	+	0	0
0	+	+	0	+	0
0	+	−	0	−	0
0	−	+	0	−	0
0	−	−	0	+	0
0	0	+	+	0	+
0	0	+	−	0	−
0	0	−	+	0	−
0	0	−	−	0	+
+	0	0	+	+	0

Experiment and Product Design

Table 9-17 Six-Variable Box-Behnken Design
(Two Blocks of 27 Points)

x_1	x_2	x_3	x_4	x_5	x_6
+	0	0	−	−	0
−	0	0	+	−	0
−	0	0	−	+	0
0	+	0	0	+	+
0	+	0	0	−	−
0	−	0	0	+	−
0	−	0	0	−	+
+	0	+	0	0	+
+	0	−	0	0	−
−	0	+	0	0	−
−	0	−	0	0	+
0	0	0	0	0	0
0	0	0	0	0	0
0	0	0	0	0	0

Table 9-18 Seven-Variable Box-Behnken Design
(Two Blocks of 31 Points)

x_1	x_2	x_3	x_4	x_5	x_6	x_7
0	0	0	+	+	−	0
0	0	0	+	−	+	0
0	0	0	−	+	+	0
0	0	0	−	−	−	0
+	0	0	0	0	+	−
+	0	0	0	0	−	+
−	0	0	0	0	+	+
−	0	0	0	0	−	−
0	+	0	0	+	0	−
0	+	0	0	−	0	+
0	−	0	0	+	0	+
0	−	0	0	−	0	−
+	+	0	−	0	0	0
+	−	0	+	0	0	0
−	+	0	+	0	0	0
−	−	0	−	0	0	0
0	0	+	+	0	0	−
0	0	+	−	0	0	+
0	0	−	+	0	0	+
0	0	−	−	0	0	−

Table 9-18 Seven-Variable Box-Behnken Design (Two Blocks of 31 Points)

x_1	x_2	x_3	x_4	x_5	x_6	x_7
+	0	+	0	−	0	0
+	0	−	0	+	0	0
−	0	+	0	+	0	0
−	0	−	0	−	0	0
0	+	+	0	0	−	0
0	+	−	0	0	+	0
0	−	+	0	0	+	0
0	−	−	0	0	−	0
0	0	0	0	0	0	0
0	0	0	0	0	0	0
0	0	0	0	0	0	0
0	0	0	+	+	+	0
0	0	0	+	−	−	0
0	0	0	−	+	−	0
0	0	0	−	−	+	0
+	0	0	0	0	+	+
+	0	0	0	0	−	−
−	0	0	0	0	+	−
−	0	0	0	0	−	+
0	+	0	0	+	0	+
0	+	0	0	−	0	−
0	−	0	0	+	0	−
0	−	0	0	−	0	+
+	+	0	+	0	0	0
+	−	0	−	0	0	0
−	+	0	−	0	0	0
−	−	0	+	0	0	0
0	0	+	+	0	0	+
0	0	+	−	0	0	−
0	0	−	+	0	0	−
0	0	−	−	0	0	+
+	0	+	0	+	0	0
+	0	−	0	−	0	0
−	0	+	0	−	0	0
−	0	−	0	+	0	0
0	+	+	0	0	+	0
0	+	−	0	0	−	0
0	−	+	0	0	−	0
0	−	−	0	0	+	0
0	0	0	0	0	0	0
0	0	0	0	0	0	0
0	0	0	0	0	0	0

The required number of runs (i.e., number of experimental points) is

No. independent variables, p	3	4	5	6	7
No. points, n	15	27	46	54	62

Constraints of time or money may dictate that the test be limited to a smaller number of independent variables than listed to enable fewer runs. Tables 9-17 and 9-18 list the individual experimental points using coded $(-, 0, +)$ levels for the independent variables. One must consider whether the points should be run in separate blocks (if the design allows orthogonal blocking). Blocking is desirable if external conditions are likely to be more uniform within a block than between blocks. The table shows dashed horizontal lines to separate orthogonal blocks.

The next step involves randomizing the sequence for running the experimental points within each block. Sometimes it is very costly to do the points in strictly random order; in such cases it may be necessary to compromise the random order somewhat. It is desirable to have the repeated center points scattered throughout the experiment. Then write out the *experimental schedule" in decoded (physical) units and in the randomized order;* *for example:*

Run	Temperature	Pressure	Concentration
1	80	70	400
2	80	50	600
3	100	70	600

Review each run to verify that it will be operable. It may be necessary to adjust some factor levels or to move individual points slightly to achieve operability.

An important class of response surface experiments concerns those involved with mixtures. In these experiments, factors cannot be varied independently of each other. Many rubber or plastics products are manufactured by mixing together several ingredients or components. In mixture experiments the response is an "intensive" property, that is, a property whose value does not depend on the quantity of the mixture. The propor-

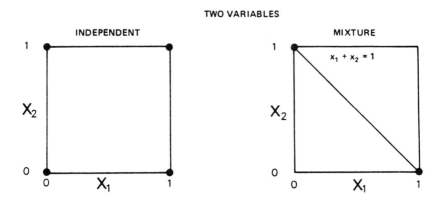

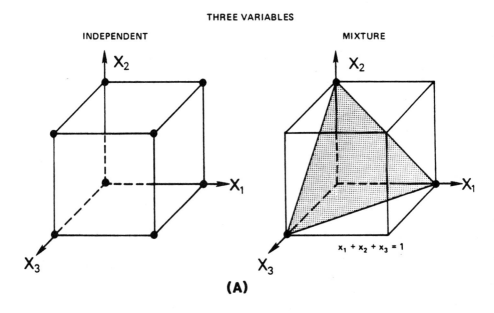

Figure 9-13. (A) Three-component mixture design; (B) tetrahedron, four-component simplex.

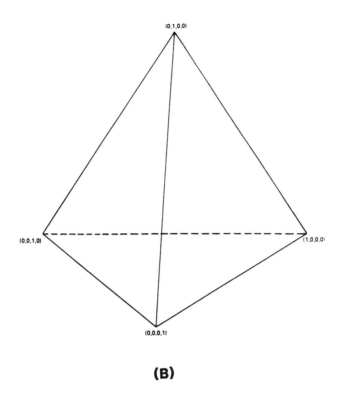

(B)

tion of each component x_j ($j = 1, 2,, q$) in the mixture must lie between 0 and 1.0, and the sum of the proportions of all the components must equal 1.0.

$$\sum_{j=1}^{q} x_j = 1 \qquad 0 \leq x_j \leq 1.0$$

Thus the level of the qth component can be calculated from the levels of the $q - 1$ remaining components:

$$x_q = 1.0 - \sum_{q=1}^{j-1} x_j \qquad (21)$$

Factorial and response surface designs cannot be used directly to study the response of a multicomponent mixture system because of the constraint that the sum of all the components must be unity. Mixture data are, however, generated using an experimental design, whereby a polynomial model is fitted to the data, and the response surface contours are examined to find the region of the best values of the response.

Figure 9-13A illustrates the geometry of mixture factor spaces for two and three components. With two components the rectangular factor space (at the left) in ordinary independent factors $x_1 x_2$ reduces to a single diagonal line when the mixture constraint is imposed (at the right). In three variables the cuboidal space reduces to the triangular space at the right. In every case the reduced space has dimension one less than the number of factors. The shapes of the reduced spaces are referred to as "simplexes," and mixture experiments are often called simplex experiments. Figure 9-13B shows the tetrahedron, which is the four-component simplex.

When the mixture constraint is applied to a quadratic model in two variables, a reduced model with only three terms is obtained:

$$y = b_0 + b_1 x_1 + b_2 x_2 + b_{11} x_1^2 + b_{22} x_2^2 \tag{22}$$

The constraints on this model are

$$x_1 + x_2 = 1 \tag{23a}$$

$$x_1^2 = x_1 x_1 = x_1(1 - x_2) = x_1 - x_1 x_2 \tag{23b}$$

$$x_2^2 = x_2 x_2 = x_2(1 - x_1) = x_2 - x_1 x_2 \tag{23c}$$

Constraints (23b) and (23c) are a consequence of constraint (23a).

The quadratic mixture and model has the form

$$y = b_1^* x_1 + b_2^* + b_{12}^* x_1 x_2 \tag{24}$$

where

$$b_1^* = b_0 + b_1 + b_{11}$$

$$b_2^* = b_0 + b_2 + b_{22}$$

$$b_{12}^* = b_{12} - b_{11} - b_{22}$$

Application of this procedure for other models results in a set of reduced models called the Scheffé canonical forms for mixture models (Scheffé, 1958). This is also discussed in detail by Marquardt and Snee (1974).

Experiment and Product Design

In the general case of q components the linear Scheffé model is

$$y = \sum_{j=1}^{q} b_j x_j$$

$$b_1 x_1 + b_2 x_2 + \cdots b_q x_q \tag{26}$$

The quadratic model is

$$y = \sum_{j=1}^{q} b_j x_j + \sum_{i=1}^{q-1} \sum_{j=i+1}^{q} b_{ij} x_i x_j \tag{27}$$

This describes many systems in which the components do not blend linearly, and in some instances the special cubic model having the form

$$y = \sum_{i=1}^{q} b_i x_i + \sum_{i=1}^{q-1} \sum_{j=i+1}^{q} b_{ij} x_i x_j + \sum_{i=1}^{q-2} \sum_{j=i+1}^{q-1} \sum_{k=j+1}^{q} b_{ijk} x_i x_j x_k \tag{28}$$

is needed to allow an adequate representation of the response surface. Scheffé also defines a full cubic model form.

For a three-component system the models are as follows:

Linear:

$$y = b_1 x_1 + b_2 x + b_3 x_3 \tag{29}$$

Quadratic:

$$y = b_1 x_1 + b_2 x_2 + b_3 x_3 + b_{12} x_1 x_2 + b_{23} x_2 x 3 \tag{30}$$

Special cubic:

$$y = b_1 x_1 + b_2 x_2 + b_3 x_3 + b_{12} x_1 x_2 + b_{13} x_1 x_3 + b_{23} x_2 x_3 \\ + b_{123} x_1 x_2 x_3 \tag{31}$$

Cubic:

$$y = b_1 x_1 + b_2 x_2 + b_3 x_3 + b_{12} x_1 x_2 + b_{13} x_1 x_3 + C_{12} x_1 x_2 (x_1 - x_2) \\ + C_{13} x_1 x_3 (x_1 - x_3) + C_{23} x_2 x_3 (x_2 - x_3) + b_{123} x_1 x_2 x_3 \tag{30}$$

Note that there are no constant (b_0) and squared terms ($b_{ii}x_i^2$). Also, since the components cannot be varied independently as interaction terms but describe the nonlinear blending of components i and j.

The proper mixture design should have three characteristics:

1. Provide good estimates of the coefficient (b's).
2. Provide an estimate of the experimental error.
3. Provide a measure of the lack of fit of the model.

The simplex design is a popular design for studying the response of a mixture system. For the linear model only the pure components need to be run. The ith coefficient (b_i) in the model is simply the average response determined for the ith pure component ($b_i = Y_i$). Additional runs would have to be added to the design to obtain a measure of lack of fit. As a general rule of thumb, it is best to start with a design that will permit the estimation of at least the quadratic coefficients (b_{ij}) and perhaps the special cubic coefficient (b_{ijk}). Such a design for three components is given in Table 9-19.

In a graphical design the first seven points in the design are used to estimate the coefficients in the special cubic response surface model. The remaining three points are used to check the accuracy of the prediction of the model. The coefficients in the special cubic model are estimated using the following formulas.

$$b_1 = Y_1$$

$$b_2 = Y_2$$

$$b_3 = Y_3$$

$$b_{12} = 4Y_{12} - 2(Y_1 + Y_2) \tag{33}$$

$$b_{13} = 4Y_{13} - 2(Y_1 + Y_3)$$

$$b_{23} = 4Y_{23} - 2(Y_2 + Y_3)$$

$$b_{123} = 27Y_{123} - 12(Y_{12} + Y_{13} + Y_{23}) + 3(Y_1 + Y_2 + Y_3)$$

After the coefficients are calculated, it is necessary to check the prediction of the model to see if it provides an adequate representation of the response surface.

Experiment and Product Design

Table 9-19 Three-Component Simplex Design

Run	x_1	x_2	x_3	Response, Y	
1	1	0	0	Y_1	
2	0	1	0	Y_2	
3	0	0	1	Y_3	
4	1/2	1/2	0	$Y_{1,2}$	
5	1/2	0	1/2	$Y_{1,3}$	
6	0	1/2	1/2	$Y_{2,3}$	
7	1/3	1/3	1/3	$Y_{1,2,3}$	
8	2/3	1/6	1/6	$Y_{1,1,2,3}$	Check
9	1/6	2/3	1/6	$Y_{1,2,2,3}$	points

The lack of fit of the model is estimated by examining the difference between the observed and predicted responses at each of the checkpoints, where the lack of fit variance is

$$s_{LF}^2 = \frac{1}{r} \sum_{i=1}^{r} (\bar{Y}_{oi} - \hat{Y}_i)^2 \tag{34}$$

where $\bar{Y}_{oi}$ is the observed average response at ith checkpoint, $\hat{Y}_i$ is the predicted response at ith checkpoint, based on a model derived without using checkpoint data, and r is the number of checkpoints. The variance for lack of fit is compared to the error variance (degrees of freedom = u) using an F ratio, $F = s_{LF}^2/s_{error}^2$.

If the F-ratio does not exceed the tabulated F-value for r and u degrees of freedom at the desired probability level, the model can be assumed to provide a presentation of the response surface within the experimental error of the data.

APPLICATION TO PRODUCT DEVELOPMENT

The paper by Benefield and Ruiz (1988) embodies many of the concepts explored in this chapter with specific importance to elastomer product development. The role of statistical experimental design in product development is well illustrated in their work, some of which is reviewed

Table 9-20 Experimental Design Variables

Independent variables	Low level	High level	
Taguchi L8 Design			
Unsaturation (Diene)	0	5	$C = C/1000°C$
Mooney viscosity (Mooney)	40	70	ML (1 + 4) at 125°C
Molecular weight distribution (MWD)	Narrow	Broad	
Taguchi L4 Design			
Ethylene content (Ethylene)	60	80	Mol %
Mooney viscosity (mooney)	25	55	ML (1 + 4) at 125°C

here. Let's consider a product development program for ethylene propylene rubber. The major product characteristics that can be varied during the polymerization of this elastomer are the diene content, ethylene content, Mooney viscosity (which provides a measure of molecular weight), and the molecular weight distribution (MWD). Each of these characteristics can play an important role in end-use performance and must be considered when choosing an EP polymer for a particular application.

In the study by Benefield and Ruiz (1988) an eight-run experiment was executed, aimed at establishing the dominant polymer properties affecting a variety of end-use performance variables, such as tensile strength, hardness, Gardner impact, spiral flow rate, die swell, and shear sensitivity. The test construction is shown in Table 9-20 and 9-21 and in Figure 9-14 and 9-15.

Each measured characteristic contains these pieces of related information concerning the suspected significant variables. For tensile at yield, it was found that by changing the diene content form 0 to 5% resulted in an increase in 1.2 MPa.

Table 9-21 EP(D)M Polymer Characteristics

EP(D)M[a]	Controlled factors			Trial Order
	Diene	Mooney	MWD	
Taguchi L8 Design				
Epsyn 7506	5	70	Broad	1
Polymer A	0	70	Broad	2
EPsyn 70A	5	70	Narrow	3
EPsyn 4506	5	40	Broad	4
Polymer B	0	40	Broad	5
Polymer C	0	40	Narrow	6
EPsyn 40A	5	40	Narrow	7
Polymer D	0	70	Narrow	8

Taguchi L4 Design	Controlled factors		Trial Order
	Ethylene	Mooney	
Polymer E	80	25	1
EPsyn 5009	80	55	2
Polymer F	60	25	3
Polymer G	60	55	4

[a]Polymers A, B, E: experimental EP(D)M polymers designed for this study; polymers C, D, F, G: propietary EPM polymers.

Three-Factor, Two-Level Design Matrix

Trial Order	Controlled Factors and Interactions						
	Diene A	Mooney B	AB	MWD C	AC	BC	ABC
1	+	+	−	+	−	−	+
2	−	+	+	+	+	−	−
3	+	+	−	−	+	+	−
4	+	−	+	+	+	+	+
5	−	−	−	−	−	−	−
6	−	−	−	−	+	−	+
7	+	−	+	−	+	−	+
8	−	+	+	−	−	+	+

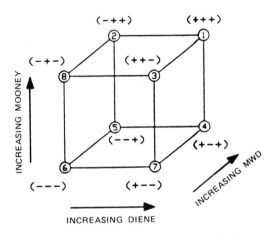

Figure 9-14. Taguchi L8 design matrix and design cube.

The following is a summary of the variable effects identified in this study, which contributed to at least half of experimental variation with confidence levels of at least 95%. The positive/negative indicates direction of effect—not necessarily desirable or undesirable.

Increase diene content:

Positive effect on tensile at yield
Positive effect on tensile at break
Positive effect on flexural modulus

Experiment and Product Design

Two-Factor, Two-Level Design Matrix

Trial Order	Controlled Factors and Interactions		
	Ethylene A	Mooney B	AB
1	+	−	+
2	+	+	−
3	−	−	−
4	−	+	+

Two-Factor, Two Level Design

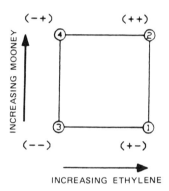

Figure 9-15. Taguchi L4 design matrix and design cube.

Increasing ethylene content:

Positive effect on tensile at yield
Positive effect on tensile at break
Positive effect on elongation
Positife effect on hardness
Positive effect on spiral flow (high shear)

Increasing Mooney viscosity:

Positive effect on Gardner impact at −30°C
Negative effect on die swell
Positive effect on shear sensitivity

Although molecular weight distribution was not determined to be significant as a single variable, it appears to play an important role in several interaction effects with Mooney viscosity. The data further indicate that these interactions significantly affect rheological properties such as flow rate, spiral flow, and shear sensitivity.

Let's examine a final example, where we wish to supply an EPDM rubber to customers in the form of friable bales. A friable bale is one that will break apart rather easily and hence helps to promote fast, easy mixing in a Banbury operation, along with better dispersion of compound ingredients in the rubber mass. To assist the plant in making such a product form for a particular grade, a two-level factorial designed experiment in a pilot plant was to run to determine (1) to what extent polymerization conditions and resulting polymer variables affect bale friability, and (2) how best to control the reactor and rubber baling process operations. In the experimental program, polymers were synthesized over a range of process conditions; in particular, $1.0 <$ base/catalyst < 2.5 moles/mole and $3 < H_2$/catalyst < 7.5 moles/mole. The base/catalyst ratio controls the level of branching in the EPDM, and the H_2/catalyst ratio helps to control the molecular weight distribution. Since we are interested in making a particular grade in friable form, the ethylene and diene contents, as well as weight average molecular weight (i.e., Mooney viscosity) are kept within specs. This means that reactor residence time and temperatures must be kept constant.

There are a number of empirical test methods that can be used to study friability. For example, the force required to break apart a friable rubber sample can readily be measured on an Instron tensile tester. From the plant's viewpoint, relaxation testing is a quick and effective quality control test to implement. Therefore, one approach is to relate the friability data to a QC stress relaxation test.

Table 9-22 summarizes the two-level factorial design used for this study. The independent variables, base/catalyst and H_2/catalyst, were varied over a broad experimental range to define clearly if these variables have a statistically significant effect on the dependent variables of interest. The latter included polymer MWD, crystallinity, branching, and bale friability. Seven polymers were included in the design, including three center-point replicates, to the target Mooney ethylene and diene values shown. The center points were included in the design so that one could estimate the magnitude of error in the measurement of the dependent variables and hence determine if the apparent effects of the independent variables were truly statistically significant at the 95% confidence level.

Experiment and Product Design

Table 9.22 Two-Level Factorial Design

Polymerization conditions (held constant during study):
35C
9.5-min residence time
8 AL/V, M/M

Product targets:
M_L (1 + 4) at 125C   60 ± 5
C_2^{2-}, wt % (IR)   70 ± 1
ENB, wt % (RI)   5 ± 0.3

	Code		Actual			Catalyst Efficiency
Trial	Base/cat.	Hz/cat.	Base/cat.	H$_2$/cat.	H$_2$ (ppm)	(W/W)
1	−	−	1.0	2.96	70	683
2	+	−	2.5	2.96	70	655
3	−	+	1.0	7/63	105	1133
4	+	+	2.5	7.27	100	1078
5	0	0	1.8	5.19	90	916
6	0	0	1.8	5.19	90	920
7	0	0	1.8	5.19	90	897
Plus "controls"						
		B	1.9	7.8	—	960
		N	2.3	12.8	179	1050
		V	2.2	4.2	63	1055
		C				

The following schematic representation of the values of the independent variables is used. A two-part symbol is used to represent the "corner" points of the experimental design. The first part represents the coded value of the first independent variable, the base/catalyst molar ratio; the second, the coded value of the second independent variable, the H$_2$/catalyst molar ratio.

As a result of conducting the two-factorial designed experiment, a number of "dependents" variables can be tested for statistical significance at the 95% confidence level, that is, to determine if a simple linear relationship in terms of the independent variables and H$_2$/catalyst and their cross-product) could predict the value of the dependent variable. This was done and the results are summarized in Table 9-23. The study

Table 9-23 Summary of Tests for Statistical Significance

Dependent variable	Is the variable a function of base/cat., H_2/cat., or product of the two at 95% confidence limit?
MWD:	
M_w/M_n (DRI)	Yes
M_z/M_w (LALLS)	Yes
Stress relaxation (at constant M_L)	No
Crystallinity:	
DSC, heat of fusion	No
DSC, peak temperature	No
Branching:	
GPC branching index	Yes
FTIR branching index	No
Reactor C_3^{2-} conversion	Yes

shows that polymer molecular weight distribution and branching are both controlled by the two independent variables. Polymer crystallinity is not. It was found that base/catalyst ratio changes had a minor impact on polymer Mooney (ML). In general, ML increased slightly with an increase of base/catalyst at a constant H_2/catalyst ratio. This slight increase (and absence of any decrease) suggests that polymer made at the lowest base/catalyst ratio was not significantly more branched than polymer made at the highest ratio.

Both independent variables were found to have a statistically significant impact on polymer MWD, that is, MW/MN from GPC (gel permeation chromatography) and stress relaxation (at constant ML). In general, it appears that the high ends of the distribution, the MZ/MW ratio, correlates better with the polymer's stress relaxation than does the MW/MN polydispersity ratio. To stabilize variations in polymer ethylene content, stress relaxation control via H_2/catalyst changes is preferred over base/catalyst changes. For illustration purposes, Table 9-24 shows the worksheets for analysis of the data.

Table 9-24 Tests for Statistical Significance: DOE Calculation Worksheet, (2 × 2) at 95% Confidence, 3 Center Points, M_w/M_n as Dependent Variable

Run	Mean	X_1 Base/cat.	X_1 M/M	X_2 H$_2$/Cat.	X_2 M/M	x_1x_2	Deep var. M_w/M_n
1	+	−		−		+	3.02
2	+	+		−		−	2.32
3	+	−		+		−	2.56
4	+	+		+		+	2.29
5	0		0		0	0	2.43
6	0		0		0	0	2.51
7	0		0		0	0	2.27
Sum(+)	10.19		4.61		4.85	5.31	
Sum(−)	0		5.58		5.34	4.88	
Sum	10.19		10.19		10.19	10.19	
Difference	10.19		−0.97		−0.49	0.43	
Effect	2.5475		−0.485		−0.245	0.215	
95% CL	+/0.262734		0.371562		0.371562	0.371562	0.144166
Significant?	Yes		Yes		No	No	0.401333
							Curv.
							No

Center Point   Average = 2.403333   Std = 0.122202

$$\frac{M_w}{M_n} = 2.55 - 0.24\left(\frac{\text{base/cat} - 1.75}{0.75}\right)$$

Table 9-24 *(continued)*

Run	Mean	X₁ Base/cat.	X₁ M/M	X₁ H₂/Cat.	X₂ M/M	x₁x₂	Deep var. M_w/M_n
1	+	−		−		+	3.57
2	+	+		−		−	1.76
3	+	−		+		−	1.96
4	+	+		+		+	1.73
5	0		0		0	0	2.14
6	0		0		0	0	1.89
7	0		0		0	0	1.86
Sum(+)	9.02		3.49		3.69	5.3	
Sum(−)	0		5.53		5.53	3.72	
Sum	9.02		9.02		9.02	9.02	
Difference	9.02		−2.04		−1.64	1.58	Curv.
Effect	2.5475		−1.02		−0.82	0.79	0.29166
95% CL	+/0.330522		0.467529		0.467429	0.467429	0.504881
Significant?	Yes		Yes		Yes	Yes	No

Center Point   Average = 1.963333   Std = 0.153731

$$\frac{M_z}{M_w} = 2.26 - 0.51\left(\frac{\text{base/cat} - 1.75}{0.75}\right) + 0.40\left(\frac{H_2/\text{cat.} - 5.25}{2.25}\right) = 2.26 - 0.51\left(\frac{\text{base/cat} - 1.75}{0.75}\right)\left(\frac{H_2/\text{cat.} - 5.25}{2.25}\right)$$

Experiment and Product Design

Run	Mean	X₁ Base/cat.	X₁ M/M	X₂ H₂/Cat.	X₂ M/M	$x_1 x_2$	Deep var. M_w/M_n
1	+	−		−		+	175
2	+	+		−		−	30.5
3	+	−		+		−	54.6
4	+	+		+		+	28
5	0				0	0	49.6
6	0				0	0	34.8
7	0				0	0	34.7
Sum(+)	288.1		58.5		82.6	203	
Sum(−)	0		229.6		205.5	85.1	
Sum	288.1		288.1		288.1	288.1	
Difference	288.1		−171.1		−122.9	117.9	Curv.
Effect	72.025		−85.5		−61.45	58.95	32.325
95% CL	±18.43366		26.06913		26.06913	26.06913	28.15788
Significant?	Yes		Yes		Yes	Yes	Yes

Center Point   Average = 39.7   Std = 8.573797

$$\text{Stress relaxation} = 72 - 43\left(\frac{\text{base/cat} - 1.75}{0.75}\right) - 31\left(\frac{H_2/\text{cat.} - 5.25}{2.25}\right) + 29\left(\frac{\text{base/cat} - 1.75}{0.75}\right)\left(\frac{H_2/\text{cat.} - 5.25}{2.25}\right)$$

Table 9-24 (continued)

Run	Mean	X_1		X_2		x_1x_2	Deep var. M_w/M_n
		Base/cat.	M/M	H$_2$/Cat.	M/M		
1	+	−		−		+	0.59
2	+	+		−		−	0.76
3	+	−		+		−	0.73
4	+	+		+		+	0.75
5	0		0		0	0	0.68
6	0		0		0	0	0.72
7	0		0		0	0	0.73
Sum(+)	2.83		1.51		1.48	1.34	
Sum(−)	0		1.32		1.35	1.49	
Sum	2.83		2.83		2.83	2.83	
Difference	2.83		0.19		0.13	−0.15	Curv.
Effect	0.7075		0.095		0.065	−0.075	−0.0025
95% CL	+/0.056883		0.080445		0.080445	0.080445	0.086891
Significant?	Yes		Yes		No	No	No
Center Point			Average = 0.71		Std = 0.026457		

$$\text{GPC Branching Index} = 0.71 + 0.05\left(\frac{\text{base/cat} - 1.75}{0.25}\right)$$

NOTATION

c	number of center points
k	number of replicates
ML	Mooney viscosity
M_{sn}	number-average molecular weight
M_w	weight-average molecular weight
M_z	z-average molecular weight
m	number of + signs in the column (= 2^{p-1} for factor effect columns; 2^p for the mean column)
n	number of observation
q	number of unassigned factor effects
r	number of checkpoints
s	pooled standard deviation of a single response observation
t	value of Student's t statistic
x, x	independent variables
y, y	dependent variables

Greek Symbols

α	parameter used to specify probability levels
Δ	size of the factor experiment
μ	population mean
σ	standard deviation

REFERENCES AND SUGGESTED READINGS

Bacon, D.W., *Ind. Eng. Chem.*, **62**, 27-34 (1970).

Benefield, R.E., and O.A. Ruiz, Statistical Design of Rubber Modified Thermoplastics, Part IV: Effects of EP(D)M Polymerization Variables on Properties of Olefinic TPEs, paper presented at the 134th meeting of Rubber Division, American Chemical Society, Cincinnati, Ohio, October 18-21, 1988.

Box, G.E.P., and D.W. Behnken, *Technometrics*, **2**, 455-475 (1960).

Box G.E.P., and J.S. Hunter, *Technometrics*, **3**, 311-351 (1961).

Box, G.E.P., and K.B. Wilson, *J. R. Stat. Soc., Ser. B*, **23**, 1-45, (1951).

Cheremisinoff, N.P., *Practical Statistics for Engineers and Scientists,* Technomic Publishing, Lancaster, Pa. 1987.

Cochran, W.G., and G.M. Cox, *Experimental Designs,* 2nd ed., Wiley, New York, 1957.

Davies, O.L., ed, *The Design and Analysis of Industrial Experiments,* 2nd ed., Hafner Press, New York, 1967.

Doehlert, D.H., *J. R. Stat. Soc., Ser C*, **19**, 231-239 (1970)

Hald, A., *Statistical Theory with Engineering Applications,* Wiley, New York, 1952.

Hicks, C.R., *Fundamental Concepts in the Design of Experiments,* Holt, Rinehart and Winston, New York, 1964.
Hoerl, A.E., and R.W. Kennard, *Technometrics, 12,* 55–67, 69–82, (1970).
Kennard, R.W., and L.A. Stone, *Telchnometrics, 11,* 137–148, (1969).
Kurotori, I.S., *Ind. Quality Control, 22,* 592–596 (1966).
Lindgren, B.W., and G.W. McElrath, *Introduction to Probability and Statistics,* Macmillan, New York, 1959.
Lucas, J.M., *Technometrics, 16,* 561–567, (1974).
Marquardt, D.W., *Chem. Eng. Prog., 55,* 65–70 (1959).
Marquardt, D.W., *J. Soc. Ind. Apl. Match., 11,* 431–441 (1963).
Marquardt, D.W., Least Squares Estimation of Nonlinear Parameters, Fortran Computer Program, *IBM Share Library 3094.01,* 1966.
Marquardt, D.W., *Technometrics, 12,* 591–612, (1970).
Marquardt, D.W., and R.D. Snee, *Technometrics, 16,* 533–537, (1974).
Marquardt, D.W., and R.D. Snee, *Am. Stat., 29,* 1–20 (Feb., 1975).
Marquardt, D.W., R.G. Bennett, and E.J. Burrell, *J. Mol. Spectross., F,* 269–279 (1961).
McLean, R.A., and V.L. Anderson, *Technometrics, 8,* 447–456 (1966).
Natrella, M.G., Experimental Statistics, *National Bureau of Standards Handbook 91,* U.S. Government Printing Office, Washington, D.C. 1963.
Plackett, R.L., and J.P. Burman, *Biometrika, 33,* 305–325 (1946).
Scheffé, H., *J. R. Stat. Soc., Ser. B, 20,* 344–360 (1958).
Scheffé, H., *J. R. Stat. Soc., Ser. B, 25,* 235–263 (1963).
Snee, R.D., *J. Qual. Technol., 5.,* 67–79 (1973a).
Snee, R.S., *Technometrics, 15,* 517–528 (1973b).
Snee, R.D., and D.W. Marquardt, *Technometrics, 16,* 399–408 (1974).
Wheeler, R.E., *Technometrics, 16,* 193–201 (1974).
Winkel, B.F., and T.A. Cooper, *Chem. Technol., 1,* 660–663 (1971).

APPENDIX A
Glossary of Plastics and Engineering Terms

A-stage: Initial or early stage in the reaction of some thermosetting resins, the material is still soluble in certain liquids and fusible; referred to as *resol*.

Acid acceptor: Chemical that acts as a stabilizer by chemically combining with an acid that may be present initially in trace quantities in a plastic; may also be formed via decomposition of the resin.

Acrylic plastics: Group of plastics based on resins generated from the polymerization of acrylic monomers (e.g., ethyl acrylate and methyl methacrylate).

Activation: Process of inducing radioactivity in a material by bombardment with other types of radiation, such as neutrons.

Adherend: A component or body held to another body by an adhesive.

Adhesion: Condition in which two surfaces are bonded together by interfacial forces caused by valence forces or interlocking forces or both. *see also* mechanical and specific Adhesion.

Adhesion, mechanical: Bonding between two surfaces caused by interlocking action of molecules.

Adhesion, specific: Adhesion between surfaces whereby valence forces predominate that are similar to those promoting cohesion.

Adhesive: Material that holds parts together by surface attachment. Examples include glue, mucilage, paste and cement. Various forms of adhesives include liquid or tape adhesives (physical type) and silicate or resin adhesives (chemical type).

Adhesive, assembly: Adhesive for bonding materials togehter, e.g., boat, airplane, furniture, etc; term commonly used in wood chemistry to distinguish between "joint glues" and veneer glues. Term applied to adhesives employed in fabricating finished goods and differs from adhesives used in fabricating sheet materials such as laminates or plywood.

Aging: The effect of exposure of plastics to the environment for a length of time. The specific effect and degree depend on the moisture in, and temperature and composition of the environment, in addition to the length of exposure.

Air vent: Small outlet for preventing gas entrapment.

Alkyd plastics: Group of plastics composed of resins based on saturated polymeric esters whereby the recurring ester groups are an integral part of the primary polymer chain and the ester groups exist in crosslinks that are present between chains.

Allyl plastics: Group ofplastics composed of resins formulated by addition polymerization of monomers containing allyl groups (e.g., diallyl phthalate).

Amino plastics: Group of plastics generated by the condensation of amines (e.g. urea and melamine with aldehydes).

Anneal: As applied to molded plastics, the process of heating material to a specified temperature and slowly cooling it to relieve stresses.

Assembly: The positioning or placing together in proper order layers of veneer or other materials, with adhesives, for purposes of pressing and bonding into a single sheet or unit.

Assembly time: Refers to the elapsed time after an adhesive is applied until applied pressure effects curing.

Autoclave: A closed vessel or reactor for chemical reaction to take place under pressure.

B-stage: Intermediate-stage reaction steps for various thermosetting resins. During this stage, the material swells when in contact with certain liquids and becomes soft when heat is applied. The material may not dissolve or fuse entirely. Resin in this stage is referred to a resitol.

Back-pressure-relief port: Opening from an extrusion die used for excess material to overflow.

Backing plate: Also called support plate, it serves to back up cavity blocks, guide pins, bushings, etc.

Binder: Part of adhesive composition responsible for adhesive forces.

Blanket: Veneers laid up on a flat table. Complete assembly is positioned in a mold at one time; used primarily on curved surfaces to be molded by the flexible bag process.

Blister: Elevation of the surface of a plastic caused by trapped air, moisture, solvent; can be caused by insufficient adhesive, inadequate curint time, or excess temperature or pressure.

Blocking: Adhesion between layers of plastic sheets in contact; condition arises during storage or use when components are under pressure.

Bloom: Visible exudation or efflorescence on the surface of a plastic; caused by plasticizer, lubricant, etc.

Bolster: Spacer or filler material in a mold.

Bond: The attachment at the interface or exposed surfaces between an adhesive and an adherend; to attain materials together with adhesives.

Bulk density: Density of a molding material in loose form, such as granular, nodular, etc., with units g/cm^3 or lb/ft^3.

Bulk factor: Ratio of the volume of loose molding compound to the volume of the same amount in molded solid form; ratio of density of solid plastic component to apparent density of loose molding compound.

C-stage: Final reaction stage of various thermosetting resins. In this stage material is insoluble and infusible. Resin in fully cured thermosetting molding is in this stage and is referred to as resite.

C-veil: Thin, nonwoven fabric composed of randomly oriented and adhered glass fibers of a chemically resistant glass mixture.

Case harden: Process of hardening the surface of a piece of steel to a relatively shallow depth.

Cast film: Film generated by depositing a layer of liquid plastic onto surface and stabilizing by evaporating the solvent, by fusing after deposition or by cooling. Cast films generated from solutions or dispersions.

Catalyst: Material used to activate resins to promote hardening. For polyesters, organic peroxides are used primarily. For epoxies, amines and anhydrides are used.

Cavity: Portion of a mold that forms the outer surface of the molded product.

Cell: Single cavity caused by gaseous displacement in a plastic.

Cellular striation: Layering of cells within a cellular plastic.

Cellulosic plastics: Group of plastics composed of cellulose compounds: for example, esters (e.g., cellulose acetate) and ethers (e.g., ethyl cellulose).

Centrifugal casting: Process in which tubular products are fabricated through the application of resin and glass strand reinforcement to the inside of a mold that is rotated and heated. The process polymerizes the resin system.

Chalking: Dry, chalklike deposit on the surface of a plastic.

Chase: Main portion of the mold, containing the molding cavity, mold pins, guide pins, etc.

Chemically foamed plastic: Cellular plastic whereby the materials' structure is formed by gases generated from the chemical reaction between its constituents.

Clamping plate: Mold plate that matches the mold and is used to fasten the mold to the machine.

Closed-cell foam: Cellular plastic that is composed predominantly of noninterconnecting cells.

Cohesion: Forces binding or holding a single material together.

Cold flow: Creep: the dimensional change of a plastic under load with time followed by the instantaneous elastic or rapid deformation at room temperature; permanent deformation caused by prolonged application of stress below the elastic limit.

Cold molding: The fashioning of an unheated mixture in a mold under pressure. The article is then heated to effect curing.

Cold pressing: Bonding process whereby an assembly is subjected to pressure without applying heat.

Cold slug: First material to enter an injection mold.

Cold-slug well: Section provides opposite sprue opening of the injection mold, used for trapping cold slug.

Condensation: Chemical reaction whereby two or more molecules combine and separate out water or other substance. When polymers are formed, it is referred to as polycondensation.

Contact molding: Process whereby layers of resin-impregnated fabrics are built up one layer at a time onto the mold surface forming the product. Little or no pressure is required for laminate curing.

Consistency: Resistance of a material to flow or undergo permanent deformation under applications of shearing stresses.

Copolymer: Formed from two or more monomers. See also Polymer.

Core: Portion of the mold that forms the inner surfaces of the molded product.

Core and separator: Center section of an extrusion die.

Core pin: Pin for molding a hole.

Core-pin plate: Plate that holds core pins.

Crazing: Tiny cracks that develop on a laminate's surface. caused by mechanical or thermal stresses.

Creep: *See* Cold flow.

Cross-linking: Generation of chemical linkages between long-chain molecules; can be compared to two straight chains joined together by links. The rigidity of the material increases with the number of links. The function of a monomer is to provide these links.

Cull: Remaining material in the transfer vessel after the mold has been filled.

Cure: Process in which the addition of heat, catalyst or both, with or without pressure, causes the physical properties of the plastic to change through a chemical reaction. Reaction may be condensation, polymerization, or addition reactions.

Degradation: Deleterious change in a plastic's chemical structure.

Delamination: Separation of a laminate's layers.

Deterioration: Permanent adverse change in the physical properties of a plastic.

Diaphragm gate: Gate employed in molding tubular or annular products.

Die-adaptor: Piece of an extrusion die that serves to hold die block.

Die body: Part of an extrusion die that holds the core and forming bushing.

Dilatant: Property of a fluid whose apparent viscosity increases with increasing shear rate.

Dished: Displays a symmetrical distortion of a flat or curved section; as viewed, it appears concave.

Dispersant: In an organosol, the liquid constituent which displays solvating or peptizing action on the resin; subsequent action aids in dispersing and suspending resin.

Dispersion: Heterogeneous mixture in which finely divided material is distributed throughout the matrix of another material. Distribution of finely divided solids in a liquid or a solid (e.g., pigments, fillers).

Doctor bar: Device for regulating the amount of material on the rollers of a spreader.

Doping: Coating a mandrel or mold with a material that prevents the finished product from sticking to it.

Dowel: Pin that maintains alignment between the various sections of a mold.

Draft: Angle of clearance between the molded article and mold, allowing removal from the mold.

Dry spot: Incompleted area on laminated plastics; the region in which the interlayer and glass are not bonded.

Durometer hardness: A material's hardness as measured by the Shore Durometer.

Ejector pin: Pin or dowel used to eject molded articles from a mold.

Ejector-pin-retainer plate: Receptacle into which ejector pins are assembled.

Elasticity: Property of materials whereby they tend to retain or recover original shape and size after undergoing deformation.

Elastomer: A material under ambient conditions that can be stretched and, upon release of the applied stress, returns with force to its approximate original size and shape.

Epoxy plastics: Group of plastics composed of resins produced by reactions of epoxides or oxiranes with compounds such as amines, phenols, alcohols, carboxylic acids, and anhydrides, and unsaturated compounds.

Ethylene plastics: Group of plastics formed by polymerization of ethylene or by the copolymerization of ethylene with various unsaturate compounds.

Evenomation: Softening, discoloration, mottling, crazing, etc. Process of deterioration of a plastic's surface.

Exotherm: Indicates that heat is a given from a reaction between a catalyst and a resin.

Expandable plstics: Plastics that can be transformed to cellular structures by chemical, thermal, or mechanical means.

Extender: A material which, when added to an adhesive, reduces the amount of primary binder necessary.

Extraction: Transfer of materials from plastics to liquids with which they are in contact.

Extrusion: Process in which heated or unheated plastic compound is forced through an orifice, forming a continuous article.

Filament winding: Process in which continuous strands of roving or roving tape are wound, at a specified pitch and tension, onto the outside surface of a mandrel. Roving is saturated with liquid resin or is impregnated with partially cured resin. Application of heat may be required to promote polymerization.

Filler: Inert material that is added to a plastic to modify the finished product's strength, permanence and various other properties; an extender.

Fin: Portion of the "flash" that adheres to the molded article.

Finishing: Removal of any defects from the surfaces of plastic products.

Fisheye: A clump or globular mass that does not blend completely into the surrounding plastic.

Flash: Excess material that builds up around the edges of a plastic article; usually trimmed off.

Foamed plastic: Cellular structured plastic.

Force plate: A plate used for holding plugs in place in compression molding.

Furane plastics: Group of plastics composed of resins in which the furane ring is an integral portion of the polymer chain; made from polymerization or polycondensation of furfural, furfural alcohol, and other compounds containing furane rings; also formed by reation of furane compounds with an equal weight or less of other compounds.

Fusion: As applied to vinyl dispersions, the heating of a dispersion, forming a homogenous mixtur.

Fusion temperature: Fluxing temperature; temperature at which fusion occurs in vinyl dispersions.

Gel: State at which resin exists before becoming a hard solid. Resin material has the consistency of a gelatin in this state; initial jellylike solid phase that develops during the formation of a resin from a liquid.

Gel coat: Specially formulated polyester resin that is pigmented and contains fillers. Provides a smooth, pore-free surface for the plastic article.

Gel point: Stage at which liquid begins to show psuedoelastic properties.

Gelation: Formation of a gel.

Glass: Inorganic product of a fusion reaction. Material forms upon cooling to a rigid state without undergoing crystallization. Glass is typically hard and brittle and will fracture concoidally.

Glass transition: Transition region or state in which an amorphous polymer changed from (or to) a viscous or rubbery condition to (or from) a hard and relatively brittle one. Transition occurs over a narrow temperature region; similar to solidification of a liquid to a glassy state. This transformation causes hardness, brittleness, thermal expansibility, specific heat, and other properties to change dramatically.

Gum: Class of colloidal substances prepared from plants. Composed of complex carbohydrates and organic acids that swell in water. A number of natural resins are gums.

Halocarbon plastics: Group of plastics composed of resins generated from the polymerization of monomers consisting of a carbon and a halogen or halogens.

Hardener: Compound or mixture that, when added to an adhesive, promotes curing.

Heat treat: Refers to annealing, hardening, tempering of metals.

Hot soils: Soils having a resistivity of less than 1000 $\Omega \cdot$ cm; generally very corrosive to base steel.

Hydrocarbon plastics: Plastics composed of resins consisting of carbon and hydrogen only.

Inhibitor: Material that retards chemical reaction or curing.

Isocyanate plastics: Group of plastics produced by the condensation of organic isocyanates with other plastics. Examples are the urethane plastics.

Isotactic: Type of polymeric molecular structure that contains sequences of regularly spaced asymmetric atoms that are arranged in similar configuration in the primary polymer chain. Materials having isotactic molecules are generally in a highly crystalline form.

Isotropic: Refers to materials whose properties are the same in all directions. Examples are metals and glass mats.

Laminate: Article fabricated by bonding together several layers of material or materials.

Laminated, cross: Laminate in which some of the layers of materials are oriented at right angles to the remaining layers. Orientation may be based on grain or strength direction considerations.

Laminated, parallel: Laminate in which all layers of materials are oriented parallel with respect to grain or strongest direction in tension.

Lignin plastics: Group of plastics composed of resins formulated from the treatment of lignin with heat or by reaction with chemicals.

Line pipe: Pipeline used for transportation of gas, oil, or water; utility distribution pipeline system ranging in sizes from ⅛ to 42 in. OD inclusive. Fabricated to American Petroleum Institute (API) and American Water Works Association (AWWA) specifications.

Lyophilic: Referring to vinyl dispersions, having affinity for the dispersing medium.

Lyophobic: Referring to vinyl dispersions, no affinity or attraction for dispersing medium.

Mechanical tubing: Welded or seamless tubing manufactured in large range of sizes of varied chemical composition (sizes range from 3/16 to 10¾ in OD inclusive for carbon and alloy material); usually not fabricated to meet any specification other than application requirements; fabricated to meet exact outside diameter and decimal wall thickness.

Mechanically foamed plastic: Cellular plastic whose structure is fabricated by physically incorporated gases.

Melamine plastics: Group of plastics whose resins are formed by the condensation of melamine and aldehydes.

Metastable: Unstable state of plastic as evidenced by changes in physical properties not caused by the surroundings. Example is the temporary flexible condition some plastics display after molding.

Mold base: Assembly of all parts making up an injection mold, excluding cavity, cores, and pins.

Molding, bag: Process of molding or laminating in which fluid pressure is applied, usually by means of water, steam, air, or vacuum, to a flexible film or bag that transmits the pressure to the material being molded.

Molding, blow: Method of forming plastic articles by inflating masses of plastic material with compressed gas.

Molding, compression: Process of shaping plastic articles by placing material in a confining mold cavity and applying pressure and usually heat.

Molding, contact pressure: Method of molding or laminating whereby pressure used is slightly greater than is necessary to bind materials together during molding stage (pressures generally less than 10 psi).

Molding, high pressure: Molding or laminating with pressures in excess of 200 psi.

Molding, injection: Process of making plastic articles from powdered or granular plastics by fusing the material in a chamber under pressure with heat and forcing part of the mass into a cooler cavity where it solidifies; used primarily on thermoplastics.

Molding, low pressure: Molding or laminating with pressures below 200 psi.

Molding, transfer: Process of molding plastic articles from powdered, granular, or preformed plastics by fusing the material in a chamber with heat and forcing the mass into a hot chamber for solidification. Used primarily on thermosetting plastics.

Monomer: Reactive material that is compatible with the basic resin. Tends to lower the viscosity of the resin.

Nonrigid plastic: Plastic whose apparent modulus of elasticity is not greater than 10,000 psi at room temperature in accordance with the Standard Method of Test for Stiffness in Flexure of Plastics (ASTM Designation: D747).

Novolak: Phenolic-aldehydic resin that remains permanently thermoplastic unless methylene groups are added.

Nylon plastics: Group of plastics comprised of resins that are primarily long-chain synthetic polymeric amides. These have recurring amide groups as an integral part of principal polymer chain.

Organosol: Suspension of finely divided resin in a volatile organic slurry.

Phenolic plastics: Group of plastics whose resins are derived from the condensation of phenols (e.g., phenol and cresol, with aldehydes).

Piling pipe: Round-welded or seamless pipe for use as foundation piles where pipe cylinder acts as a permanent load-carrying member; usually filled with concrete. Used below the ground in foundation work in the construction industry for piers, docks, highways, bridges, and all types of buildings. Fabricated to ASTME piling specifications (ASTM A252).

Plastic: According to ASTM, a material containing an organic substance of large molecular weight is sold in its finished state, and at some stage in its manufacture into finished goods, it can be shaped to flow.

Plastic, semirigid: Plastic having apparent modulus of elasticity in the range 10,000 to 100,000 psi at 23°C, as determined by the Standard Method of Test for Stiffness in Flexure Plastics (ASTM Designation: D747).

Plastic welding: Joining of finished plastic components by fusing materials either with or without the addition of plastic from another source.

Plasticate: Softening by heating or kneading.

Plasticity: Property of plastics that permits the material to undergo deformation permanently and continuously without rupture from a force that exceeds the yield value of the material.

Plasticize: Softening by adding a plasticizer.

Plasticizer: Material added to a plastic to increase its workability and flexibility. Plasticizers tend to lower the melt viscosity, the glass transition temperature, and/or the elastic modulus.

Plastisol: Suspension of finely divided resin in a plasticizer.

Polyamide plastics: *See* Nylon plastics.

Polyester plastics: Group of plastics composed of resins derived principally from polymeric esters that have recurring polyester groups in the main polymer chain. These polyester groups are cross-linked by carbon-carbon bonds.

Polyethylene: Plastic or resin made by the polymerization of ethylene as the sole monomer.

Polymer: Material produced by the reaction of relatively simple molecules with functional groups that allow their combination to proceed to high molecular weights under suitable conditions; formed by polymerization or polycondensation.

Polymerization: Chemical reaction that takes place when a resin is activated.

Polypropylene: Plastic or resin derived from the polymerization of propylene as the principal monomer.

Polystyrene: Plastic derived from a resin produced by the polymerization of styrene.

Poly(vinyl acetate): Resin derived from the polymerization of vinyl acetate.

Poly(vinyl alcohol): Polymer derived from the hydrolysis of polyvinyl esters.

Poly(vinyl chloride): Resin derived from the polymerization of vinyl chloride.

Poly(vinyl chloride-acetate): Copolymer of vinyl chloride and vinyl acetate.

Pot life: Time period beginning once the resin is catalyzed and terminating when material is no longer workable; working life.

Preform: Coherent block of granular plastic molding compound or of fibrous mixture with or without resin. Prepared by sufficiently compressing material, forming a block that can be handled readily.

Prepolymer: An intermediate chemical structure between that of a monomer and the final resin.

Pressure tubing: Tubing used to convey fluids at elevated temperatures and/or pressures. Suitable for head applications, it is fabricated to exact OD and decimal wall thickness in sizes ranging from ½ to 6 in. OD inclusive and to ASTM specifications.

Primer: Coating that is applied to a surface before application of an adhesive, enamel, etc. The purpose is to improve bonding.

Promoted resin: Resin with an accelerator added but not catalyst.

Reinforced plastic: According to ASTM, those plastics having superior properties over those consisting of the base resin, due to the presence of high-strength fillers embedding in the composition. Reinforcing fillers are fibers, fabrics, or mats made of fibers.

Resin: Highly reactive material which, in its initial stages, has fluidlike flow properties. When activation is initiated, material transforms into a solid state.

Roller: A serrated piece of aluminum used to work a plastic laminate. Purpose of device is to compact a laminate and to break up large air pockets to permit release of entrapped air.

Roving: Bundle of continuous, untwisted glass fibers. Glass fibers are wound onto a roll called a "roving package."

Saran plastics: Group of plastics whose resins are derived from the polymerization of vinylidene chloride or the copolymerization of vinylidene chloride and other unsaturated compounds.

Shelf life: Period of time over which a material will remain usable during storage under specified conditions such as temperature and humidity.

Silicone plastics: Group of plastics whose resins consist of a main polymer chain with alternating silicone and oxygen atoms and with carbon-containing side groups.

Softening range: Temperature range in which a plastic transforms from a rigid solid to a soft state.

Solvation: Process of swelling of a resin or plastic. Can be caused by interaction between a resin and a solvent or plasticizer.

Standard pipe: Pipe used for low-pressure applications such as transporting air, steam, gas, water, oil, etc. Employed in machinery, buildings, sprinkler and irrigation systems, and water wells but not in utility distribution systems; can transport fluids at elevated temperatures and pressures not subjected to external heat applications. Fabricated in standard diameters and wall thicknesses to ASTM specifications, its diameters range from ⅛ to 42 in. OD.

Stress crack: Internal or external defect in a plastic caused by tensile stresses below its short-time mechanical strength.

Structural pipe: Welded or seamless pipe used for structural or load-bearing applications in aboveground installations. Fabricated in nominal wall thicknesses and sizes to ASTM specifications in round, square, rectangular, and other cross-sectional shapes.

Structural shapes: Rolled flanged sections, sections welded from plates and specialty sections with one or more dimensions of their cross section greater than 3 in. They include beams, channels, and tees, if depth dimensions exceed 3 in.

Styrene plastics: Group of plastics whose resins are derived from the polymerization of styrene or the copolymerization of styrene with various unsaturated compounds.

Styrene-rubber plastics: Plastics that are composed of a minimum of 50% styrene plastic and the remainder rubber compounds.

Syneresis: Contraction of a gel, observed by the separation of a liquid from the gel.

Thermoelasticity: Rubberlike elasticity that a rigid plastic displays; caused by elevated temperatures.

Thermoforming: Forming or molding with heat.

Thermoplastic: Reverse of thermoset. Materials that can be reprocessed by applying heat.

Thermoset: Those plastics that harden upon application of heat and cannot be reliquefied, the resin state being infusible.

Thixotropy: Describes those fluids whose apparent viscosity decreases with time to an assymptotic value under conditions of constant shear rate. Thixotropic fluids undergo a decrease in apparent viscosity by applying a shearing force such as stirring. If shear is removed, the material's apparent viscosity will increase back to or near its initial value at the onset of applying shear.

Tracer yarn: Strand of glass fiber colored differently from the remainder of the roving package. It allows a means of determining whether equipment used to chop and spray glass fibers are functioning properly and provides a check on quality and thickness control.

Urea plastics: Group of plastics whose resins are derived from the condensation of urea and aldehydes.

Urethane plastics: Group of plastics composed of resins derived from the condensation of organic isocyanates with compounds containing hydroxyl groups.

Vacuum forming: Fabrication process in which plastic sheets are transformed to desired shapes by inducing flow; accomplished by reducing the air pressure on one side of the sheet.

Vinyl acetate plastics: Group of plastics composed of resins derived from the polymerization of vinyl acetate with other saturated compounds.

Vinyl alcohol plastics: Group of plastics composed of resins derived from the hydrolysis of polyvinyl esters or copolymers of vinyl esters.

Vinyl chloride plastics: Group of plastics whose resins are derived from the polymerization of vinyl chloride and other unsaturated compounds.

Vinyl plastics: Group of plastics composed of resins derived from vinyl monomers, excluding those that are covered by other classifications (i.e., acrylics and styrene plastics). Examples include PVC, poly(vinyl acetate), poly(vinyl butyral), and various copolymers of vinyl monomers with unsaturated compounds.

Vinylidene plastics: Group known as saran plastics.

Weathering: Exposure of a plastic to outdoor conditions.

Yield value: Also called yield stress; force necessary to initiate flow in a plastic.

Appendix B
General Properties and Data on Elastomers and Plastics

Table B.1 Properties of Important Plastics and Elastomers

	Molecular mass (g/mol)	Enthalpy of melting (kJ/mol)	Glass transition Temperature		Melt temperature		Density (g/cm³)	
			°C	°F	°C	°F	Amorphous	Crystalline
Polyolefins								
Polyethylene	28.1	8.0	−78	−108.4	141	285.8	0.85	1.00
Polypropylene	42.1	1.9/10.9	−35, 26	−31, 78.8	112, 208	233.6, 406.4	0.85	0.95
1,2-Poly(1,3-butadiene) (iso)	54.1	—	−75, −30	−103, −22	2, 44	35.6, 111.2	—	0.96
1,2-Poly(1,3-butadiene) (syndio)	54.1	—	—	—	155	311	0.92	0.96
Poly(1-butene)	56.1	4.1/13.9	−45, −24	−49, −11.2	106, 142	222.8, 287.6	0.85	0.95
Polyisobutylene	56.1	12.0	−75, −30	−103, −22	2, 44	35.6, 111.2	0.84	0.94
Poly(3-methyl-1-butane)	70.1	17.3	50	122	300, 310	572, 590	0.90	0.93
Poly(1-pentane)	70.1	4/6.3	−50, 14	−58, 57.2	111, 130	231.8, 266	0.85	0.92
Poly(4-methyl-1-pentene)	84.2	11.9/19.7	22, 42	71.6, 107.6	228, 250	442.4, 482	0.84	—
Poly(1-hexene)	84.2	—	−50	−58	48	118.4	0.86	0.91
Poly(5-methyl-1-hexene)	98.2	—	−14	6.8	110, 120	230, 248	—	0.84
Poly(1-octadecane)	252.5	—	55	131	68, 80	154.4, 176	0.86	0.95
Polyoxides								
Poly(methylene oxide)	30	3.7/10	−88, −79	−126.4, −110	60, 198	140, 388.4	1.25	1.54
Poly(ethylene oxide)	44.1	7.3/12	−67, −27	−88.6, −16.6	62, 72	143.6, 161.6	1.13	1.33
Polyacetylaldehyde	44.1	—	−30	−22	165	329	1.07	1.14
Poly(tetramethylene oxide)	72.1	12.4/14.4	−88, −79	−126.4, −110	60, 198	140, 388.4	0.98	1.18
Poly(*trans*-2-butene oxide)	72.1	—	4	39.2	114	237.2	1.01	1.10
Poly(ethylene formal)	74.1	16.7	−65	−83.2	55, 74	131, 165.2	—	1.32/1.414
Poly(3-methoxypropylene oxide)	88.1	—	−62	−79.6	57	134.6	1.10	—
Poly(3-chloropropylene oxide)	92.5	—	—	—	117, 135	242.6, 275	1.37	1.46
Poly(hexene oxide)	100.2	—	−69	−92.2	72	161.6	0.92	0.97
Poly(tetramethylene formal)	102.1	14/14.7	−84	−119.2	23	73.4	—	1.23
Poly(octane oxide)	128.1	—	−70	−94	87	188.6	0.94	0.97

Data on Elastomers and Plastics

Polyhalo-olefins

Polymer							
Poly(vinyl fluoride)	46	7.5	−20, 40	200	392	1.37	1.44
Poly(vinyl chloride)	62.5	2.8/11.3	−26, 83	212, 83	413.6, 181.4	1.385	1.52
Poly(vinylidene fluoride)	64	8.9	−40, 13	137, 238	278.6, 460.4	1.74	2
Poly(vinylidene chloride)	97	1.4	−18, 15	190, 210	374, 410	1.66	1.95
Poly(tetrafluoroethylene)	100	5.7	−0.4, 59	19, 299	66.2, 750.2	2	2.35
Poly(vinyl bromide)	107	—	−113, 127	100	212	—	—

Polyvinyls

Polymer							
Poly(vinyl alcohol)	44.1	6.97/7	−171.4, 260.6	—	—	—	—
Poly(vinyl methyl ether)	58.1	—	70, 99	232, 265	449.6, 509	1.26	1.35
Poly(vinyl methyl ketone)	70.1	—	−31, −13	144, 150	291.2, 302	1.03	1.175
Poly(vinyl ether ether)	72.1	—	158, 210.2	170	338	1.12	1.216
Poly(vinyl formate)	72.1	—	−23.8, 8.6	86	186.8	0.94	0.97
Poly(methyl isopropenyl ketone)	84.1	—	—	—	—	1.35	1.49
Poly(vinyl propyl ether)	86.1	—	−43.6, −2.2	200, 240	392, 464	1.12/1.15	1.15/1.17
Poly(vinyl isopropyl ether)	86.1	—	87.8, 98.6	76	168.8	0.94	—
Poly(vinyl acetate)	86.1	—	176, 237.2	191	375.8	0.924	0.93
Poly(vinyl cyclopentane)	96.2	—	−3	—	—	1.19	1.2
Poly(vinyl butyl ether)	100.2	—	26.6	292	557/6	0.965	0.986
Poly(vinyl isobutyl ether)	100.2	—	82.4	64	147.2	0.927	0.944
Poly(vinyl sec-butyl ether)	100.2	—	167	170	338	0.93	0.94
Poly(vinyl propionate)	100.2	—	−53	170	338	0.92	0.956
Poly(vinyl hexyl ether)	128.2	—	−27, −18	—	—	1.02	—
Poly(vinyl propionate)	100.2	—	−16.6, −0.4	—	—	0.925	—
			−20			1.02	—
			−4				
			10				
			50				
			−77, −50				
			−106.6, −58				
			10				
			50				

Table B.1 (continued)

	Molecular mass (g/mol)	Enthalpy of melting (kJ/mol)	Glass transition Temperature °C	Glass transition Temperature °F	Melt temperature °C	Melt temperature °F	Density (g/cm³) Amorphous	Density (g/cm³) Crystalline
Polystyrene								
Polystyrene	104.1	8.4/10.1	90, 100	176, 212	225, 250	437, 482	1.05	1.13
Poly(2-methylstyrene)	118.2	—	136	276.8	360	680	1.027	1.07
Poly(4-methylstyrene)	118.2	—	93, 106	199.4, 222.8	—	—	1.04	—
Poly(3-phenyl-1-propene)	118.2	—	60	140	230, 240	446, 464	1.046	1.05
Poly(4-methoxystyrene)	134.2	—	89	192.2	238	460.4	—	1.12
Poly(2-chlorostyrene)	138.6	—	119	246/2	—	—	1.25	—
Poly(4-chlorostyrene)	138.6	—	110, 126	230, 258.8	—	—	—	—
Polyacrylates								
Poly(acrylic acid)	72.1	—	106	222.8	—	—	—	—
Poly(methyl acrylate)	86.1	—	8	46.4	—	—	1.22	—
Poly(ethyl acrylate)	100.1	—	−22	−7.6	—	—	1.22	—
Poly(propyl acrylate)	114.1	—	−44	−47.2	115, 162	239, 323.6	1.08	1.18
Poly(isopropyl acrylate)	114.1	5.9	−11, 11	12.2, 51.8	116, 180	240.8, 356	—	1.08/1.18
Poly(butyl acrylate)	128.2	—	−52	−61.6	47	116.6	1.00/1.09	—
Polyacrylics								
Polyacrylonitrile	53.1	4.9/.2	80, 105	176, 221	318	604.4	1.184	1.27/1.54
Polymethacrylonitrile	67.1	—	121	249.8	250	482	1.1	1.134
Polyacrylamide	71.1	—	165	329	—	—	1.302	—
Poly(methyl chloroacrylate)	120.5	—	143	289.4	—	—	1.45/1.49	—

Data on Elastomers and Plastics 533

Polyesters

Poly(glycolic acid)	58	12	38, 95	100.4, 203	223, 260	433.4, 500	1.6	1.7
Poly(*para*-hydrobenzoate)	120.1	—	147	296.6	317, 497	602.6, 926.6	1.44	1.48
Poly(ethylene succinate)	144.1	320/338	—	—	−70, −40	−94, −40	1.175	1.36
Poly(ethylene oxybenzoate)	164.2	10.5	82	179.6	202, 227	395.6, 440.6	1.34	—
Poly(ethylene adipate)	172.2	—	−70, −40	−94, −40	47, 65	116.6, 149	1.18/1.22	1.25/1.45
Poly(tetramethylene isophthalate)	220.2	—	—	—	153	307.4	1.27	1.31
Poly(ethylene azelate)	214.3	—	−45	−49	46	114.8	—	1.17/1.22
Poly(ethylene-1,5-naphthalene)	242.2	—	71	159.8	230	446	1.33	1.35
Poly(ethylene-2,6-naphthalene)	242.2	—	113, 180	235.4, 356	260, 268	500, 514.4	1.3	1.37
Poly(decamethylene terephthalate)	303.3	43.5/48.6	−5, 25	23, 77	123, 138	253.4, 280.4	—	1.022
Poly(decamethalene sebacate) 1.13	340.5	30.2/56.5	—	—	—	71, 85	159.8, 185	—

Polyamides

Poly(6-aminohexanoic acid)	113.1	17.6/23	50, 75	122, 167	214, 233	417.2, 451.4	1.08	1.23
Poly(7-aminoheptanoic acid)	127.2	—	52, −38	125.6, −36.4	217, 233	422.6, 451.4	1.095	1.21
Poly(9-aminononanoic acid)	155.2	—	51	123.8	194, 209	381.2, 408.2	1.052	1.066
Poly(11-aminoundecanoic acid)	183.3	41.4	46	114.8	182, 220	359.6, 428	1.01	1.12/1.23
Poly(hexamethylene adipamide)	226.2	36.8/46.9	45, 57	113, 134.6	250, 272	482, 521.6	1.07	1.24
Poly(hexamethylene sebacamide)	282.4	30.6/58.6	30, 50	86, 122	215, 233	419, 451.4	1.04	1.19
Poly(decamethylene azelamide)	324.5	36.3/68.2	—	—	214	417.2	1.04	—
Poly(decamethylene sebacamide)	338.5	32.7/51.1	46, 60	114.8, 140	196, 216	384.8, 420.8	1.03	1.06

Table B.2 Terminology and Properties of Important Elastomers

Type elastomer	Common name	ASTM name	ASTM D1418 designation	IUPAC trivial name	IUPAC structure-base names	SAE J20/ ASTM D200	Specific gravity D200	Durometer Range	Tensile strength (max., MPa)	Elongation (max., %)	Glass-transition temperature (K)
Fluoro (copolymer)	Fluoro-elastomer	Fluoro rubber	FKM	Poly(vinylidene fluoride-co-hexafluoropropylene)	Poly[(1,1-difluoroethylene)-co-perfluoropropylene]	FKM	1.86	60–95	20	250	255
Fluoro (terpolymer)	Fluoro-elastomer	Fluoro rubber	FKM	Poly(vinylidene fluoride-tetrafluoroethylene-co-hexafluoropropylene)	Poly[(1,1-difluoroethylene)-co-difluoromethylene-co-perfluoropropylene]	FKM	1.88–1.90	65–95	20	250	255, 270
Fluorosilicone	Fluorosilicone	Fluorosilicone	FVMQ	Poly(methyltrifluoropylsiloxane)	Poly(oxymethyl-3,3,3,-trifluoropropylsilylene)	FK	1.4	40–80	10	400	193

Data on Elastomers and Plastics

Epichloro-hydrin	Epichloro-hydrin	Polychloro-methyl oxirane	CO	Poly(epi-chloro hydrin)	Poly[oxy-(chloro-methyl) ethylene]	CH	1.36	40-90	18	350	251
Polychloro-prene	Neoprene	Chloroprene	CR	Poly(chloro-prene)	Poly(1-chloro-1-butenylene)	BC, BE	1.25	30-95	22	600	233
Chlorosul-fonated polyethylene	Chlorosul-fonated polyethylene	Chlorosul-fonylpoly-ethylene	CSM	Not applicable	Not applicable	CE	1.08–1.27	45-95	28	500	274
Chlorinated polyethylene	Chlorinated polyethylene	Chloropoly-ethylene	CM	Not applicable	Not applicable	CE	1.16–1.25	50-90	20	350	261
Halobutyl	Bromobutyl	Bromoiso-butene-isoprene	BIIR	Polyhalo-genated(iso-butylene-co-isoprene)		CA	0.93	40-90	20	850	200
Halobutyl	Chlorobutyl	Chloroiso-butene-isoprene	CIIR	Polyhalo-genated(iso-butylene-co-isoprene)		CA	0.92	40-90	20	850	200

Table B.3 Synthesis and Features of Hydrogenated Diene-Diene Copolymers

Types of starting polymers [a]	Unhydrogenated precursor		Hydrogenated polymer		Features
	Block 1	Block 2	Block 1	Block 2	
$B_{1,4}B_{1,2}$	1,4-Polybutadiene (low vinyl)	1,2-Polybutadiene (high vinyl)	Polyethylene	Polybutylene	Improved material stress-strain properties
$B_{1,4}B_{mv}$	1,4-Polybutadiene (low vinyl)	1,2-Polybutadiene (medium vinyl) (30–60%)	Polyethylene	Poly(ethylene-co-butylene)	Improved stress-strain properties
$B_{1,4}B_{mv}B_{1,4}$	1,4-Polybutadiene (low vinyl)	1,2-Polybutadiene [low, 34 (or low isopropenyl)] (see comments above)	Polyethylene	Poly(ethylene-co-butylene)	Properties dependent on composition and architecture
IBI	1,4-Polyisoprene	1,4-Polybutadiene	Poly(ethylene-co-propylene)	Polyethylene	Inverse block polymer—properties dependent on composition

[a] B = $B_{1,4}$, 1,4-polybutadiene block; $B_{1,2}$, 1,2-polybutadiene block; B_{mv}, medium vinyl (35–60%) polybutadiene block); I, 1,4-polyisoprene block.
[b] Selective hydrogenation; this block not hydrogenated.

Table B.4 Synthesis and Features of Hydrogenated Aromatic-Diene Copolymers

Types of starting polymers[a]	Unhydrogenated precursor		Hydrogenated polymer		Features
	Block 1	Block 2	Block 1	Block 2	
S-B$_{mv}$S	Polystyrene	1,2-Polybutadiene (medium vinyl)	Polystyrene[b]	Poly(ethylene-co-butylene)	Thermally and oxidatively stable
SBS (linear or star)	Polystyrene	Polybutadiene	Polystyrene[b]	Polyethylene	Thermoplastic elastomer
S-B$_{1,2}$S	Polystyrene	1,2-polybutadiene	Polystyrene[b]	Polybutylene	Enhanced compression set
S-(α-MeS)B$_{1,4}$	Polystyrene poly(α-methylstyrene)	High 1,4-polybutadiene 1,2-polybutadiene	Polystyrene (poly(α-methylpoly-styrene)[b]	Polyethylene	
S-B$_{1,2}$	Polystyrene	1,2-Polybutadiene	Polystyrene[b]	Polybutylene	Improved tensile strength and elongation
S-B$_{1,4}$	Polystyrene	1,4-Polybutadiene	Poly(vinyl cyclohexane)	Polyethylene	Hydrogenation of both blocks
S-I-S	Polystyrene	Polyisoprene	Polystyrene[b]	Poly(ethylene-co-propylene)	
S-(α-MeS)-I	Polystyrene poly(α-methylstyrene)	Polyisoprene	Polystyrene poly(α-methylstyrene)[b]	Poly(ethylene-co-propylene)	
Random SBR	Polystyrene	1,4-Polybutadiene (~20% 1,2)	Polystyrene[b]	Poly(ethylene-co-butylene)	

[a]B$_{mv}$, medium vinyl (35–60% 1,2-polybutadiene block); B = B$_{1,4}$, high 1,4-polybutadiene block; B$_{1,2}$, high 1,2-polybutadiene block; αMeS, α-methylstyrene block; SBS, poly(styrene-b-butadiene-b-styrene); SBR, poly(styrene-co-1,4-butadiene-b-styrene).
[b]Selective hydrogenation; this block not hydrogenated.

Table B.5 Hydrogenation of Functional Diene Polymers

Functionality	Backbone[a]	Method	Features
–OH	Polybutadiene	Homogeneous hydrogenation [e.g., Ni or Co/triethyl aluminum (or Ru, Rh catalysis)]; heterogeneous catalysts (Raney Ni)	Hydrogenated mixed 1,4- and 1,2-polybutadienes as heat-resistant polyurethane prepolymers
–COOH	SBR, NBR	Rh catalysis	
–COOCH$_3$	Poly(butadiene-alt-methyl methacrylate	Pt black	
–CN	SBR, NBR, IR	Rh catalysis	
	PP	[NH=NH]	
–SO$_3$Na	PP	[NH=NH]	
–PO(OCH$_3$)$_2$	PP	[NH=NH]	
–NO$_2$	NR, PIP, 1,4-BR	Raney Ni, Zn–AcOH	Partial reduction of NO$_2$ groups; complete hydrogenation of double bonds
	Poly(butadiene-b-vinyl pyridine) or poly(isoprene-b-vinyl pyridine)	Ni/Et$_3$Al catalysis	BF$_3$ and Cl$_2$AlEt used to complex vinyl pyridine to increase rate of hydrogenation; BF$_3$ released by NH$_4$OH

[a]BR, polybutadiene; SBR, poly(styrene-co-1,4-butadiene); NBR, poly(acrylonitrile-co-1,4-butadiene); PP, polypentenamer; IR, polyisoprene.

Data on Elastomers and Plastics

Table B.6 Properties of Liquid Polysulfide Polymers

	LP-31	LP-2	LP-32	LP-12	LP-3	LP-33
Viscosity (poise) (25°C)	950–1550	410–525	410–525	410–525	9.4–14.4	15–20
Mercaptan content (%)	1.0–1.5	1.5–2.0	1.5–2.0	1.5–2.0	5.9–7.7	5.0–6.5
Average molecular weight	8000	4000	4000	4000	1000	1000
Pour point (°C)	10	7	7	7	−26	−23
Cross-linking agent (%)	0.5	2.0	0.5	0.2	2.0	0.5
Specific gravity (25°C)	1.31	1.29	1.29	1.29	1.27	1.27
Average viscosity (poise) (4°C)	7400	3800	3800	3800	90	165
Stress–strain properties						
Tensile strength (MPa)	2.59	2.82	2.07	2.07	2.07	2.59
300% Modulus (MPa)	2.07	2.41	1.45	1.38	1.03	1.38
Elongation (%)	600	510	930	900	275	700
Hardness, Shore A	48	50	50	48	45	34

Recipe: LP-31, LP-32, LP-2, LP-12

Base compound:
Liquid polymer 100
M774 black 30

Curing paste:
PbO_2[a] 7.8
HP-40 (plasticizer) 0.2
Alumina (Al_2O_3) 0.2

Cure: 2 hr at 70°C in closed mold, postcure 20 hr at 23°C, 50% RH after demolding

Recipe: LP-33 and LP-3

Base compound:
Liquid polymer 100
N990 black 20

p-Quinonedioxime (GMF) 6.67
Diphenylguanidine (DPG) 0.67
Magnesium oxide 0.50

Cure: 20 hr at 77°C in mold, postcure minimum 2 hr at 23°C and 50% RH after demolding

Source: Courtesy of Morton Thiokol Inc.
[a]With LP-31, 5.0 parts of PbO_2 was used.

Table B.7 Properties of Arco Poly bd R-45 HT Urethane Composition

	Formulations						
Poly bd R-45 HT (g)	100	100	100	100	100	100	100
Isonol 100 (g)[a]	—	2.22	4.45	8.89	11.85	17.78	26.67	35.56
Catalyst T-12 (drops)[b]	4	4	4	4	4	4	4	4
CAO-14 (g)[c]	12.76	15.45	19.14	25.53	29.78	38.29	51.05	63.81
Equivalent ratio: Poly bd/Isonol 100	—	4/1	2/1	1/1	3/4	1/2	1/3	1/4
Vulcanizate properties (press cure 30 min at 80°C; postcure 64 hr at 49°C.)								
Tensile strength (MPa)	1.2	1.7	2.6	6.2	8.2	13.9	18.6	24.0
200% modulus (MPa)	—	—	1.3	2.6	5.3	10.3	14.7	19.4
Hardness								
Shore A	53	56	62	75	82	—	—	—
Shore D	—	—	—	—	—	43	51	53

Source: Courtesy of ARCO Chemical Co.
[a] N,N-bis(2-Hydroxypropyl)aniline (Upjohn Co.).
[b] Dibutyltin dilaurate.
[c] Antioxidant (Sherex Chemical Co.).
[d] Polyfunctional liquid isocyanate (Upjohn Co.)

Table B.8 Properties of CTBN-Epoxy Resin Compositions

	Formulations				
DGEBA[a]	100	100	100	100	100
Hycar CTBN 1300X8[b]	—	5	10	15	20
Piperidine	5	5	5	5	5
Physical properties (cure 16 hr at 120°C)					
Tensile strength (MPa)	65.8	62.8	58.4	51.4	47.2
Tensile modulus (GPa)	2.8	2.5	2.3	2.1	2.2
Elongation (%)	4.8	4.6	6.2	8.9	12.0
Fracture surface energy (kJ/m^2)	0.18	2.63	3.33	4.73	3.33
Gardner impact (J)	6	8	8	8	25
Heat-distortion temperature (°C)	80	76	74	71	69

Source: Courtesy of the B.F. Goodrich Co.
[a]Diglycidyl ether of bisphenol A.
[b]ACN content 18 wt%, functionality 1.8, $\bar{M}_n$ 3600.

Table B.9 Properties of Unfilled Thermoplastic Compositions

Composition number	Resin type/parts per 100 parts of rubber (phr)[a]	Sulfur (phr) Y	Method of preparation[b]	Cross-link density, $r/2$ (moles × 10^5 per ml of rubber)	Rubber particle size (μm) d_n	d_w	Shore D hardness	Young's modulus (MPa)	Stress at 100% strain (MPa)	Tens. str. (MPa)	Ult. elong. (%)	Tens. set (%)
1	Polypropylene/66.7	2.0	S	16.4	72	750	43	97	8.2	8.6	165	—
2	Polypropylene/66.7	2.0	S	16.4	39	200	41	102	8.4	9.8	215	22
3	Polypropylene/66.7	2.0	S	16.4	17	96	41	105	8.4	13.9	380	22
4	Polypropylene/66.7	2.0	S	16.4	5.4	30	42	103	8.4	19.1	380	22
5	Polypropylene/66.7	2.0	D	16.4	About 1 to 2		42	58	8.0	24.3	530	16
6	Polypropylene/66.7	1.0	D	12.3	—	—	40	60	7.2	18.2	490	17
7	Polypropylene/66.7	0.05	D	7.8	—	—	39	61	6.3	15.0	500	19
8	Polypropylene/66.7	0.25	D	5.4	—	—	40	56	6.7	15.8	510	19
9	Polypropylene/66.7	0.125	D	1.0	—	—	35	57	6.0	9.1	407	27
10	Polypropylene/66.7	0.00	—	0.0	—	—	22	72	4.8	4.9	190	66
11	Polypropylene/33.3	1.00	D	12.3	—	—	29	13	3.9	12.8	490	7
12	Polypropylene/42.9	2.00	D	16.4	—	—	34	22	5.6	17.9	470	9
13	Polypropylene/53.8	2.00	D	16.4	—	—	36	32	7.6	25.1	460	12
14	Polypropylene/81.8	2.00	D	16.4	—	—	43	82	8.5	24.6	550	19
15	Polypropylene/122	2.00	D	16.4	—	—	48	162	11.3	27.5	560	31
16	Polypropylene/233	5.00	D	14.5	—	—	59	435	13.6	28.8	580	46
17	None[c]/0.00	2.00	S	16.4	—	—	11	2.3	1.5	2.0	150	1
18	Polypropylene[c]/∞	0	—	—	—	—	71	854	10.2	28.5	530	—
19	Polyethylene/66.7	2.00	D	12.3	—	—	35	51	7.2	14.8	440	18
20	Polyethylene/66.7	0.0	—	0.0	—	—	21	46	4.1	3.5	240	24

[a] Polypropylene is Profax 6723 and polyethylene is Marlex EHM 6006.
[b] S, static; D, dynamic. [c] Compositions 17 and 18 control compositions purely of cured rubber or polypropylene.

Data on Elastomers and Plastics

Table B.10 True Stress at Break of Selected Melt-Mixed Rubber-Plastic Blends[a]

Rubber	Plastic	True stress at break (MPa)
IIR	Polypropylene	26
EPDM	Polypropylene	26
NR	Polypropylene	16
NBR	Polypropylene	23
EPDM	Polyethylene	27
NBR	Polyethylene	13
NBR	Poly(tetramethylene terephthalate)	27
IIR	Polystyrene	2.3
EPDM	Polystyrene	3.7
BR	Polystyrene	4.1
IIR	SAN	5.0
NBR	SAN	10.5
IIR	PMMA	3.6

[a]The compositions were 60:40 and 50:50 rubber-plastic weight ratio. The stress at the break is the product of the ultimate extension ratio.

Table B.11 Properties of Various Types of Elastomer Compositions

	Partially vulcanized EPDM/Polypropylene blend[a]	Completely vulcanized EPDM/Polypropylene blend[b]	Neoprene vulcanizate	Ester-ether copolymer thermoplastic elastomer[c]
Shore A hardness	77	80	80	92
Tensile strength (MPa)	6.6	9.7	9.7	25.5
Ultimate elongation (%)	200	400	400	450
Volume swell in ASTM No. 3 Oil (74 hr at 100°C) (%)	Disintegrated	50	35	30
Compression set (method B: 22 hr at 100°C) (%)	70	39	35	33
Use temperature (°C)	100	125	110	125
Type of processing[d]	TP	TP	CV	TP

[a]TPR-1700 (Uniroyal).
[b]Santoprene (Monsanto).
[c]Hytrel (DuPont).
[d]TP, thermoplastic; CV, conventional vulcanizate.

Table B.12 Nonextended Polymers with Unsaturated Center Block

Polymer[a]	Solprene or Finaprene					Kraton		
	406	414	411	416	418	1101	1102	1107
Diolefin/styrene ratio	60:40	60:40	70:30	70:30	85:15[b]	70:30	72:28	86:14[b]
Molecular weight	High	Low	High	Low	—	—	—	—
Type structure	Radial	Radial	Radial	Radial	Radial	Linear	Linear	Linear
Specific gravity	0.95	0.95	0.94	0.94	0.92	0.94	0.94	0.92
Melt flow (200°C/5 kg)[c]	0	4	0	3	3	<1	6	9
300% modulus (MPa)	4.1	4.1	2.1	2.9	1.0	2.8	2.8	0.7
Tensile (MPa)	26.9	27.6	19.3[d]	20.0	16.5	31.7	31.7	21.4
Elongation (%)	700	750	700	720	1050	880	880	1300
Shore A hardness	93	90	78	68	34	71	62	37

[a]Partial listing only.
[b]Isoprene-styrene; others are butadiene-styrene.
[c]ASTM D1238.
[d]Resin additives result in higher tensile strength—retains tensile better than polymers of lower molecular weight on extension.

Table B.13 Some Commercial Macroglycols That Have Been Used to Make TPU Elastomers

Abbreviation	Common chemical name	Structure
PTAd	Poly(tetramethylene adipate) glycol[a]	$HO-[(CH_2)_4-OCO(CH_2)_4-COO]-(CH_2)_4-OH$
PCL	Poly(ε-captrolactone)glycol[b]	$H-[O-(CH_2)_5CO-]_x CO-ORO-[CO(CH_2)_5-O]_y-H$
PHC	Poly(hexamethylenecarbonate)glycol[c]	$HO-[(CH_2)_6-OCOO]_n-(CH_2)_6-OH$
PTMO	Poly(oxytetramethylene)glycol[d]	$HO-[(CH_2)_4-O]_n-H$
PPG	Poly(1,2-oxypropylene)glycol[d]	$HO-[CH-CH_3-CH_2-O]_n-CH_2-CHCH_3-OH$

[a] Polyester(carboxylate).
[b] Polyester(lactone).
[c] Polyester(carbonate).
[d] Polyether.

Table B.14 TPU Product Comparison Chart

Property	ASTM method	Units	BFG Estane Polyester 58206	BFG Estane Polyester 58137	BFG Estane Polyether 58300	BFG Estane Polyether 58810	Dow (Upjohn) Polyester 2102-85A	Dow (Upjohn) Polyester 2355-65D	Pellethane Polyether 2103-80A	Pellethane Polyether 2103-90A	Mobay Texin Polyester 480A	Mobay Texin Polyester 902E
Shore hardness	D2240	degrees	85A	70D	80A	90A/42D	87A±5	63D±4	80A±5	90A/47D	86A	75D
Specific gravity	D792	g/cm³	1.20	1.23	1.13	1.15	1.18	1.22	1.13	1.14	1.20	1.21
Ultimate tensile strength	D412	Mpa	45	40	32	44	43	41	41	43	40	41–48
		psi	6500	5800	4600	6400	6300	5900	6000	6250	5800	6–7000
Tensile stress	D412	MPa	8.5	22	4.8	9	8	22	6	11	5	37
At 100% elongation		psi	800	3200	700	1300	1100	3200	800	1530	700	5400
		Mpa	10	33	6.9	17	15	31	12	24	11	—
At 300% elongation		psi	1500	4800	1000	2400	2100	4500	1675	3430	1600	—
Ultimate elongation	D412	%	550	440	700	590	600	450	550	475	520	150–175
Compression set[a] (22 hr at 70°C)	D395	%	64	80	78	64	30	50	25–40	25–40	55	—
Flex modulus at 23°C	D790	Mpa	—	331	—	—	—	269	—	—	55	1068
		psi	—	48,000	—	—	—	39,000	—	—	8000	155,000
Vicat softness temperature (method B)	D1525	°C	85	149	76	111	—	—	—	—	91	182
		°F	185	300	169	232	—	—	—	—	196	360

Property										
Taber abrasion	D1044 mg/1000 cycles									
CS17 wheel, 1-kg load		3	12	—	—	—	—	—	—	2.7
H18 wheel, 1-kg load		36	119	36	70	—	—	—	—	—
H22 wheel, 1-kg load		—	—	—	—	50	15	20	10	—
Tear resistance, die C	D624 kN/m	88	228	70	88	88	245	83	95	94
	pli	500	1300	400	500	500	1400	475	540	535
Mold shrinkage[b]	— in./in.	0.012	0.005	0.016	0.614	[c]	[c]	—	—	0.008[d]
Price (list)	$/lb (T/L qty)	2.13	2.11	2.70	2.75	2.42	2.04	2.53	2.50	2.19	2.19

Source: Estane thermoplastic polyurethane product comparison data (1985).

[a]BFG samples unannealed, Dow (Upjohn) and Mobay samples annealed 16 hr at 240°F.
[b]Mold shrinkage determined on 0.125 × 3 × 6 in molded plaques; actual shrinkage will vary with part size and design.
[c]Mold shrinkage values for pellethane:

Part thickness (in.)	80A	55D
0.125	0.011–0.015	0.008–0.011
0.250	0.015–0.020	0.010–0.015
>0.250	0.020–0.030	0.015–0.020

[d]0.120-in. wall thickness.

Table B15 Physical Properties of 1,2-Polybutadiene

Properties	Testing methods	Measured values		
		JSR RB810	JSR RB820	JSR RB830
Density (kg/m^3)	Density-gradient tube method	901	906	909
Crystallinity (%)	Density-gradient tube method	~15	~25	~29
Microstructure 1,2-unit content (%)	Infrared ray spectrum (Morero method)	90	92	93
Refractive index, n_{25}^d	ASTM D542	1.513	1.515	1.517
Melt flow index (g/10 min) 150°C, 2160 g	ASTM D1238	3	3	3
Thermal properties				
Vicat softening point (°C)	ASTM D1525 (DSC method)	39	52	66
Melting point (°C)[a]		75	80	90
Brittle point (°C)	JIS K 6301	−40	−37	−35
Tensile properties	JIS K6301			
300% Modulus (MPa)		4.0	6.0	8.0
Tensile strength (MPa)		6.5	10.5	13.5
Elongation (%)		750	700	670
Hardness				
Shore D (degrees)	ASTM D1706	32	40	47
JIS A (degrees)	JIS K6301	79	91	95
Izod impact (kg · m/m)	ASTM D256	Not broken	Not broken	Not broken
Light transmittance (%)	JIS K6714	91	89	82
Haze (%)[b]	JIS K6714	2.6	3.4	8.0

Source: Reprinted with permission from Japan Synthetic Rubber Co. Ltd.
[a] Endothermic peak temperature according to the differential scanning calorimeter method. (Speed of temperature rise: 20°C/min.)
[b] The figures apply to 2-mm-thick sheet injection molded with cylinder temperature of 150°C and mold temperature 20°C.

Table B.16 Applications and features of 1,2-Polybutadiene

Application	Features
Application as a thermoplastic resin	
Films: stretch film, laminated film, shrinkable film	Safe for food packages, transparency, self-tack, pliability, shrinkability at low temperature, puncture strength resistance, heat sealability at low temperature, gas permeability
Various footwear soles: unit soles, inner soles, and outer soles by injection molding	Light weight, hardness, rubbery feeling, no deformation snappiness, reproducibility of mold pattern, coating performance, adhesive properties, crack resistance
Tubes and hoses: liquid food transfer tubes, tubes for medical practice	Safe for food transfer, safe for medical practice, transparency, flexibility
Other: blow moldings, injection moldings, resin modifier	Flexibility, safe for food packages
Application as a rubber	
Various sponges: microcellular sponge, hard sponge, semihard sponge, soft sponge, crepe-tone sponge	One-step vulcanization, wide range of curing conditions, high loading elasticity, snappiness, no deformation, weatherability, ozone resistance, heat resistance, tear resistance, coating performance, adhesive properties, skid resistance, abrasion resistance
Various high-hardness rubber goods: footwear, solid tires, industrial goods, dock fenders, sporting goods, sundries	Elongation, tensile strength, hardness, snappiness, good flowability, easy vulcanization, weatherability, ozone resistance, heat resistance, skid resistance, abrasion resistance
Injection cured goods: footwear, solid tires, industrial goods, rubber gloves	Flowability, injection molding processability easy vulcanization, weatherability, ozone resistance, heat resistance, skid resistance, abrasion resistance, snappiness
Rubber modifier: various rubber goods	Green strength, flowability, extrudability injection molding processability, weatherability, ozone resistance, heat resistance snappiness

Table B.16

Application	Features
Application as a rubber	
Other: transparent cured rubber goods	Transparency, safe for food application, weatherability, heat resistance
Other applications	
Adhesive (hot melt type): adhesive for various woven or unwoven cloth, paper, leather, and wooden board	Low melting point, flowability
Reaction accelerator: cross-linking accelerator for polyolefins	Easy cross-linking, reduction in use of cross-linking agents
Photo-sensitive polymer: printing plates, photo-sensitive paint, etching-resistant material	Photo sensitivity (photo curing), flowability, low solution viscosity
Thermo-setting resin: electrical insulation material	Chemical resistance, heat resistance, electric properties
Others: fibers, modifiers for composite resins, photo-degradable polymer	

Source: Reprinted with permission from Japan Synthetic Rubber Co. Ltd.

Table B17 Chemical and Oil Resistance of Silicone Rubber

	Percent volume change	
Chemical or medium	Fluorosilicone rubber (FVMQ)	Silicone rubber (VMQ)
Acid (7 days, 24°C)		
10% Hydrochloric	+1	+2
Hydrochloride	+8	+15
10% Sulfuric	Nil	+5
10% Nitric	+1	+8
Alkali (7 days, 24°C)		
10% Sodium hydroxide	Nil	Nil
50% Sodium hydroxide	+1	+9
Solvent (24 hr, 24°C)		
Acetone	+180	+15
Ethyl alcohol	+5	+9
Xylene	+20	Over 150
JP4 fuel	+10	Over 150
Butyl acetate	Over 150	Over 150
Oils		
ASTM No. 3 (7 days, 149°C)	+6	+20 +50
Turbo oil 15 (Mil L07808) (1 day, 177°C)	+8	30
Dimethyl siloxane, 500 cS (14 days, 205°C)	Nil	Swells, deteriorates

Table B.18 Summary of Solid EP and EPDM Worldwide Products[a]

Manufacturer	Trade name	Type elastomer	Grade designation	Oil content (phr)	Mooney viscosity (1' + 8') at 127°C	Ethylene content	Diene type	Diene content	Molecular weight distribution classification
Exxon Chemical	Vistalon	Copolymers	V-504	—	25	Low	—	—	Medium
			V-404	—	33	Very low	—	—	Very broad
			MD-78-5	—	26	Very low	—	—	Narrow
			MD-80-6	—	29	High	—	—	Narrow
			V-808	—	40	Very high	—	—	Narrow
			V-606	—	57	Low	—	—	Medium
			MD805-1	—	33	Very high	—	—	Narrow
		Terpolymers	MD-7504-1	—	30	Low	ENB	Medium	Narrow
			V-2504	—	24	Low	ENB	Medium	Very broad
			V-2200	—	30	Low	ENB	Low	Broad
			V-4507	—	45	Low	ENB	Medium	Medium
			V-7507	—	50	Low	ENB	Medium	Narrow
			V-2555	—	50	Medium	ENB	Low	Broad
			V-3708	—	50	High	ENB	Medium	Broad
			V-7000	—	55	High	ENB	Medium	Narrow
			V-4608	—	60	Low	ENB	Medium	Broad
			V-7509	—	68	Low	ENB	Medium	Narrow
			V-5600	—	70	Medium	ENB	Medium	Broad
			V-5630	30	35	Medium	ENB	Medium	Broad
			V-3666	75	52	Medium	ENB	Medium	Medium
			V-3777	75	45	High	ENB	Medium	Medium
			V-6505	—	46	Low	ENB	High	Narrow
			V-6630	30	33	Medium	ENB	High	Broad

Data on Elastomers and Plastics

Manufacturer	Type	Subtype	Grade					
Huels	Buna AP	Copolymers	AP201	—	22	Low	—	—	Broad
			AP 407	—	51	High	—	—	NA
			AP 301	—	45	Low	—	—	NA
		Terpolymers	AP 321	—	40	Medium	DCPD	High	NA
			AP 421	—	60	Medium	DCPD	High	NA
			AP 324	50	40	Medium	DCPD	High	NA
			AP 521	—	70	Medium	DCPD	High	NA
			AP 147	—	22	High	ENB	Medium	Narrow
			AP 241	—	25	Low	ENB	Medium	Narrow
			AP 331	—	41	Low	ENB	Low	Narrow
			AP 341	—	41	Low	ENB	Medium	Narrow
			AP 437	—	52	High	ENB	Low	Narrow
			AP 447	—	52	High	ENB	Medium	Narrow
			AP 238	30	30	High	ENB	Low	Narrow
			AP 248	30	30	Medium	ENB	Medium	Narrow
			AP 541	—	70	Low	ENB	Medium	Narrow
			AP 344	50	43	Low	ENB	Medium	Narrow
			AP 244	50	33	Low	ENB	Medium	Narrow
			AP 251	—	26	Low	ENB	High	Narrow
			AP 451	—	56	Low	ENB	High	Narrow
			AP 258	30	31	High	ENB	High	Narrow
			AP 338	25	50	High	ENB	High	Broad
			AP 451 H	—	51	Low	ENB	Very high	Narrow
			AP 454	50	50	Low	ENB	Very high	NA
Montedison	Dutral	Copolymers	CO 034	—	23	Medium	—	—	Narrow
			CO 054	—	29	Low	—	—	Broad
			CO 038	—	56	High	—	—	Narrow
			CO 059	—	60	Low	—	—	Narrow
			CO 554	100	24	Very low	—	—	Narrow
			CS 15/84	—	60	Low	—	—	Narrow

Table B.18 Summary of Solid EP and EPDM Worldwide Products[a]

Manufacturer	Trade name	Type elastomer	Grade designation	Oil content (phr)	Mooney viscosity (1' + 8') at 127°C	Ethylene content	Diene type	Diene content	Molecular weight distribution classification
			CS 16/84	—	23	Low	—	—	Narrow
		Terpolymers	044 E	—	25	Low	ENB	Low	Narrow
			054 E	—	27	Low	ENB	Medium	Narrow
			045 E	—	32	Low	ENB	Medium	Narrow
			048 E	—	60	Low	ENB	Medium	Narrow
			058 E	—	58	Low	ENB	Medium	Narrow
			038 E	—	57	High	ENB	Medium	Narrow
			334 E	40	23	High	ENB	Medium	Narrow
			535 E	100	20	Medium	ENB	Medium	Narrow
			436 X	65	50	High	ENB	High	Narrow
			346	45	32	Medium	ENB	High	Narrow
			TS 03/84	—	45	High	ENB	Medium	NA
			TS 12/84	—	55	High	ENB	Medium	NA
			TS 04/84	—	56	Low	ENB	Medium	NA
			TS 09/84	—	20	High	ENB	Medium	NA
			TS 11/84	—	70	High	ENB	Medium	NA
			0436	24	36	Low	ENB	Medium	Narrow
			TS 2084	—	30	Low	ENB	Medium	Medium
			TS 10/84	—	17	High	ENB	Medium	NA
			046 E3	—	40	Medium	ENB	Very high	Narrow
			235 E2	30	30	Medium	ENB	High	Narrow
			537 E2	100	46	Medium	ENB	High	NA
			TS 13/84	—	30	Medium	ENB	High	NA
			TS 14/84	—	75	Medium	ENB	High	NA

Data on Elastomers and Plastics

Manufacturer	Family	Grade				Diene		Distribution
DSM	Keltan	Terpolymers	520	—	45	Medium	DCPD	Medium	Broad
			720	—	57	Medium	DCPD	Medium	Broad
			520 x 58	50	40	Medium	DCPD	Medium	Broad
			480 x 100	100	34	High	DCPD	Medium	Broad
			540	—	40	Medium	ENB	Medium	Broad
		Tetrapolymers	312	—	31	Low	ENB/DCPD	High	Broad
			512	—	37	Low	ENB/DCPD	High	Broad
			578	—	46	High	ENB/DCPD	High	Broad
			712	—	50	Medium	ENB/DCPD	Medium	Broad
			778	—	58	High	ENB/DCPD	High	Broad
		Terpolymers	740	—	60	Medium	DCPD	Low	Narrow
		Tetrapolymers	812	—	67	Low	ENB/DCPD	High	Broad
		Terpolymers	4802	—	73	Low	ENB	Medium	Narrow
			802	—	64	Low	ENB	Medium	Broad
		Tetrapolymers	512 x 50	50	33	Low	ENB/DCPD	High	Broad
		Terpolymers	700 x 15	15	60	High	ENB	Medium	Narrow
			808	—	81	High	ENB	Medium	Medium
		Tetrapolymers	738	—	32	High	ENB/DCPD	High	Narrow
			514	—	40	Medium	ENB/DCPD	Very high	Broad
			714	—	58	Medium	ENB/DCPD	Very high	Broad
			DR 90	—	36	Medium	ENB/DCPD	Very high	Broad
		Terpolymers	4703	—	56	Low	ENB	High	Narrow
			509 x 100	100	40	High	ENB	High	Broad
Polysar	Polysar EP	Copolymers	306	—	20	Medium	—	—	Narrow
			405	—	22	Low	—	—	Narrow
			807	—	52	High	—	—	Narrow
		Terpolymers	P 227	—	12	High	ENB	Low	Narrow
			345	—	20	Medium	ENB	Medium	Narrow
			346	—	18	Medium	ENB	Medium	Narrow
			545	—	27	Low	ENB	Medium	Narrow

Table B.18 Summary of Solid EP and EPDM Worldwide Products[a]

Manufacturer	Trade name	Type elastomer	Grade designation	Oil content (phr)	Mooney viscosity (1' + 8') at 127°C	Ethylene content	Diene type	Diene content	Molecular weight distribution classification
DuPont	Nordel	Terpolymers	965	—	90	Low	ENB	High	Narrow
			6463	50	50	Medium	ENB	Medium	Narrow
			5465	100	55	Medium	ENB	Medium	Narrow
			585	—	32	Medium	ENB	Very high	Narrow
			5875	100	44	High	ENB	Very high	Narrow
			1320	—	18	Low	1,4-HD	Low	Narrow
			2522	—	21	Medium	1,4-HD	Medium	Narrow
			2722	—	21	Very high	1,4-HD	Medium	Narrow
			1040	—	35	Low	1,4-HD	Low	Very broad
			1440	—	33	Low	1,4-HD	Medium	Very broad
			1145	—	38	Medium	1,4-HD	Low	Narrow
			2744	—	40	High	1,4-HD	Medium	Very broad
			1660	—	52	Medium	1,4-HD	Medium	Narrow
			1070	—	62	Low	1,4-HD	Low	Very broad
			1070E	50	26	Low	1,4-HD	Low	Very broad
			1470	—	62	Low	1,4-HD	Medium	Very broad
			2670	—	54	High	1,4-HD	Medium	Narrow
Copolymer	Epsyn	Copolymers	4006	—	40	Medium	—	—	Narrow
			3007	—	35	Medium	ENB/VNB	Very low	NA
			2308	—	23	High	ENB/VNB	Low	Narrow
			2506	—	25	Medium	ENB/VNB	Medium	Medium
		Tetrapolymers	P 337	30	35	High	ENB/VNB	Medium	Narrow
			P 558	50	25	High	ENB/VNB	Medium	Medium

Data on Elastomers and Plastics

	Grade							
Uniroyal Terpolymers	5009	—		50	High	VNB	Very low	Narrow
	4506	—		35	Low	VNB	Low	Medium
	40A	—		38	Low	ENB	Low	Medium
	5508	—		55	High	ENB	Medium	Narrow
	5509	—		50	Very high	ENB	Medium	Narrow
	5206	—		50	Medium	ENB	Low	Medium
	7506	—		62	Low	ENB	Medium	Broad
	70 A	—		66	Low	ENB	Low	Narrow
	N 597 (P597)	100		50	50	Medium	ENB	Medium
Narrow			N 557 (P557)	50	Medium	ENB	Medium	Narrow
	4986	—		45	Medium	ENB	High	Narrow
	55	—		50	Low	ENB	High	Narrow
	N 997	—	100	50	Medium	ENB	High	Narrow
Royalene Terpolymers	301T	—		31	Medium	DCPD	Medium	Medium
	375	—		48	High	DCPD	Medium	Medium
	359	—		53	Low	DCPD	Medium	Medium
	400	—	100	36	Medium	DCPD	High	High
	X2753	—		40	Low	ENB	Low	Broad
	X2794	—		25	Low	ENB	Medium	Broad
	521	—		25	Low	ENB	Medium	Medium
	501	—		30	Low	ENB	Medium	Narrow
	580HT	—		33	Low	ENB	Low	Medium
	512	—		52	Medium	ENB	Medium	Medium
	X2817	—		50	Low	ENB	Low	Broad
	502	—		53	Low	ENB	Medium	Narrow
	552	—		50	High	ENB	Medium	Narrow
	X2859	—		53	Low	ENB	Medium	Broad
	539	—		68	High	ENB	Medium	Broad
	611	40		24	High	ENB	High	Medium
	622	40		37	High	ENB	Medium	Narrow

Table B.18 Summary of Solid EP and EPDM Worldwide Products[a]

Manufacturer	Trade name	Type elastomer	Grade designation	Oil content (phr)	Mooney viscosity (1' + 8') at 127°C	Ethylene content	Diene type	Diene content	Molecular weight distribution classification
Japan Synthetic Rubber Co.	JSR	Copolymers	X2899	75	37	High	ENB	Medium	NA
			X2935	25	43	Medium	ENB	Low	Broad
			505	—	42	Low	ENB	High	Broad
			509	—	57	High	ENB	High	Broad
			525	—	60	Low	ENB	High	Medium
			X2914	75	60	High	ENB	High	Medium
		Terpolymers	EP 01P	—	10	High	—	—	Very narrow
			EP 02P	—	12	High	—	—	Narrow
			EP 07P	—	32	High	—	—	Narrow
			EP 11	—	21	Low	—	—	Very broad
			911P	—	10	Very high	—	—	Very narrow
			912P	—	6	High	—	—	Very narrow
			EP 75F	—	50	Medium	DCPD	High	Medium
			EP 86	—	25	Medium	DCPD	Medium	Broad
			EP 87	—	31	Medium	DCPD	Low	Broad
			7001DE	100	40	Medium	DCPD	Low	Very broad
			EP 21	—	20	Medium	ENB	High	Medium
			EP 22	—	22	Low	ENB	Medium	Broad
			EP 24	—	35	Low	ENB	Medium	Broad
			EP 27	—	65	Low	ENB	Medium	Medium
			EP 43	—	26	Low	ENB	Low	Broad
			EP 51	—	19	High	ENB	Medium	Broad
			EP 57C	—	58	High	ENB	Medium	Medium
			EP 93	—	28	Low	ENB	Low	Narrow
			EP 96	50	58	High	ENB	High	Medium
			EP 25	—	49	Medium	ENB	Medium	Broad
			EP 98	75	63	High	ENB	Medium	Narrow

Data on Elastomers and Plastics

Company	Type	Grade	Oil	Mooney	Level	Diene	Level2	Distribution
Sumitomo Chemical Co.	Esprene Terpolymer	EP 103	—	80	Low	ENB	Medium	Narrow
		EP 33	—	22	Low	ENB	High	Broad
		EP 35	—	56	Low	ENB	High	NA
		EP 37C	—	55	Low	ENB	High	Broad
		EP 65	—	34	Low	ENB	High	Medium
		301	—	30	High	DCPD	NA	Broad
		305	—	35	High	DCPD	NA	Broad
		400	100	39	High	DCPD	NA	Broad
		501A	—	24	Low	ENB	Medium	Broad
		507	—	35	Low	ENB	Low	Broad
		522	—	50	Low	ENB	Medium	Narrow
		502	—	50	Low	ENB	Medium	Narrow
		512F	—	52	High	ENB	Medium	Medium
		532	—	65	Low	ENB	Medium	Narrow
		601F	70	52	Medium	ENB	Medium	Narrow
		600F	100	34	High	ENB	Medium	Broad
		505A	—	26	Medium	ENB	Very high	Broad
		505	—	40	Medium	ENB	Very high	Medium
		606	40	47	High	ENB	High	Broad
Mitsui Chemical Co.	Mitsui Copolymer	0045	—	17	Low	—	—	Broad
		1035	—	21	Medium	DCPD	NA	Broad
		1045	—	21	Medium	DCPD	NA	Broad
		1070	—	35	Medium	DCPD	NA	Broad
		1071	—	45	Medium	DCPD	Medium	Broad
	Terpolymers	3045	—	19	Medium	ENB	Medium	Broad
		3070	—	38	Medium	ENB	Medium	Narrow
		3091	—	50	Medium	ENB	Medium	Broad
		3075E	40	47	High	ENB	Medium	Narrow
		X-3092P	—	50	High	ENB	Medium	Narrow
		4021	—	11	Medium	ENB	High	Broad
		4045	—	21	Medium	ENB	High	Broad
		4070	—	36	Medium	ENB	High	Broad

[a]These data for typical properties of EP polymers are either as measured or as advertised by respective manufacturers. This table is not intended to be definitive either in terms of the total grade slate or the specific data reported for each producer. Note that the molecular weight distribution data are based on a qualitative comparison of GPC curves. Mooney viscosities are reported for final product form (i.e., in the case of oil-extended rubbers, the viscosity is that of the EP plus oil.

Index

ABS polymer melt rheology, 209
Accelerators, 51
Acceptance charts, 431
Acetylene blacks, 33
Activation energy, 207
Activation energy of flow, 21, 249
Addition polymerization, 2
Additives, 7
Adhering fillers, 41
Adhesives testing, 157
Aggregate breakdown, 43
Allowance design, 352
Amorphous polymers, 116
Amorphous state, 9
Analysis of variance, 345
Angular velocity, 19
Antioxidants, 8
Apparent modulus of elasticity, 166
Apparent shear rate, 285
Apparent viscosity, 14, 204, 208
Arrhenius equation, 144, 270
Attribute inspections, 397
Avogadro's number, 93

Bagley correction factor, 18, 220-223
Balance equations, 198
Banbury mixing, 45
Bending formulas, 166
Bias error, 437
Bingham plastic, 204
Binomial coefficients, 378
Black incorporation time, 277
Blocking in experimental design, 341
Blooming, 51
Blowing agents, 8
Box-Behnken designs, 484-491
Branched polystyrene, 106
Branching, 117, 118, 247
British standards, 149
Brookfield viscometer, 19, 233
Bulk viscosity, 20

Capillary dies, 284
Capillary rheometry, 241

Capillary tube flow, 215
Capillary viscometer, 17, 99, 213, 220
Carbon black, aggregates, 40
    dispersion, 55
    elastomer applications, 34-35
    incorporation, 55
    incorporation time, 57
    loadings, 39
    structures, 32
    transfer, 44
    wetting, 55
Carbon black-rubber interactions, 40
Chain stiffness, 117
Chain structure, 36
Channel blacks, 33
Charpy test, 171
Chemical potential, 86
Chromatographic processes, 103
Cold crystallization, 137
Cold-feed extrusion, 246
Colora apparatus, 142
Commercial carbon blacks, 40
Commercial plastics, 189
Compatibility, 42, 43
Complex modulus, 182
Complex shear modulus, 184
Compression forces, 63
Compression molding, 52
Compression ratio, 63
Compression stress-strain properties, 162
Compressive forces, 162
Condensation polymerization, 3
Cone-and-plate viscometer, 18, 19, 234
Confidence intervals, 449
Continuous vulcanization, 52, 53
Contour mapping, 49
Control charts, 367, 400-430
Control limits, 402
Copolymers, 6
Couette flow, 210, 211

Couette viscometer, 19
Critical molecular weight, 16
Cross-linked polymers, 3
Cross-linking, 117, 155
    agents, 8, 51
Crystalline melting behavior, 119
Crystalline polymers, 116
Crystallinity-temperature dependency, 135
Crystallite perfection, 120
Crystallization, 43, 120
Crystallization exotherms, 136
Crystallization process, 134
Cumulative molecular weight distribution, 110
Curemeter, 308
Curing, 278
    exotherms, 132
    temperature, 52
Customer window model, 325

Debora number, 207
Degree of crystallinity, 134
Degrees of freedom, 446
Density, 125
Die swell, 225
Diene, 23
Differential molecular weight distribution, 111
Differential power calorimetry, 130
Differential refractometer, 103
Differential scanning calorimetry, 129
Differential thermal analysis (DTA), 42, 125-129, 139
Dilute-solution viscosity, 99
Dispersion, 43
Dispersive mixing, 265
Divinyl benzene, 103
Doolittle equation, 22
Drag-induced plug flow, 278
Duality principle, 371

Index

Dynamic mechanical properties, 181
Dynamic mechanical test apparatus, 183
Dynamic property behavior, 175
Dynamic strain amplitudes, 188

Effective natural frequency, 179
Elastic modulus, 123
Elastic properties, 226
Elastic recovery, 5
Elastomer blends, 42
Electrical cable, 2
Electrical resistance, 89
Ellis model, 208
Elongation, 44
Elongation at break, 153, 155
Elongation of plastic films, 159
Engine mountings, 185
Entanglement molecular weight, 16
Enthalpy, 6, 116, 133, 140
Entrance corrections, 249
EPDM, applications, 51
    melting behavior, 136
    rubber, 25
EPM copolymers, 50
Equilibrium theory of Gibbs, 123
Ethylene content, 29
Ethylene-propylene elastomers, 22, 23
Experimental design, 436
Experimentation strategy, 443
External lubricants, 8
Extruder mixing devices, 281
Extrusion, 60, 61, 63, 365
Extrusion, performance, 287
    pressures, 63
    productivity, 294
    venting, 282

Factorial designs, 443, 450-473
Failure envelope, 156
Failures, 359
Falling weight apparatus, 173

Fatigue limit, 190
Fatigue testing, 187
Ferranti viscometer, 19
Filler loadings, 48
Filler-polymer adhesion, 41
Filler-reinforced elastomers, 58
Fillers, 7, 38, 197
Flame-ionization detectors, 103
Flex resistance, 31
Flexural stress-strain properties, 162
Flory temperature, 100
Flow activation energy, 201
Focal point leadership, 69
Force at break, 58
Force at failure, 153
Fourier's law, 202
Fraction of free volume, 124
Fractionation, 342
Fractionation methods, 113
Free surface energy, 137
Free volume, 118
Free-radical curing, 50
Free-volume theory, 117, 124, 125
Frequency curves, 384
Frequency distribution histograms, 366
Frequency distributions, 397
F-statistic, 449

Garvey die, 305
Gas law constant, 201
Gas-solid chromatography, 103
Gel permeation chromatography (GPC), 42, 102
Gelation reactions, 274
Geometric probability, 392
German standards, 151
Gibbs free energy, 122
Glass dilatometers, 122
Glass transition temperature, 116, 118, 119, 121-123, 127
Glassy transition, 117
GPC, calibration, 106
    chromatograms, 108

columns, 104

Hardness, 47, 148
  number, 150
  scales, 151
  test, miscellaneous, 152
Heat capacity, 127, 129, 132
Heat of fusion, 122
Heat of vaporization, 142
Heat resistance of EPDM, 28
Heat setting, 138
Hidden replication, 454
High-frequency vibration testing, 187
High-structure blacks, 42, 44
Homopolymers, 118
Hookean behavior, 161
Hookean elastic solid, 199
Hooke's law, 160, 164, 176
Hydrodynamic volume, 102, 106, 107
Hysteresis for time-dependent fluids, 206

Impact testing, 169
Inherent viscosity, 97
Injection molding, 52
Instron capillary rheometer, 17
Interaction geometry, 456
Interfacial bonding, 43
Intrinsic viscosity, 98, 100, 257
ISO standards, 148
Izod impact testing, 172

Kinetic theory of elasticity, 117

Lampblacks, 33
Light intensity, 113
Light intensity ratio, 94
Light scattering, 91-93, 95
Light-scattering photometers, 90, 91, 113

Linear polymers, 137
Long-chain branching, 294
Loss factor, 183
Loss function, 347
Low-functionality rubber, 43
Low-structure blacks, 41
Lubricants, 197

Malaysian Rubber Producers' Association, 176
Market research, 317, 319
Mark-Houwink correlation, 101
Mark-Houwink exponent, 248
Mechanical behavior, 117
Mechanical fatigue, 192
Mechanical spectrometer, 235
Melt flow index tester, 239, 240
Melt viscosity, 21
Melting point, 116
Melting temperature, 8, 121
Melting thermograms, 136
Membrane osmometry, 84, 85, 87
Mill mixing, 59, 60
Milling, 59
Mineral fillers, 47
Miniaturized internal mixer (MIM), 264
Mixer rotor configurations, 267
Mixing energy, 245
Modulus, 47
Modulus of rigidity, 164
Molecular diffusivity, 203
Molecular motion, 132
Molecular size, 96, 97, 100
Molecular weight, 4, 6, 13, 80, 100, 118
Molecular weight distribution, 29, 80, 102, 197
Molecular-structure parameters, 116
Molten polymers, 197
Mooney viscometer, 238
Mooney viscosity, 23, 29, 30, 50, 244
MPI tester, 227

Index

Newtonian behavior, 13
Newtonian viscosity, 201
Newton's law of viscosity, 200
Nonlinear elastic materials, 178
Non-Newtonian flow curves, 220
Non-Newtonian flow index, 16
Non-Newtonian fluids, 203
Normal distribution curve, 395
Normal frequency distribution, 365
Normal probability plot, 343
Normal stress differences, 237
Normal stresses, 14
Null set, 369
Number average molecular weight, 82

Oil types, 48
Oil-extended elastomers, 59
Oligomers, 87
Onset of glass transition, 117
Oscillating disc rheometer, 307
Oscillating stresses, 182
Ostwald viscometer, 98
Ostwald-de Waele model, 204
Oxygenated solvents, 36

Packing density, 44
Parallel disk viscometer, 236, 237
Parallel-plate viscometer, 235
Pelletization, 30
Plackett-Burman design, 473-478, 480
Plastic viscosity, 204
Plasticity, 205
Plasticizer, 8, 38, 83, 118, 205
Plastics testing, 159
Plate count method (GPC), 105
Polyethylene, 61, 101
Polyfunctional monomers, 4
Polynomial models, 479
Polypropylene, 101
Polyvinyl chloride, 118
Poly(methyl) methacrylate, 118
Power law model, 15, 204

Precision ratio, 459
Probability, 364
   distributions, 381, 390
   function, 391
   surface, 388
   theorems, 368
Processability, 9, 45
Processability testing, 138, 197
Processing behavior, 38
Processing maps, 54
Processing temperature, 307
Property method, 369
Pseudoplastic materials, 217
Pseudoplasticity index, 204
PVC fusion, 272

Quality control, 147, 313
   principles, 332
   techniques, 392
Quality improvement schemes, 327

Rabinowitsch-Mooney equation, 218
Radius of gyration, 96, 163
Random coil polymers, 81
Random error, 438
Random experiments, 371
Random variables, 381
Randomization, 424
Rate of crystallization, 120
Recovery time, 229
Reduced viscosity, 96
Refractive index, 113
Reinforcement, 8, 43
Reinforcing properties, 45
Relative stiffness, 37
Relative viscosity, 96, 97
Relaxation phenomenon, 12
Relaxation time, 14, 207
Relaxed die swell, 251
Resilience, 37
Response surface concepts, 479
Rheological responses, 12
Rheopectic fluids, 205

Rockwell hardness, 149, 150
Roster method, 369
Rotating cylinder viscometers, 229, 230
Rotational fatigue testing, 191
Rubber bearings, 178
Rubber modulus, 44, 161
Running die swell, 291

Sampling inspection, 428-430
Sandwich viscometer, 236
Screening designs, 472
Screw configurations, 63
Segmental motion, 117
Semicrystalline polymers, 155
Semipermeable membranes, 83
Shear flow geometries, 214
Shear modulus, 199, 200
Shear strength, 164
Shear stress, 19
Shear stress-strain properties, 162
Shear thinning behavior, 205
Shore D hardness, 148
Shore durometers, 148, 149
Short-term stress-strain properties, 175
Signal-to-noise ratio, 353
Silica, 47
Single-screw extrusion, 277, 279
Slippage, 225
Slit rheometer, 238
Soft clays, 45
Softness number, 149
Solubility parameter, 117
Solution polymerization, 139
Specific viscosity, 96
Specific volume, 6, 125
Specific volume curves, 133
Sponge weatherstrips, 357
Spring-mass combinations, 178
Stability testing, 273
Stabilizers, 8
Standard deviation, 415
Static property behavior, 175
Static shear moduli, 177

Statistical experimental design, 335-355
Statistical process control, 351
Statistics, 364
Steelyard test machine, 157
Stirling's approximation, 378
Stored energy, 14, 15
Strain amplitude, 185
Stress, 176
Stress relaxation testing, 229
Stress rupture tests, 193
Stress-strain, behavior, 153
    curves, 154
    measurements, 11
    properties, 12, 147, 156
    properties tests, 151
Stress/strain ratio, 182
Structure-property relationships, 11
Sulfur curing, 52
Surface activity, 45
Surface aesthetics, 289
Surface appearance rating, 298
Surface rating classification, 293
Surface smoothness, 301
Swelling, 11, 43

Taguchi method, 350, 354, 355
Talc, 47
Taylor series, 479
Tear resistance, 40
Tear testing, 169
Tear testing apparatus, 170
Tensile impact apparatus, 174
Tensile machines, 163
Tensile modulus of elasticity, 157
Tensile strain, 151
Tensile strength, 8, 153, 155
Tensile stress, 151
Tensile stress-strain testing, 162
Thermal analysis, 116
Thermal blacks, 33
Thermal conductivity, 14, 116, 120, 122, 125, 139
Thermal degradation, 116
Thermal failure, 191

Index

Thermal properties, 116
Thermochemical analysis, 42
Thermodynamic first order transition, 119
Thermodynamic properties, 138
Thermodynamic transitions, 123
Thermogravimetric analysis, 143-144
Thermoset materials, 261
Thermosets, 5
Three-factorial designed experiment, 339
Time-dependent fluids, 203
Time-independent fluids, 203
Torque balance, 210
Torque rheometry, 261, 288
Total shear energy, 286
Total work energy, 271
Totalized torque, 271
Toughness, 169
Transfer molding, 52
Tree diagrams, 376, 377
True shear stress, 285
Turbidimetric titration, 111
Twin-screw extruders, 64
Two-level factorial designs, 339, 463

Ultimate elongation, 153
Ultimate stretch ratio, 58
Ultracentrifugation, 102
Ultraviolet light stabilizers, 8
Unfilled elastomers, 181
Unsaturation, 43

Vapor-phase osmometry, 87
Variance, 445
Velocity profiles, 217
Venn diagrams, 369, 370
Vibration spectrum, 181
Vibration testing, 178-180
Virial coefficients, 86
Viscoelastic behavior, 9, 12, 189
Viscoelastic properties, 198
Viscoelasticity, 189
Viscometers, 209-241
Viscosity, 13, 43
    at zero shear rate, 204
    number, 96
    pressure coefficient, 201
    ratio, 96
Viscous heating, 242
Volume dilatometers, 123
Vulcanizate properties, 30
Vulcanization, 50
Vulcanization rate, 309

Weight average molecular weight, 90-93
Weissenberg rheogoniometer, 235

Yerzley test, 37
Yield stress, 14, 204
Young's modulus, 161

Zero-shear viscosity, 16
Zimm plot, 95